Department of Trade and Industry

Digest of United Kingdom Energy Statistics 2000

D1514129

Production editor: Mari Scullion

A National Statistics publication

London: The Stationery Office

ISBN 011 5154973
ISSN 0307 0603

Digest of United Kingdom Energy Statistics

Enquiries about statistics in this publication should be made to the contact named at the end of the relevant chapter. Brief extracts from this publication may be reproduced provided the source is fully acknowledged. General enquiries about the publication, and proposals for reproduction of larger extracts, should be addressed to the Production Editor, Mari Scullion, at the address given in paragraph XXV of the Introduction.

The Department of Trade and Industry reserves the right to revise or discontinue the text or any table contained in this Digest without prior notice.

About The Stationery Office's Standing Order Service

The Standing Order Service, open to all Stationery Office account holders, allows customers to automatically receive the publications they require in a specified subject area, thereby saving them the time, trouble and expense of placing individual orders, also without handling charges normally incurred when placing ad-hoc orders.

Customers may choose from over 4,000 classifications arranged in 250 sub groups under 30 major subject areas. These classifications enable customers to choose from a wide variety of subjects, those publications which are of special interest to them. This is a particularly valuable service for the specialist library or research body. All publications will be dispatched immediately after publication date. A Standing Orders Handbook describing the service in detail and a complete list of classifications may be obtained on request. Write to The Stationery Office, Standing Order Department, PO Box 29, St Crispins, Duke Street, Norwich, NR3 1GN, quoting reference 12.01.013; Alternatively telephone 0870 600 5522 and select the Standing Order Department (option 2); fax us on 0870 600 5533; or finally e-mail us at book.standing.orders@theso.co.uk.

National Statistics

National Statistics are produced to high professional standards set out in the National Statistics Code of Practice. They undergo regular quality assurance reviews to ensure that they meet customer needs. They are produced free from any political interference.

You can find a range of National Statistics on the Internet – www.statistics.gov.uk

---

Cover design by Michelle Franco, ONS Graphic design Unit

Photographs courtesy of British Petroleum (2nd and 3rd from top)
2nd from top: BP Exploration - Wytch Farm project, Dorset Gathering centres at Wytch Heath
3rd from top: Production platform in BP's Magnus oilfield, north-east of Shetland

Contents

A list of tables

Annex A Energy, commodity balances, calorific values and conversion factors

Annex B Energy and the environment

Annex C United Kingdom oil and gas resources

Introduction

I   This issue of the Digest of United Kingdom Energy Statistics continues a series which commenced with the Ministry of Fuel and Power Statistical Digest for the years 1948 and 1949, published in 1950. The Ministry of Fuel and Power Statistical Digest was previously published as a Command Paper, the first being that for the years 1938 to 1943, published in July 1944 (Cmd. 6538).

II   The current issue brings up to date the figures given in the Department of Trade and Industry's *Digest of United Kingdom Energy Statistics 1999*, published in July 1999.

III   This issue consists of nine chapters, the first of which deals with overall energy. The other chapters cover the specific fuels, combined heat and power, renewable sources of energy, trade, and prices. The period covered in detail in the fuel chapters is 1997 to 1999, whilst the prices chapter covers 1990 to 1998 however most chapters also contain some detailed information covering 1995 to 1999 as well as data on long term trends. As in previous years there is an annex dealing with energy and the impact of its production and consumption on the environment, and annexes on major events in the energy industries, calorific values and conversion factors, a glossary of terms and further sources of information. An annex on energy consumption by industry and the service sector has also been added in this edition.

IV   The series available on a long term trend basis go back to 1960 in most cases, although for some series data are not readily available for all the years. However, sufficient data have been included to provide a comprehensive view of developments in energy supply and consumption over the last thirty-eight years.

V   Where necessary, data have been converted or adjusted to provide consistent series, however, in some cases changes in methods of data collection have affected the continuity of the series. The presence of remaining discontinuities is indicated in the chapter text or in footnotes to the tables.

VI   The first chapter covers general energy statistics and includes tables showing energy consumption by final users and an analysis of energy consumption by main industrial groups. Fuel production and consumption statistics are derived mainly from the records of fuel producers and suppliers. In general the statistics have the same coverage as those in other chapters and where this is so the explanatory notes in these chapters are applicable. Statistics in the Foreign trade chapter are derived largely from returns made to HM Customs and Excise and published in the *Overseas Trade Statistics of the United Kingdom*, by HM Customs and Excise, although some of the data shown in this Digest may contain unpublished revisions and estimates of trade from additional sources.

VII   Chapters 6 and 7 summarise the results of surveys conducted by ETSU on behalf of the Department of Trade and Industry and the Statistical Office of the European Communities. These estimate the contribution made by combined heat and power (CHP) and renewable energy sources to the United Kingdom's energy during the period from 1995 to 1999.

VIII   In Chapters 2, 3, 4, 5 and 7 production and consumption of individual fuels are presented using *commodity balances*. A commodity balance shows the flows of an individual fuel through from production to final consumption, showing its use in transformation and energy industry own use. Commodity balances are presented for 1999, 1998 and finally 1997. Further details of commodity balances and their use are given in the Annex A, paragraphs A.7 to A.41.

IX   The individual commodity balances are combined in an *energy balance*. The energy balance is not the same as a commodity balance; it shows the interactions between different fuels in addition to illustrating their consumption. The energy balance thus gives a fuller picture of the production, transformation and use of energy showing all the flows. The energy balance forms the basis of Chapter 1, Energy, where the United Kingdom energy balance is presented for 1999, then 1998 and finally 1997. Expenditure on energy is also presented in energy balance format in Chapter 1. Further details of the energy balance and its use are given in Annex A, paragraphs A.42 to A.57.

Definitions

X   The text at the beginning of each chapter explains the main features of the tables. Technical notes and definitions, given at the end of this text, provide detailed explanations of the figures in the tables and how they are derived. Explanations of the logic behind an energy balance and for commodity balances are given in Annex A.

XI   Most chapters contain some information on 'oil' or 'petroleum'; these terms are used in a general sense and vary according to usage in the field examined. In their widest sense they are used to include all mineral oil and related hydrocarbons (except methane) and any products derived therefrom.

XII   An explanation of the terms used to describe electricity generating companies is given in Chapter 5, paragraphs 5.57 to 5.58.

XIII   Data in this issue have been prepared on the basis of the Standard Industrial Classification (SIC) 1992 as far as is practicable. For further details of classification of consumers see Chapter 1, paragraphs 1.74 to 1.78.

XIV   Where appropriate, further explanations and qualifications are given in footnotes to the tables.

Geographical coverage

XV   The geographical coverage of the statistics is indicated on each table. Almost all the tables relate to the United Kingdom; the main exceptions being the tables on temperatures (Tables 1.15 and 1.16), and manufacturing industry fuel prices (Tables 9.11 and 9.12) which cover Great Britain, and the tables in Annex C on oil and gas resources (Tables C.1 to C.6) which give data for the United Kingdom Continental Shelf. Exports to the Channel Islands and the Isle of Man from the United Kingdom are not classed as exports, and supplies of solid fuel and petroleum to these islands are therefore included as part of United Kingdom inland consumption or deliveries.

Periods

XVI   Data in this Digest are for calendar years or periods of 52 weeks, depending on the reporting procedures within the fuel industry concerned. Actual periods covered are given in the notes to the individual fuel sections. Where no periods are shown, the data are for calendar years, except in Chapter 1, where data are for periods of 52 weeks.

Revisions

XVII   The tables contain revisions to some of the previously published figures, and where practicable the revised data have been indicated by an 'r'.

Energy data on the Internet

XVIII   The digest is also available on the internet at [http://www.dti.gov.uk/epa/dukes]. Information on DTI energy publications can also be found on the energy statistics Internet site at http://www.dti.gov.uk/epa.

Other sources

XIX   The Department is also publishing with this issue of the Digest *'UK Energy in Brief'*, a booklet summarising the main statistics presented in this Digest.

XX   Short term statistics are published by the Department of Trade and Industry in the monthly statistical bulletin *Energy Trends,* and by the Office for National Statistics in the *Monthly Digest of Statistics (The Stationery Office)*. Provisional monthly figures are published in advance of *Energy Trends* via the Department's *'Advance Energy Statistics'* press notice. To subscribe to *Energy Trends,* to obtain copies of the press notice, or for free copies of *'UK Energy in Brief'* or *'Energy Sector Indicators'* booklet please contact Gillian Purkis or Shopna Zaman at the address given at paragraph XXV

Table numbering

XXI   The numbering and order of some tables in this issue differ from that in earlier issues. Pages 9 and 10 contain a list showing the tables in the order in which they appear in this issue, and their corresponding numbers and pages in previous issues.

Symbols used

XXII   The following symbols are used in this Digest:-

..	not available
-	nil or negligible (less than half the final digit shown)
r	Revised since the previous edition

Rounding convention

XXIII   Individual entries in the tables are rounded independently and this can result in totals which are different from the sum of their constituent items.

Acknowledgements

XXIV   Acknowledgement is made to the main coal producing companies, the electricity companies, the oil companies, the gas pipeline operators, the gas suppliers, the Institute of Petroleum, the Coal Authority, the United Kingdom Iron and Steel Statistics Bureau, the National Environmental Technology Centre, ETSU, the Department of the Environment, Transport and the Regions, the HM Customs & Excise, the Office for National Statistics, and other contributors to the enquiries used in producing this publication.

Contacts

XXV   For general enquiries on energy statistics contact Gillian Purkis on 020-7215 2697 (E-mail: Gillian.Purkis@dti.gsi.gov.uk), or Shopna Zaman on 020-7215 2698 (E-mail: Shopna.Zaman@dti.gsi.gov.uk) or at:-

> Gillian Purkis/Shopna Zaman
> Department of Trade and Industry
> Bay 1128
> 1 Victoria Street
> London SW1H 0ET
> Fax: 020-7215 2723

Enquirers with hearing difficulties can contact the Department on the DTI Textphone: 020-7215 6740.

XXVI   For enquiries concerning particular data series or chapters contact those named on page 8 or at the end of the relevant chapter.

Mari Scullion, Editor
July 2000

Contact list

The following people in the Department of Trade and Industry may be contacted for further information about the topics listed:

Chapter	Contact	Telephone	E-mail
		020-7215	
Energy	Chris Bryant	5183	Chris.Bryant@dti.gsi.gov.uk
	Caroline Barrington	5187	Caroline.Barrington@dti.gsi.gov.uk
Coal and other solid fuels	Mike Janes	5186	Mike.Janes@dti.gsi.gov.uk
	James Achur	2717	James.Achur@dti.gsi.gov.uk
Oil and gas resources	Kevin Williamson	5184	Kevin.Williamson@dti.gsi.gov.uk
	Clive Evans	5189	Clive.Evans@dti.gsi.gov.uk
North Sea profits, operating costs and investments	Suhail Siddiqui	5262	Suhail.Siddiqui@dti.gsi.gov.uk
	Philip Beckett	5260	Philip.Beckett@dti.gsi.gov.uk
Petroleum (downstream)	Kevin Williamson	5184	Kevin.Williamson@dti.gsi.gov.uk
	Clive Evans	5189	Clive.Evans@dti.gsi.gov.uk
Gas supply (downstream)	Mike Janes	5186	Mike.Janes@dti.gsi.gov.uk
	Natasha Chopra	2718	Natasha.Chopra@dti.gsi.gov.uk
Electricity	Mike Janes	5186	Mike.Janes@dti.gsi.gov.uk
	Joe Ewins	5190	Joe.Ewins@dti.gsi.gov.uk
Combined heat and power	Mike Janes	5186	Mike.Janes@dti.gsi.gov.uk
Prices and values	Lesley Petrie	2720	Lesley.Petrie@dti.gsi.gov.uk
Domestic prices	Sharon Young	6531	Sharon.Young@dti.gsi.gov.uk
Industrial, international and oil prices	Sara Atkins	6532	Sara.Atkins@dti.gsi.gov.uk
Foreign trade	Caroline Barrington	5187	Caroline.Barrington@dti.gsi.gov.uk
Renewable sources of energy	Mike Janes	5186	Mike.Janes@dti.gsi.gov.uk
Energy and the environment	Chris Bryant	5183	Chris.Bryant@dti.gsi.gov.uk
	Kevin Williamson	5184	Kevin.Williamson@dti.gsi.gov.uk
Calorific values and conversion factors	Caroline Barrington	5187	Caroline.Barrington@dti.gsi.gov.uk
	Kevin Williamson	5184	Kevin.Williamson@dti.gsi.gov.uk
General enquiries (energy helpdesk)	Gillian Purkis	2697	Gillian.Purkis@dti.gsi.gov.uk
	Shopna Zaman	2698	Shopna.Zaman@dti.gsi.gov.uk

All the above can be contacted by fax on 020-7215 2723

Tables as they appear in this issue and their corresponding numbers in the previous four issues

Section	1996	1997	1998	1999	2000
ENERGY	1	1	1	-	-
	3	3	1.3	1.1	1.1
	2	2	1.2	1.2	1.2
	-	-	-	1.3	1.3
	4	4	1.4	1.4	1.4
	-	-	-	1.5	1.5
	-	-	-	1.6	1.6
	5	5	1.5	-	-
	6	6	1.6	-	-
	7	7	1.7	-	-
	8	8	1.8	-	-
	9	9	1.9	1.7	1.7
	-	10	1.10	1.8	1.8
	A1	13	1.13	1.9	1.9
	A2	14	1.14	1.10	1.10
	A3	15	1.15	1.11	1.11
	A4	16	1.16	1.12	1.12
	A5	17	1.17	1.13	1.13
	-	-	-	1.14	1.14
	10	11	1.11	1.15	1.15
	A19	12	1.12	1.16	1.16
SOLID FUELS & DERIVED GASES	-	-	-	2.1	2.1
	-	-	-	2.2	2.2
	-	-	-	2.3	2.3
	-	-	-	2.4	2.4
	-	-	-	2.5	2.5
	-	-	-	2.6	2.6
	11	18	2.1	2.7	2.7
	12/14	19/21	2.2-4	2.8	2.8
	15	22	2.5	-	-
	17	24	2.7	-	-
	-	-	-	2.9	2.9
	A6	25	2.8	2.10	2.10
	A7	26	2.9	2.11	2.11
PETROLEUM	-	-	-	3.1	3.1
	-	-	-	3.2	3.2
	-	-	-	3.3	3.3
	-	-	-	3.4	3.4
	-	-	-	3.5	3.5
	-	-	-	3.6	3.6
	28	37	4.4	3.7	3.7
	29	38	4.2	-	-
	31	39	4.3	-	-
	30	40	4.1	3.8	-
	33	41	4.5	-	-
	32	42	4.6	-	-
	34	43	4.7	-	-
	37	44	4.8	3.9	3.8
	35	45	4.9	-	-
	-	-	-	3.10	-
	36	46	4.10	3.11	-
	38	47	4.11	3.12	3.9
	39	48	4.12	3.13	3.10
	A8	49	4.13	3.14	3.11
	A9	50	4.14	3.15	3.12
NATURAL GAS	-	-	-	4.1	4.1
	-	-	-	4.2	4.2
	40	51	5.1	-	-
	41	52	5.2	-	-
	42	53	5.3	4.3	4.3
	A10	57	5.7	4.4	4.4
	43	54	5.4	A9.1	A9.1
	44	55	5.5	-	-
	45	56	5.6	-	-
ELECTRICITY	-	-	-	5.1	5.1
	46	58	6.1	5.2	5.4
	-	-	-	5.3	5.2
	47	59	6.2	5.4	5.5
	48	60	6.3	-	5.3
	-	-	-	-	5.6

Section	1996	1997	1998	1999	2000
ELECTRICITY continued	49	61	6.4	A9.1	A9.1
	50	62	6.5	5.5	5.7
	51	63	6.6	5.6	5.8
	52	64	6.7	5.7	5.9
	A13	65	6.8	5.8	5.10
	A12	67	6.10	5.9	5.11
	A11	66	6.9	5.10	5.12
COMBINED HEAT AND POWER	-	-	-	-	6.1
	-	68	7.1	6.1	6.2
	C7	69	7.2	6.2	6.3
	C4	-	-	-	-
	C6	-	-	-	-
	-	70	7.3	6.3	6.4
	-	71	7.4	6.4	6.5
	-	72	7.5	6.5	6.6
	-	73	7.6	6.6	6.7
	C8	-	-	-	-
	C9	-	-	-	-
	C10	-	-	-	-
	C3/C5	74	7.7	6.7	6.8
	-	75	7.8	6.8	6.9
	C1	7.2	7B	6A	6A
	-	-	-	6B	6B
	-	-	-	6C	6C
	C2	7.1	7A	6D	6D
	-	-	-	6E	6E
	-	-	-	6F	6F
RENEWABLE SOURCES	-	-	-	7.1	7.1
	-	-	-	7.2	7.2
	-	-	-	7.3	7.3
	B2	96	10.2	7.4	7.4
	-	10.1	10A	7.5	7.5
	B1	95	10.1	7.6	7.6
FOREIGN TRADE	64	90	9.1	8.1	8.1
	A17	94	9.2	8.2	8.2
	65	91	9.3	8.3	8.3
	66	92	9.4	8.4	8.4
	16/67	23/93	2.6/9.5	8.5	8.5
PRICES AND VALUES	53	76	8.1	9.1	9.1
	54	77	8.2	9.2	9.2
	55	78	8.3	9.3	9.3
	63	-	-	-	-
	A14	79	8.4	9.4	9.4
	-	-	-	9.5	9.5
	-	-	-	9.6	9.6
	-	-	-	9.7	9.7
	-	-	-	9.8	9.8
	-	-	-	9.9	9.9
	58	80	8.5	-	-
	-	81	8.6	-	-
	62	82	8.7	9.10	9.10
	59	83	8.8	9.11	9.11
	60	84	8.9	9.12	9.12
	61	-	-	-	-
	-	85	8.10	9.13	9.13
	56	87	8.12	9.14	9.14
	A15	88	8.13	9.15	9.15
	57/A16	89	8.14	9.16	9.16
	-	86	8.11	9.17	9.17
	-	-	-	9.18	9.18
	-	-	-	9.19	9.19
	43/49	54/61	5.4/6.4	A9.1	A9.1
ANNEX B: ENERGY AND THE ENVIRONMENT	D1	A.1	A.1	A.1	B.1
	-	-	A.2	A.2	B.2
	D2	A.2	A.3	A.3	B.3
	-	-	A.4	A.4	B.4
	D3	A.3	A.5	A.5	B.5
	D4	A.4	A.6	A.6	B.6
	D5	A.5	A.7	A.7	B.7
	-	A.6	A.8	A.8	B.8

Tables as they appear in this issue and their corresponding numbers in the previous four issues (cont.)

Chapter 1
Energy

Introduction

1.1   This chapter presents figures on overall energy production and consumption. Figures showing the flow of energy through from production, transformation and energy industry use through to final consumption are presented in the format of an energy balance based on the individual commodity balances presented in Chapters 2 to 5 and 7. The tables included in this chapter have been rationalised to avoid duplication of information as far as possible.

1.2   The chapter begins with aggregate energy balances covering the last three years (Tables 1.1, 1.2 and 1.3) starting with the latest year, 1999. Energy value balances then follow this for the same years (Tables 1.4, 1.5 and 1.6). Table 1.7 covers final energy consumption by the main industrial sectors over the last five years followed by Table 1.8 which shows the fuels used for electricity generation by these industrial sectors. Tables 1.9 to 1.14 present long-term trends for energy production, consumption and expenditure on energy as well as analyses such as the relationship between energy consumption and the economy of the UK. The final tables (Tables 1.15 and 1.16) present figures for average temperatures. The explanation of the principles behind the energy balance and commodity balance presentations are in Annex A.

The energy industries

1.3   The energy industries in the UK play a central role in the economy by producing, transforming and supplying energy in its various forms to all sectors. They are also major contributors to the UK's Balance of Payments through the exports of crude oil and oil products. The box below summarises the energy industries' contribution to the economy:

- 5 per cent of GDP

- 6 per cent of total investment

- 36 per cent of industrial investment

- Value added per head 5 times the industrial average

- 170 thousand people directly employed (4 per cent of industrial employment)

- Many others indirectly employed (e.g. an estimated 360,000 in support of UK Continental Shelf activities)

- Trade surplus in fuels of £5 billion.

Aggregate energy balance (Tables 1.1, 1.2 and 1.3)

1.4   These tables show the flows of energy in the United Kingdom from production to final consumption through conversion into secondary fuels such as coke, petroleum products and secondary electricity. The principles behind the presentation used and how this links with the figures presented in the other chapters are explained in Annex A. The figures are presented on an energy supplied basis, in tonnes of oil equivalent, see paragraphs 1.45 to 1.48.

1.5   In 1999 the primary supply of fuels was 243 million tonnes of oil equivalent, hardly any change from 1998 and an increase of 2 per cent compared to 1997. However indigenous production in 1999 was over 5 per cent higher than in 1998 or 1997. Chart 1.1 illustrates the figures for the production and consumption of individual primary fuels in 1999. In 1999, as in the previous 6 years, overall primary fuel consumption was fully met by indigenous production, with the trade balances for petroleum and its products and gas more than offsetting net imports of coal, manufactured fuels and electricity.

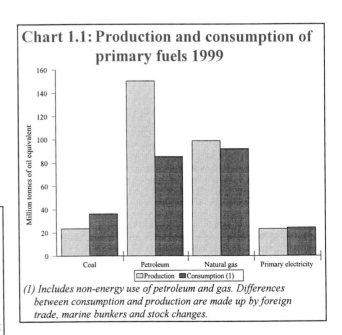

Chart 1.1: Production and consumption of primary fuels 1999

(1) Includes non-energy use of petroleum and gas. Differences between consumption and production are made up by foreign trade, marine bunkers and stock changes.

1.6   Total primary energy demand was just under 1 per cent lower in 1999 than in 1998, but just under 1 per cent higher than in 1997. Chart 1.2 shows the composition of primary demand in 1999.

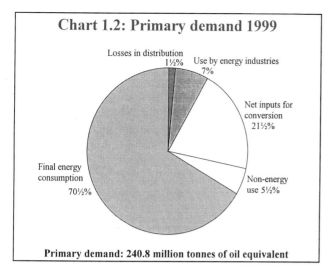

Chart 1.2: Primary demand 1999

Losses in distribution 1½%

Use by energy industries 7%

Net inputs for conversion 21½%

Final energy consumption 70½%

Non-energy use 5½%

Primary demand: 240.8 million tonnes of oil equivalent

1.7   The transfers row in Tables 1.1, 1.2 and 1.3 should ideally sum to zero with transfers from primary oils to petroleum products amounting to a net figure of zero. Similarly the manufactured gases and natural gas transfers should sum to zero; the net difference is due to transfers of coke to breeze.

1.8   The transformation section of the energy balance shows, for each fuel, the net inputs for transformation uses. For example 6,163 thousand tonnes of oil equivalent of coal feeds into the production of 5,448 thousand tonnes of oil equivalent of coke, representing a loss of 715 thousand tonnes of oil equivalent in the manufacture of coke in 1999. In 1999 energy losses during the production of electricity and other secondary fuels amounted to 51,314 thousand tonnes of oil equivalent, shown in the transformation row in Table 1.1.

1.9   The next section of the table represents use of fuels by the energy industries themselves. This section also includes consumption by those parts of the iron and steel industry which behave like an energy

industry i.e. are involved in transformation processes (see paragraph A.28 of Annex A). In 1999 energy industry use amounted to 16,686 thousand tonnes of oil equivalent of energy, a decrease of 2 per cent on 1998 and a 2 per cent increase on 1997.

1.10   Losses presented in the energy balance include distribution and transmission losses in the supply of manufactured gases, natural gas, and electricity. These losses have fallen by 8 per cent in 1999, following a fall of 14 per cent in 1998.

1.11   Total final consumption, which includes non-energy use of fuels, in 1999 was 169,944 thousand tonnes of oil equivalent, an increase of less than 1 per cent on 1998 and 2 per cent increase on 1997. Final energy consumption in 1999 was mainly accounted for by the transport sector (32 per cent), the domestic sector (27 per cent), industry (21 per cent) and non-energy use (8 per cent). These figures are illustrated in Chart 1.3. Recent trends in industrial consumption are shown in Table 1.7 and discussed in paragraphs 1.18 to 1.20.

1.12   The main fuels used by final consumers in 1999 were petroleum products (45 per cent), natural gas (34 per cent) and electricity (16 per cent). Of the petroleum products consumed by final users 8 per cent was for non-energy purposes; for natural gas 1 per cent was consumed for non-energy purposes.

1.13   Non-energy use of fuels includes use as chemical feedstocks and other uses such as lubricants. In 1999 non-energy use of fuels was as shown in Table 1A. Further details of non-energy use are given in Chapter 3, paragraphs 3.64 to 3.70 and Chapter 4, paragraphs 4.17.

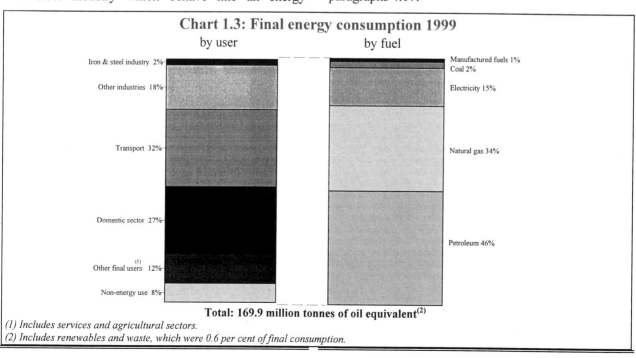

Chart 1.3: Final energy consumption 1999

by user

Iron & steel industry 2%
Other industries 18%
Transport 32%
Domestic sector 27%
Other final users [1] 12%
Non-energy use 8%

by fuel

Manufactured fuels 1%
Coal 2%
Electricity 15%
Natural gas 34%
Petroleum 46%

Total: 169.9 million tonnes of oil equivalent[2]

(1) Includes services and agricultural sectors.
(2) Includes renewables and waste, which were 0.6 per cent of final consumption.

Table 1A: Non-energy use of fuels 1999

Thousand tonnes of oil equivalent

	Petroleum	Natural gas
Petrochemical feedstocks	8,496	1,147
Other	3,526	-
Total	**12,022**	**1,147**

Value balance of traded energy (Tables 1.4, 1.5 and 1.6)

1.14   Tables 1.4 to 1.6 provide a new presentation of the value of traded energy in a similar format to the energy balances. The balance shows how the value of inland energy supply is made up from the value of indigenous production, trade, tax and margins (profit and distribution costs). The lower half of the table then shows how this value is generated from the final expenditure on energy through transformation processes and other energy sector users as well as from the industrial and domestic sectors. The balances only contain values of energy which is traded i.e. where a transparent market price is applicable. Further technical notes are given in paragraphs 1.56 to 1.61.

1.15   Total expenditure by final consumers in 1999 is estimated at £60,900 million, (£60,675 million shown as actual final consumption and £225 million of coal consumed by the iron and steel sector in producing coke for their own consumption). Of the final consumption, 27 per cent represents the basic value of primary fuels, i.e. the value of the fuels at the pit or landing terminal. Whilst a further 50 per cent was accounted for by Duty and VAT. Distribution costs and margins accounted for the remainder.

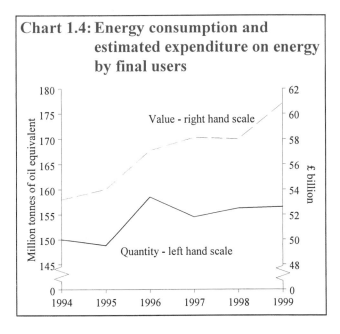

Chart 1.4: Energy consumption and estimated expenditure on energy by final users

1.16   This balance provides a guide on how the value chain works in the production and consumption of energy. For example in 1999, £10,910 million of crude oil were indigenously produced of which £7,155 million were exported and £3,280 million were imported. Allowing for stock changes this provides a total value of inland crude oil supply of £7,005 million. This fuel was then completely consumed within petroleum in the process of producing £7,835 million of petroleum products. Again some external trade and stock changes took place before arriving at a basic value of petroleum products of £6,810 million. In supplying the fuel to final consumers distribution costs were incurred and some profit was made amounting to £1,730 million whilst duty and tax meant a further £29,680 million was added to the basic price to arrive at the final market value of £38,220 million. This was the value of petroleum products purchased of which industry purchases £755 million, domestic consumers for heating £425 million, with the vast majority purchased within the transportation sectors, £35,015 million.

1.17   Of the total final expenditure on energy in 1999 (£60,900 million) the biggest share, 58 per cent fell to the transport sector. Of the remaining 42 per cent industry purchased around a quarter or £5,750 million with the domestic sector purchasing nearly three quarters or £13,650 million.

Energy consumption by main industrial groups (Table 1.7)

1.18   This table presents final energy consumption for the main industrial sub-sectors over the last 5 years.

1.19   So far as is practicable, the user categories have been grouped on the basis of the 1992 Standard Industrial Classification (see paragraphs 1.74 to 1.78). However, some data suppliers have difficulty in classifying consumers to this level of detail and the breakdown presented in these tables must therefore be treated with caution.   The groupings used are consistent with those used in Table 1.8 which shows industrial sectors' use of fuels for generation of electricity (autogeneration).

1.20   In 1999, 35 million tonnes of oil equivalent were consumed by the main industrial groups. The largest consuming groups were chemicals (20 per cent), iron and steel and non-ferrous metals (15 per cent), metal products, machinery and equipment (12 per cent), food, beverages and tobacco (11 per cent), and paper, printing and publishing (7 per cent). The remaining groups accounted for 34 per cent of total final energy consumption by industry. The figures are illustrated in Chart 1.5.

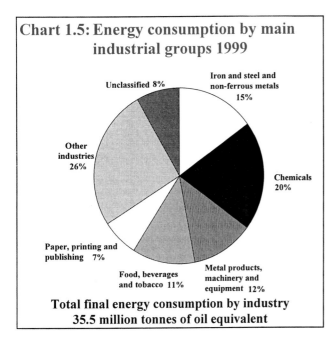

Chart 1.5: Energy consumption by main industrial groups 1999

Unclassified 8%

Iron and steel and non-ferrous metals 15%

Other industries 26%

Chemicals 20%

Paper, printing and publishing 7%

Food, beverages and tobacco 11%

Metal products, machinery and equipment 12%

Total final energy consumption by industry 35.5 million tonnes of oil equivalent

Fuels consumed for electricity generation by main industrial groups (autogeneration) (Table 1.8)

1.21 This table gives details of the amount of each fuel consumed by industries in order to generate electricity for their own use. Fuel consumption is consistent with the figures given for "other generators" in Table 5.4 of Chapter 5. The term autogeneration is explained further in paragraphs 1.52 and 1.53. Electricity produced via autogeneration is included within the figures for electricity consumed by industrial sectors in Table 1.7. Table 1.8 has been produced using the information currently available and shows the same sector detail as Table 1.7, data cannot be given in as much detail as in the individual commodity balances and the energy balance because it could disclose information about individual companies. Table 1.8 allows users to allocate the fuel used for autogeneration to individual industry groups in place of the electricity consumed. Further information on the way Table 1.8 links with the other tables is given in paragraph 1.53.

Long term trends

Inland consumption of primary fuels (Table 1.9)

1.22 The trends for inland consumption of primary fuels for energy use are illustrated in Chart 1.6.

1.23 Overall consumption for energy use increased steadily at an average rate of 2 per cent a year up to 1973, when the oil price rise following the Arab-Israeli war of that year led to a major change in patterns of fuel consumption. Having reached a level of over 220 million tonnes of oil equivalent in 1973, energy use fell, but by 1979 had returned to a similar level to that

in 1973. After the outbreak of another Middle East war, consumption fell back to less than 200 million tonnes of oil equivalent in the years 1981 to 1984. It has since grown again, and by 1996 had exceeded the peak levels of 1973 and 1979. Although overall consumption in 1999 has fallen by 1 per cent compared to 1996 it is still above the peaks seen in 1973 and 1979.

1.24 The changing trend in overall energy consumption was affected by petroleum consumption, which had continued to grow in the period 1960 to 1973 despite the strong growth in consumption of natural gas and primary electricity, mainly nuclear. After 1973 petroleum consumption declined for ten years, following much the same pattern as coal use. Over the last ten years petroleum consumption has risen again, although it fell back in 1995 and again in 1997, 1998 and 1999. Over the same period the decline in coal consumption has continued at an increasing rate, whilst the consumption of natural gas has continued to grow with natural gas consumption now exceeding petroleum consumption. Primary electricity consumption has been growing strongly in recent years. The increase in total consumption in 1996 was mainly due to the increased use of natural gas during the cold winter weather and it can be seen that total consumption fell back in 1997. Although there was a slight increase in 1998, consumption in 1999 has fallen back to its 1997 level.

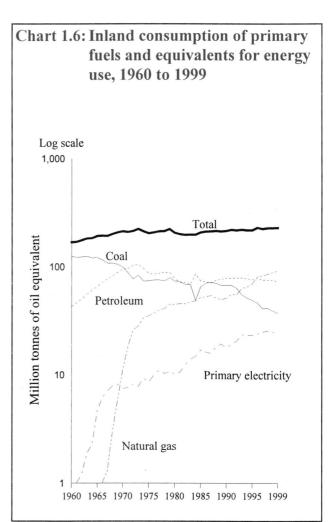

Chart 1.6: Inland consumption of primary fuels and equivalents for energy use, 1960 to 1999

Log scale

Million tonnes of oil equivalent

Total

Coal

Petroleum

Primary electricity

Natural gas

1960 1965 1970 1975 1980 1985 1990 1995 1999

Availability and consumption of primary fuels and equivalents (Table 1.10)

1.25 An overall view of energy presented in the form of energy balances is given in Table 1.10. It is based on Table 1.1 with the time series extended back to 1965. Supplies and uses of energy are expressed on an energy-supplied basis in tonnes of oil equivalent, and are balanced by fuel and for total energy. More details on the derivation of these balances and on the calculation of energy contents are given in paragraphs 1.47 to 1.48. Calorific values of fuels are shown in Annex A.

1.26 Trends in the production of primary fuels in the United Kingdom are illustrated in Chart 1.7.

1.27 In 1965 total energy production was around 129 million tonnes of oil equivalent with coal accounting for some 96 per cent. Total production declined to around 110 million tonnes of oil equivalent in the early 1970s, at about the time when natural gas from the North Sea started to be produced in substantial quantities. From 1975, petroleum production also grew rapidly to peak at over 139 million tonnes of oil equivalent in 1985 when it accounted for 55 per cent of total energy production of 253 million tonnes of oil equivalent. By 1991, temporary production problems had reduced petroleum production to less than 100 million tonnes of oil equivalent. Since then petroleum production has steadily recovered, reaching a record level of 150 million tonnes of oil equivalent in 1999. In 1999, production of petroleum was 51 per cent of total energy production. Gas production was at a record level of 99 million tonnes of oil equivalent in 1999, 33 per cent of total energy production. At the same time, coal accounted for 8 per cent of energy production and nuclear and hydro electricity together 8 per cent.

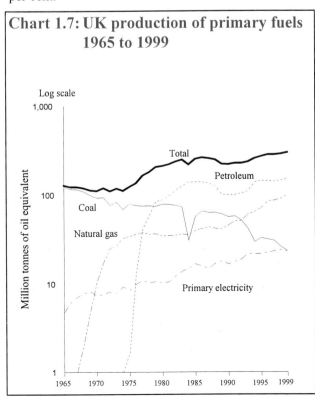

Chart 1.7: UK production of primary fuels 1965 to 1999

Comparison of net imports of fuel with total consumption of primary fuels and equivalents (Table 1.11)

1.28 In Table 1.11 gross consumption in the United Kingdom, including non-energy use and international marine bunkers, is compared with net imports of fuel to show net import dependency or net exports.

1.29 A comparison of the overall trends in energy production (from Table 1.10) and total consumption (from Table 1.11) is given in Chart 1.8.

1.30 Chart 1.8 shows United Kingdom primary energy production and consumption and illustrates the degree to which the United Kingdom was dependent on energy imports prior to North Sea oil and gas becoming available. In the early 1970s energy imports accounted for over 50 per cent of United Kingdom consumption, but in 1983 the United Kingdom was a net exporter at a level equivalent to 18 per cent of inland consumption. After 1986 net exports declined. Following temporary production losses in the North Sea, the United Kingdom became a small net importer of energy between 1989 and 1992. Since then North Sea production has recovered and the United Kingdom has become a net exporter again. Net exports represented 16 per cent of inland consumption in 1997, 17 per cent in 1998 and 21 per cent in 1999.

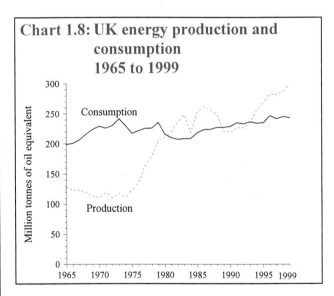

Chart 1.8: UK energy production and consumption 1965 to 1999

Energy ratio (Table 1.12)

1.31 The relationship between energy consumption and economic activity at the aggregate level can be gauged by comparing a country's temperature corrected inland primary energy consumption with its gross domestic product (GDP). This approach is simple and comprehensive but it has a number of drawbacks which were discussed in articles in the August 1976, May 1981 and May 1989 issues of *Economic Trends* (The Stationery Office).

1.32 The columns in Table 1.12 show the United Kingdom's temperature corrected inland primary energy consumption and GDP at constant prices since 1950, both expressed in absolute units (millions of tonnes of oil equivalent and billions of 1995 pounds sterling respectively). Dividing energy consumption by GDP yields the energy ratio, which is expressed in column C of the table as energy consumed per million pound of GDP and in column D as an index number based on 1995=100. For GDP at constant prices the published measure of GDP at market prices at 1995 prices has been used. The GDP figures used are now on the European System of Accounts (ESA 95) basis, consistent with the UK national accounts.

1.33 Energy consumption, GDP and the energy ratio over the period 1950 to 1999 (indexed to 1950 = 100) are illustrated in Chart 1.9.

1.34 Chart 1.9 shows that the energy ratio fell to 80 per cent of its 1950 level by 1973 and to 50 per cent by 1999, an average decrease of over 1 per cent per annum. It rose slightly during the early 1990s, but started falling again in 1993. The strong downward trend since 1950 is explained by at least four factors: improvements in energy efficiency; saturation in the ownership levels of the main domestic appliances; the unresponsiveness of certain industrial uses, like space heating, to long run output growth; and a structural shift away from energy intensive activities (such as steel making) towards low energy industries (such as services).

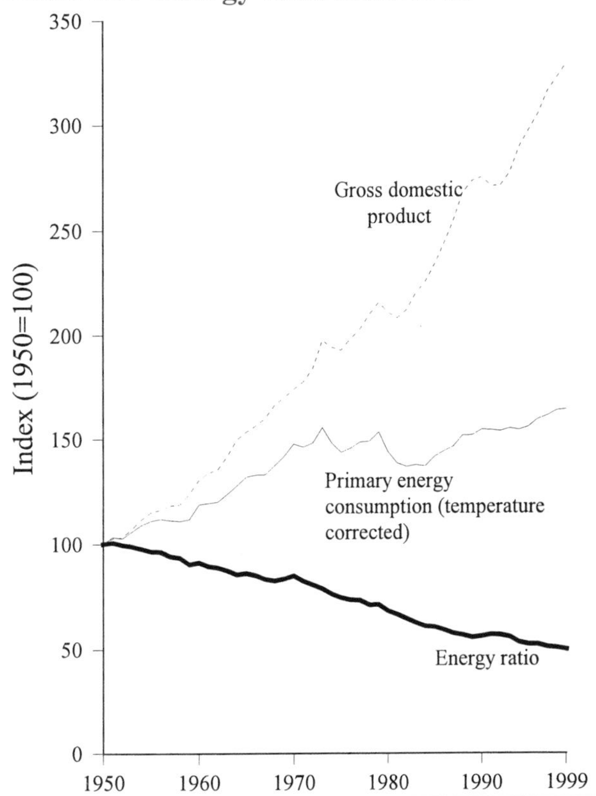

Energy consumption by final user (Table 1.13)

1.35 Figures for consumption of fuel for energy uses by category of final user are given in Table 1.13. This table excludes non-energy use. Final users' consumption is net of the fuel industries' own use and conversion, transmission and distribution losses, but it includes conversion losses by final users. The user categories are industry (including iron and steel), transport (including coastal shipping), domestic and other final users (public administration, agriculture, commerce and other sectors), see paragraphs 1.74 to 1.78.

1.36 Up to 1986 data for final consumption of electricity include acquisitions from public supply, output of industrial nuclear stations, and amounts produced by transport undertakings and industrial hydropower for final consumption. From 1987 onwards, all consumption of electricity, whether produced by major power producers or by other generators is included. There is a corresponding change in treatment, between 1986 and 1987, for other fuels used in electricity generation (see paragraph 1.55).

1.37 Overall consumption by final users followed the same pattern as overall primary energy consumption since 1965, accounting for around 70 per cent of the total consumption throughout the period.

1.38 From 1960 to 1973 industry (including iron and steel) was the sector with the greatest level of consumption, with between 41 and 43 per cent of total final consumption over the period. However, since 1973 this sector has steadily reduced its consumption so that it now stands at 22½ per cent of total final consumption for energy use. This share is now less than that of the domestic sector which, at 29½ per cent, has about the same share as in 1960, though this did fall in the intervening years, reaching 24 per cent in 1973. Greatest growth has been in the transport sector; this had a share of 17 per cent in 1960, which has risen to 34½ per cent in 1999.

1.39 A comparison of energy consumption for energy purposes by final users in 1960 and 1999 is shown in Chart 1.10.

Chart 1.9: Energy ratio since 1950

Index (1950=100)

Gross domestic product

Primary energy consumption (temperature corrected)

Energy ratio

1950 1960 1970 1980 1990 1999

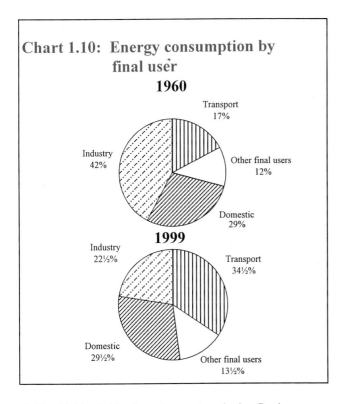

Chart 1.10: Energy consumption by final user

1960

- Transport 17%
- Industry 42%
- Other final users 12%
- Domestic 29%

1999

- Industry 22½%
- Transport 34½%
- Domestic 29½%
- Other final users 13½%

Expenditure on energy by final user (Table 1.14)

1.41   Total expenditure on fuels is presented in Table 1.14 from 1970; and figures for recent years are illustrated in Chart 1.4. Data for the latest years are taken from the value balances (Tables 1.4 to 1.6) whilst earlier years are taken from their forerunner tables of estimated values of energy purchases by sector. The total fuels series is simply the sum of fuels presented in the table and so is slightly different from the value presented in the value balances as other fuel (which accounted for around 0.1 per cent of total final expenditure in 1999) is excluded but coal purchased by the iron and steel sector is included as a final purchase of coal.

1.42   Overall final expenditure on energy rose by around £2,820 million (nearly 5 per cent) in 1999 compared to 1998. The level of £60,825 million represents a 12½ per cent rise on 1995 and 32 per cent more than in 1990. The final expenditure fell for all fuels in 1999 with the exception of petroleum where expenditure rose by 12 per cent, reflecting increases in petroleum product prices in 1999.

1.43   The make up of total expenditure has changed through time reflecting structural or long term changes in fuel and shorter term price and consumption effects. In 1970, expenditure on coal and coke accounted for around 15 per cent of total final expenditure but was down to 1½ per cent in 1999. By contrast, the general increase in the consumer price of petroleum (where duty is a major component) has meant petroleum has risen from 45 per cent of all expenditure in 1970 to 60 per cent in 1999. Electricity, despite seeing over a 50 per cent increase in volume consumed since 1970, still accounts for roughly the same share of total expenditure, 30 per cent in 1970, 26½ per cent in 1999, as prices have seen significant real term falls.

1.40   Table 1.13 also shows trends in final energy consumption for individual fuels. In 1960, consumption of coal and other solid fuels accounted for over 60 per cent of final energy consumption, but this share has declined steadily, first as the level of petroleum consumption increased but since 1973 as natural gas usage increased at the expense of both solid fuel and petroleum consumption. Electricity consumption has made steady progress over the last thirty years, rising from 7 per cent of the total in 1960 to 17½ per cent in 1999. A comparison of final energy consumption for individual fuels in 1960 and 1999 is shown in Chart 1.11.

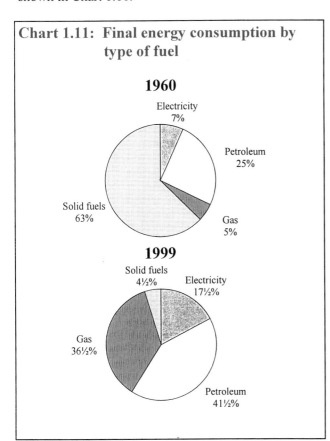

Chart 1.11: Final energy consumption by type of fuel

1960

- Electricity 7%
- Petroleum 25%
- Solid fuels 63%
- Gas 5%

1999

- Solid fuels 4½%
- Electricity 17½%
- Gas 36½%
- Petroleum 41½%

Mean air temperatures (Tables 1.15 and 1.16)

1.44   These tables give the average air temperatures in Great Britain between 1961 and 1990 by year, part year and month. Deviations from these means are presented for January 1994 to May 1999. Average monthly temperatures back to 1960 are also given in Table 1.16. These temperature deviations are used to provide the temperature corrected consumption series in Table 1.12. The average temperature in 1996 was marginally lower than the long term mean whereas in 1995, 1997, 1998 and 1999 it was 1 degree Celsius or more above the average.

Technical notes and definitions

I Units and measurement of energy

Units of measurement

1.45 The original units of measurement appropriate to each fuel are used in the individual fuel chapters. A common unit of measurement, the tonne of oil equivalent (toe), which enables different fuels to be compared and aggregated, is used in Chapter 1. For consistency with the International Energy Agency and with the Statistical Office of the European Communities, the tonne of oil equivalent is defined as follows:

$$1 \text{ tonne of oil equivalent} = 10^7 \text{ kilocalories}$$
$$= 396.8 \text{ therms}$$
$$= 41.87 \text{ Gigajoules (GJ)}$$
$$= 11{,}630 \text{ kWh}$$

1.46 This unit should be regarded as a measure of energy content rather than a physical quantity. There is no intention to represent an actual physical tonne of oil, and indeed actual tonnes of oil will normally have measurements in tonnes of oil equivalent which differ from unity.

Thermal content - energy supplied basis of measurement

1.47 Tables 1.1, 1.2, 1.3, 1.7, and 1.9 to 1.13 are compiled on an energy-supplied basis. Detailed data for individual fuels are converted from original units to tonnes of oil equivalent using gross calorific values and conversion factors appropriate to each category of fuel. The results are then aggregated according to the categories used in the tables. Gross calorific values represent the total energy content of the fuel, including the energy needed to evaporate the water present in the fuel (see also paragraph 1.72).

1.48 Estimated gross calorific values for 1999 are given on page 248. Calorific values are reviewed each year in collaboration with the fuel industries, and figures for earlier years can be found in Table A.2 on page 249. To construct energy balances on an energy supplied basis; calorific values are required for production, trade, and stocks, as follows:

Coal The weighted average gross calorific value of all indigenous coal consumed is used to derive the thermal content of coal production and undistributed stocks. Thermal contents of imports and exports allow for the quality of coal. Thermal contents of changes in coal stocks at secondary fuel producers are the average calorific values of indigenous coal consumed.

Petroleum Work was carried out in 1997 to revise calorific values for petroleum products. It has not been possible to find any recent work on the subject. In the absence of such work, the gross calorific values, included in Annex A, and used in the construction of these energy balances from 1990 onwards have been derived using a formula derived by the US Bureau of Standards. This formula estimates the gross calorific value of products according to their density. This formula is as follows:

$$Gj = 51.83 - 8.78 \times d^2$$

where d is the density of the product in terms of kilograms per litre.

For crude petroleum and refinery losses, the weighted average calorific value for all petroleum products from UK refineries is used. A notional figure of 43.3 GJ per tonne is used for non-energy petroleum products (industrial and white spirits, lubricants, bitumen, petroleum coke, waxes and miscellaneous products).

Gases Although the original unit for gases is the cubic metre, figures for gases are generally presented in the fuel sections of this Digest in gigawatt hours (GWh), having been converted from cubic metres using gross calorific values provided by the industries concerned. Conversion factors between units of energy are given on page 247.

Electricity Unlike the other fuels, the original unit used to measure electricity - GWh - is a measure of energy. The figures for electricity can therefore be converted directly to toe using the conversion factors on page 247.

Primary electricity Hydro electricity and net imports of electricity are presented in terms of the energy content of the electricity produced (the energy supplied basis). This is consistent with international practice. Primary inputs for nuclear electricity assume the thermal efficiencies at nuclear stations given in Chapter 5, Table 5.9 (36.8 per cent in 1999). (See Chapter 5, paragraphs 5.27 and 5.65.)

Temperature corrected primary fuel consumption (Tables 1.12)

1.49 The temperature corrected series of total inland fuel consumption given in Table 1.12 indicates what annual consumption might have been if the average temperature during the year had been the same as the average for the years 1961 to 1990. This average is given, with annual deviations, in Table 1.15 whilst Table 1.16 shows average temperatures for each month from 1960. The corrections used to increase demand per degree Celsius above average are:

Coal	2.1 per cent
Petroleum	0.7 per cent (June - August)
	1.8 per cent (September - May)

1.50 Figures for natural gas are corrected using a method developed by British Gas Transco. Prior to 1990, the annual temperature adjustment applied by the DTI differed from that applied by British Gas due to the effect of seasonal adjustment of the monthly data. From 1990 onwards, the DTI's annual adjustment to the gas figures for temperature is the same as that applied by BG Transco. Nuclear, hydro and net imports of electricity are not corrected for temperature.

Non-energy uses of fuel

1.51 Energy use of fuel mainly comprises use for lighting, heating, motive power and power for appliances. Non-energy use includes for example use as chemical feedstocks, solvents, lubricants, and road making material. The non-energy use of natural gas as a chemical feedstock was separately identified for the first time in the 1994 edition of the Digest. It should be noted that the estimated amounts of non-energy gas included in the Digest are very approximate. Non-energy uses of petroleum and gas are now included in the figures for final energy consumption following the move over to the presentation of energy data in the format of commodity and energy balances in this edition. Further discussion of non-energy uses of lubricating oils and petroleum coke appears in Chapter 3, paragraphs 3.64 to 3.70.

Autogeneration of electricity

1.52 Autogeneration is defined as the generation of electricity by companies whose main business is not electricity generation, the electricity being produced mainly for that company's own use. Estimated amounts of fuel used for thermal generation of electricity by such companies, the output of electricity and the thermal losses incurred in generation are included within the Transformation sector in the energy balances shown in Tables 1.1, 1.2 and 1.3. Electricity used within the power generation process by autogenerators is shown within the Energy industry use sector. Electricity consumed by industry and commerce from its own generation is included as part of Final consumption. This treatment is in line with the practice in international energy statistics.

1.53 Figures for the total amount of fuel used and electricity generated by autogenerators along with the amount of electricity they consume themselves are shown in Tables 1.8, 5.1, 5.3, and 5.5. Table 1.8 summarises the figures according to broad industrial sector. Much of the power generated is from combined heat and power (CHP) plants and data from Chapter 6 are included within Table 1.8. Differences will occur where CHP plants are classified to major power producers, and this mainly affects the chemicals sector. The method of allocating fuel used in CHP plants between electricity production and heat production is described in paragraph 6.36 of Chapter 6. This method conforms with international practice although countries adopt different percentages for the assumed efficiencies of heat-only boilers. However it can give rise to high implied conversion efficiencies in some sectors, most notably in the iron and steel sector. A large number of revised figures appear for autogeneration this year reflecting work undertaken to improve the quality of these data.

Final consumption, deliveries, stock changes

1.54 Figures for final consumption relate to deliveries, if fuels can be stored by users and data on actual consumption are not available. Final consumption of petroleum and solid fuel is on a deliveries basis throughout except for the use of solid fuel by the iron and steel industry. Figures for domestic use of coal are based on deliveries to merchants. Figures for stock changes in Tables 1.1 to 1.3 cover stocks held by primary and secondary fuel producers, major distributors of petroleum products, and stocks of coke and breeze held by the iron and steel industry. Figures for stock changes in natural gas represent the net amount put into storage by gas companies operating pipelines.

1.55 Figures for final consumption of electricity include sales by the public distribution system and consumption of electricity produced by generators other than the major electricity producing companies. Thus electricity consumption includes that produced by industry and figures for deliveries of other fuels to industry exclude amounts used to generate electricity (except for years prior to 1987 - see paragraph 1.36).

Valuation of energy purchases
(Tables 1.4, 1.5, 1.6, 1.14)

1.56 In common with the rest of the chapter, the tables covering energy expenditure now follow a balance format. Whilst a user may derive data on a similar basis as that previously published, the balance table allows for more varied use and interpretation of traded energy value data. That said the table continues to only show values for energy that has to be purchased and therefore does not include estimated values of a sectors internal consumption, such as coal used in the process of coal extraction.

The balance

1.57 The table balances around **market value of inland consumption** with the lower half of the table showing the total value of consumption by end users, sub divided into energy sector users and final users both for energy and non energy use. The top half of the table shows the supply components that go to make up the final market value of inland consumption, namely upstream cost of production, imports, taxes and the margins and costs of delivering and packaging the fuel for the final consumer. The total final consumers value of energy consumption is represented by the lines 'total non energy sector use' and iron and steel sectors purchases of coal for use in solid fuel manufacture.

Fuel definitions in value balances

1.58 **Crude oil** includes NGLs and refinery feedstocks. **Natural gas** does not include colliery methane. **Electricity** only includes electricity delivered via the public distribution system and therefore does not value electricity produced and consumed by autogenerators, but the input fuels are included in transformation. **Manufactured solid fuels** includes coke, breeze and other solid manufactured fuels, mainly products from patent fuel and carbonisation plants. **Other fuels** includes all other fuels not listed, where they can be clearly considered as traded and some reasonable valuation can be made. Fuels mainly contributing to this year's values are wood, coke oven and colliery methane gases sold on to other industrial users and some use of waste products such as tyres.

Valuation
1.59 All figures are estimates and have been rounded to the nearest £5 million.

Energy end use
1.60 Values represent the cost to the final user including transportation of the fuel. They are derived, except where actual values are available, from the traded element of the volumes presented in aggregate energy balance and end user prices collected from information supplied by users or energy suppliers. The **energy sector** consists of those industries engaged in the production and sale of energy products, but values are not given for consumption of self generated fuels e.g. coke oven gas used by coke producers. Many of the processes in the **iron and steel** industry are considered to be part of the energy sector in the energy balances, but for the purposes of this economic balance their genuine purchases are treated as those of final consumers, except for purchases of coal directly used in coke manufacture, which is shown separately as part of manufacture of solid fuel. Coal used directly in or to heat blast furnaces is shown as iron and steel final use. **Transformation** are those fuels used directly in producing other fuels e.g. crude oil in petroleum products. **Electricity generators** keep and use significant stocks of coal and the stocks used in consumption each year are shown separately. The value and margins for these being assumed to be the same as other coal purchased in the year. **Road transport** includes all motor spirit and DERV use. **Commercial and other users** includes public administration and miscellaneous uses not classified to the industrial sector.

Supply
1.61 The supply side money chain is derived using various methods. **Indigenous production** represents the estimated basic value of in year sales by the upstream producers. This value is gross of any taxes or cost they must meet. The valuation problems in attributing network losses in gas and electricity between upstream and downstream within this value chain, means any costs borne are included in the production value. **Imports and exports** are valued in accordance with Chapter 8. However crude oil is treated differently where the value is formed from price data taken from a census survey of refiners and volume data taken from Tables 3.1 to 3.3. These values are considered to reflect the complete money chain more accurately than Tables 8.1 to 8.4. **Stock**

changes are those for undistributed stocks except for coal where coke oven and generators stocks are included. A stock increase takes money out of the money chain and is therefore represented as a negative. **Distribution costs** are arrived at by removing an estimate of producers value along with any taxes from the end user values shown. For most fuel the estimate of producer value is derived from the consumption used for end use and the producer price taken from survey of producers e.g. for gas the landed price from Table 9.10. For electricity the Pool Purchase Price is used to value public distribution supply. No sector breakdown is given for gas and electricity margins because it is not possible to accurately measure delivery costs for each sector. **Taxes** include VAT where not refundable and duties paid on downstream sales. Excluded are the gas and fossil fuel levies, petroleum revenue tax and production royalties and licence fees. The proceeds from the fossil fuel levy are redistributed across the electricity industry, whilst the rest are treated as part of the production costs.

II Energy balances (Tables 1.1, 1.2 and 1.3)

1.62 Tables 1.1, 1.2 and 1.3 show the energy flows as the primary fuels are processed (or used) and as the consequent secondary fuels are used. The net inputs to transformation are shown in the transformation rows and hence outputs from transformation processes into which primary fuels are input (such as electricity generation or petroleum refining) appear as positive figures in the transformation rows under the secondary product's heading in the tables. Similarly the net inputs are shown as negative figures under the primary fuel headings.

1.63 Readers should note that following the consultation on the 1998 edition of this publication, and the responses received from that, that the energy balance presentation changed from the 1999 edition, as has the presentation of individual fuel figures throughout this publication. Annex A explains in detail the principles behind the presentation.

III Measurement of energy consumption

Primary fuel input basis
1.64 Energy consumption is usually measured in one of three different ways. The first, known as the primary fuel input basis, assesses the total input of primary fuels and their equivalents. This measure includes energy used or lost in the conversion of primary fuels to secondary fuels (for example in power stations and oil refineries), energy lost in the distribution of fuels (for example in transmission lines) and energy conversion losses by final users. Primary demands as in Table 1.1, 1.2 and 1.3 is on this basis.

Final consumption - energy supplied basis
1.65 The second method, known as the energy supplied basis, measures the energy content of the fuels, both primary and secondary, supplied to final users. Thus it is net of fuel industry own use and conversion, transmission and distribution losses, but it

includes conversion losses by final users. The final consumption figures are presented on this basis throughout Chapter 1.

1.66 Although this is the usual and most direct way to measure final energy consumption, it is also possible to present final consumption on a primary fuel input basis. This can be done by allocating the conversion losses, distribution losses and energy industry use to final users. This approach can be used to compare the total primary fuel use for which each sector of the economy is responsible. Table 1C presents shares of final consumption on this basis.

Final consumption - useful energy basis

1.67 Thirdly, final consumption may be expressed in the form of useful energy available after deduction of the losses incurred when final users convert energy supplied into space or process heat, motive power or light. Such losses depend on the type and quality of fuel and the equipment used and on the purpose, conditions, duration and intensity of use. Statistics on useful energy are not sufficiently reliable to be given in this Digest; there is a lack of data on utilisation efficiencies and on the purposes for which fuels are used.

Shares of each fuel in energy supply and demand

1.68 The relative importance of the energy consumption of each sector of the economy depends on the method used to measure consumption. Shares of final consumption on an energy supplied basis (that is in terms of the primary and secondary fuels directly consumed) in 1999 are presented in Table 1B. For comparison, Table 1C presents shares of final consumption on a primary fuel input basis.

Table 1B: Primary and secondary fuels consumed by final users in 1999 - energy supplied basis

	Industry	Transport	Domestic	Others	Total
			percentage of each fuel		
Solid fuels	56	-	38	6	100
Petroleum	10	82	5	4	100
Gas	28	-	54	18	100
Secondary electricity	34	3	35	29	100
All fuels	23	34	30	14	100

	Solid fuels	Petroleum	Gas	Secondary electricity	Total
			percentage of each sector		
Industry	12	17	45	26	100
Transport	-	99	-	1	100
Domestic	6	7	66	21	100
Others	2	12	49	37	100
All users	5	42	36	17	100

Table 1C: Total primary fuel consumption by final users in 1999 - primary input basis

	Industry	Transport	Domestic	Others	Total
			percentage of each fuel		
Coal	45	2	32	21	100
Petroleum	10	80	6	5	100
Gas	30	1	48	22	100
Primary electricity	33	3	35	29	100
All fuels	26	27	30	17	100

	Coal	Petroleum	Gas	Primary electricity	Total
			percentage of each sector		
Industry	29	12	45	14	100
Transport	1	97	1	1	100
Domestic	18	6	64	12	100
Others	21	9	52	18	100
All users	17	32	40	11	100

1.69 In 1999, every 1 toe of secondary electricity consumed by final users required, on average, 1.1 toe of coal, 0.8 toe of natural gas, 0.9 toe of primary electricity (nuclear, natural flow hydro and imports) and 0.1 toe of oil and renewables combined. The extent of this primary consumption is hidden in Table 1B, which presents final consumption only in terms of the fuels directly consumed. When all such primary consumption is allocated to final users, as in Table 1C, the relative importance of fuels and sectors changes; the transport sector, which uses very little electricity, declines in importance, whilst the true cost of final consumption in terms of coal use can now be seen.

1.70 Another view comes from shares of users' expenditure on each fuel (Table 1D based on Table 1.4). In this case the importance of fuels which require most handling by the user (solids and liquid fuels) is slightly understated, and the importance of uses taxed at higher rates (transport) is overstated in the All users line.

Table 1D: Value of fuels purchased by final users in 1999

	Solid fuels	Petroleum	Gas	Secondary electricity	Total
			percentage of each sector		
Industry	8	13	17	62	100
Transport	-	99	-	1	100
Domestic	4	3	38	55	100
Others	-	6	15	79	100
All users	**2**	**60**	**12**	**27**	**100**

Systems of measurement - international statistics

1.71 The systems of energy measurement used in various international statistics differ from the methods of this Digest as follows.

21

Net calorific values

1.72 Calorific values (thermal contents) used internationally are net rather than gross. The difference between the net and gross thermal content is the amount of energy necessary to evaporate the water present in the fuel or formed during the combustion process. The differences between gross and net values are taken to be 5 per cent for liquid and solid fuels (except for coke and coke breeze where there is no difference), 10 per cent for gases (except for blast furnace gas, 1 per cent), 15 per cent for straw, and 16 per cent for poultry litter. The calorific value of wood is highly dependent on its moisture content. In Annex A the gross calorific value is given as 10 GJ per tonne at 50 per cent moisture content and this rises to 14.5 GJ at 25 per cent moisture content and 19 GJ for dry wood (equivalent to a net calorific value).

IV Definitions of fuels

1.73 The following paragraphs explain what is covered under the terms "primary" and "secondary" fuels.

Primary fuels

Coal - Production comprises all grades of coal, including slurry.

Primary oils - This includes crude oil, natural gas liquids (NGLs) and feedstock.

Natural gas liquids - Natural gas liquids (NGLs) consist of condensates (C_5 or heavier) and petroleum gases other than methane C_1, that is ethane C_2, propane C_3 and butane C_4, obtained from the onshore processing of associated and non-associated gas. These are treated as primary fuels when looking at primary supply but in the consumption data presented in this chapter these fuels are treated as secondary fuels, being transferred from the primary oils column in Tables 1.1, 1.2 and 1.3.

Natural gas - Production relates to associated or non-associated methane C_1 from land and the United Kingdom sector of the Continental Shelf. It includes that used for drilling production and pumping operations, but excludes gas flared or re-injected. It also includes colliery methane piped to the surface and consumed by collieries or others.

Nuclear electricity - Electricity generated by nuclear power stations belonging to the major power producers. See paragraphs 5.57 to 5.58.

Natural flow hydro-electricity - Electricity generated by public supply and industrial natural flow hydroelectric power stations. Pumped storage stations are not included (see under secondary electricity below).

Renewable energy sources - In this chapter figure are presented for renewables and waste in total.

Further details, including a detailed breakdown of the commodities covered are in Chapter 7.

Secondary fuels

Manufactured fuel - This heading includes manufactured solid fuels such as coke and breeze, other manufactured solid fuels, liquids such as benzole and tars and gases such as coke oven gas and blast furnace gas. Further details are given in Chapter 2, Tables 2.4, 2.5 and 2.6.

Coke and breeze - Coke oven coke and hard coke breeze (Tables 2.4, 2.5 and 2.6).

Other manufactured solid fuels - Manufactured solid fuels produced at low temperature carbonisation plants and other manufactured fuel and briquetting plants (Tables 2.4, 2.5 and 2.6).

Coke oven gas - Gas produced at coke ovens, excluding low temperature carbonisation plants. Gas bled or burnt to waste is included in production and losses (Tables 2.4, 2.5 and 2.6).

Blast furnace gas - Blast furnace gas is mainly produced and consumed within the iron and steel industry (Tables 2.4, 2.5 and 2.6).

Petroleum products - Petroleum products produced mainly at refineries, together with inland deliveries of natural gas liquids.

Secondary electricity - Secondary electricity is that generated by the combustion of another fuel, usually coal, natural gas or oil. The figure for outputs from transformation in the electricity column of Tables 1.1, 1.2 and 1.3 is the total of primary and secondary electricity, and the subsequent analysis of consumption is based on this total.

V Classification of consumers

1.74 This issue of the Digest has been prepared, as far as is practicable, on the basis of the *Standard Industrial Classification (SIC) 1992* (The Stationery Office 1991). However, not all consumption/disposals data are on this basis, and where they are, there are sometimes constraints on the detail available. Between 1986 and 1994 data in the Digest were prepared on the basis of the previous classification, SIC 1980. The exceptions are Tables, 2.8, and 3.11 in this Digest and the corresponding tables in previous editions which have been prepared largely on the basis of SIC 1968. The main differences between the 1968 SIC (which was used as the basis for most data published for years prior to 1984) and the 1980 SIC were described in the 1986 and 1987 issues of the Digest. The differences between SIC 1980 and SIC 1992 are relatively minor. At the time of the change from the 1980 SIC to the 1992 SIC the main difference was that under the former showrooms belonging to the fuel supply industries were classified to the energy sector, whilst in the latter they are in the commercial sector. Since privatisation few gas, coal and electricity companies have retained showrooms and the difference is therefore minimal.

1.75 Table 1E shows the categories of consumer together with their codes in SIC 1992. The coverage varies between tables (e.g. in some instances the 'other' category is split into major constituents, whereas elsewhere it may include transport). This is because the coverage is dictated by what data suppliers can provide. The table also shows the disaggregation available within industry. This disaggregation forms the basis of virtually all the tables that show a disaggregated industrial breakdown. There are a few notable exceptions that are detailed in paragraph 1.74.

Table 1E: SIC 1992 classifications

Fuel producers	10-12, 23, 40

Final consumers:

Industrial

Unclassified	See text below
Iron and steel	27, *excluding* 27.4, 27.53, 27.54
Non-ferrous metals	27.4, 27.53, 27.54
Mineral products	14, 26
Chemicals	24
Mechanical engineering and metal products	28, 29
Electrical and instrument engineering	30-33
Vehicles	34, 35
Food, beverages & tobacco	15, 16
Textiles, clothing, leather, & footwear	17-19
Paper, printing & publishing	21, 22
Other industries	13, 20, 25, 36, 37, 41
Construction	45
Transport	60-63

Other final users

Domestic	Not covered by SIC 1992.
Public administration	75, 80, 85
Commercial	50-52, 55, 64-67, 70-74
Agriculture	01, 02, 05
Miscellaneous	90-93, 99

1.76 There is also an 'unclassified' category in the industry sector (see Table 1E). Wherever the data supplier is unable to allocate an amount between categories, but the Department of Trade and Industry has additional information, not readily available to readers, with which to allocate between categories, then this has been done. Where such additional information is not available the data are included in the 'unclassified' category, enabling the reader to decide whether to accept a residual, pro-rate, or otherwise adjust the figures. The 'miscellaneous' category also contains some unallocated figures for the services sector.

1.77 In Tables 6.7 and 6.8 of Chapter 6 the following abbreviated grouping of industries, based on SIC 1992, is used in order to prevent disclosure of information about individual companies:

Table 1F: Abbreviated grouping of Industry

Iron and steel and non-ferrous metal	27
Chemicals	24
Oil refineries	23.2
Paper, printing & publishing	21, 22
Food, beverages & tobacco	15, 16
Metal products, machinery and equipment	28, 29, 30, 31, 32, 34, 35
Extraction, mining and agglomeration of solid fuels	10, 11
Other industrial branches	12, 13, 14, 17, 18, 19, 20, 23.1, 23.3, 25, 26, 33, 36, 37, 40.1, 40.2, 45
Transport, commerce, and administration	1, 2, 5, 50 to 99 (except 90, 92)
Other	40.3, 90, 92

1.78 In Tables 1.7 and 1.8 the list above is further condensed and includes only manufacturing industry and construction as follows:

Table 1G: Abbreviated grouping of Industry for Tables 1.7 and 1.8

Iron and steel and non-ferrous metals	27
Chemicals	24
Paper, printing & publishing	21, 22
Food, beverages & tobacco	15, 16
Metal products, machinery and equipment	28, 29, 30, 31, 32, 34, 35
Other (including construction)	12, 13, 14, 17, 18, 19, 20, 23.1, 23.3, 25, 26, 33, 36, 37, 45

VI Monthly and quarterly data

1.79 Monthly data on energy production and consumption (including on a seasonally adjusted and temperature corrected basis) split by fuel type are provided in Tables 1 and 2 of the DTI's monthly statistical bulletin *Energy Trends*. Quarterly data on final energy consumption broken down by fuel type and by broad consuming sector are provided in Table 3. Monthly data on average temperature and deviation from the long term mean temperature are provided in Table 24. See Annex F for more information about *Energy Trends*.

Contact: *Roger Barty*
   Caroline Barrington
   0207-215 5187

1.1 Aggregate energy balance 1999

	Coal	Manufactured fuel (1)	Primary oils	Petroleum products	Natural gas (2)	Renewable & waste (3)	Primary electricity	Electricity	To
Supply									
Indigenous production	23,543	-	150,308	-	98,930	2,375	23,217	-	298,3
Imports	14,404	306	49,026	13,408	1,106	-	-	1,247	79,4
Exports	-580	-194	-100,489	-23,495	-7,260	-	-	-23	-132,0
Marine bunkers	-	-	-	-2,469	-	-	-	-	-2,4
Stock change (4)	-577	250	-214	636	-524	-	-	-	-4
Primary supply	**36,791**	**361**	**98,630**	**-11,920**	**92,253**	**2,375**	**23,217**	**1,225**	**242,9**
Statistical difference (5)	+450	-23	+100	+1,227	+220	-	-	+135	+2,1
Primary demand	**36,340**	**385**	**98,530**	**-13,147**	**92,033**	**2,375**	**23,217**	**1,089**	**240,8**
Transfers	-	+106	-1,654	+1,697	-44	-	+537	537	+1
Transformation	**-32,145**	**2,928**	**-96,485**	**94,700**	**-26,947**	**-1,434**	**-22,680**	**30,750**	**-51,3**
Electricity generation	-25,090	-854	-	-1,363	-26,947	-1,434	-22,680	30,750	-47,6
Major power producers	-24,146	-	-	-420	-24,187	-193	-22,680	28,358	-43,2
Autogenerators	-944	-854	-	-943	-2,760	-1,241	-	2,393	-4,3
Petroleum refineries	-	-	-96,485	96,342	-	-	-	-	-1
Coke manufacture	-6,163	5,448	-	-	-	-	-	-	-7
Blast furnaces	-378	-2,125	-	-279	-	-	-	-	-2,7
Patent fuel manufacture	-514	460	-	-	-	-	-	-	-
Other	-	-	-	-	-	-	-	-	
Energy industry use	**7**	**1,124**	**391**	**6,178**	**6,586**	**-**	**-**	**2,400**	**16,6**
Electricity generation	-	-	-	-	-	-	-	1,501	1,5
Oil and gas extraction	-	-	391	-	5,568	-	-	35	5,9
Petroleum refineries	-	-	-	5,960	302	-	-	460	6,7
Coal extraction	7	-	-	-	34	-	-	117	1
Coke manufacture	-	548	-	-	1	-	-	-	5
Blast furnaces	-	553	-	160	55	-	-	82	8
Patent fuel manufacture	-	24	-	-	-	-	-	-	
Pumped storage	-	-	-	-	-	-	-	75	
Other	-	-	-	57	626	-	-	131	8
Losses	-	166	-	-	384	-	-	2,433	2,9
Final consumption	**4,188**	**2,128**	**-**	**77,072**	**58,072**	**941**	**-**	**27,544**	**169,9**
Industry	**1,919**	**1,643**	**-**	**6,109**	**15,916**	**536**	**-**	**9,330**	**35,4**
Unclassified	-	231	-	2,080	13	536	-	-	2,8
Iron and steel	7	1,310	-	89	1,886	-	-	851	4,1
Non-ferrous metals	146	102	-	38	477	-	-	507	1,2
Mineral products	568	-	-	230	1,236	-	-	621	2,6
Chemicals	555	-	-	386	4,401	-	-	1,707	7,0
Mechanical engineering etc	21	-	-	200	873	-	-	760	1,8
Electrical engineering etc	6	-	-	64	339	-	-	516	9
Vehicles	68	-	-	142	911	-	-	483	1,6
Food, beverages etc	180	-	-	355	2,494	-	-	1,032	4,0
Textiles, leather, etc	45	-	-	84	599	-	-	323	1,0
Paper, printing etc	114	-	-	90	1,321	-	-	938	2,4
Other industries	209	-	-	1,808	1,228	-	-	1,470	4,7
Construction	-	-	-	541	141	-	-	123	8
Transport	**-**	**-**	**-**	**53,038**	**-**	**-**	**-**	**727**	**53,7**
Air	-	-	-	10,708	-	-	-	-	10,7
Rail	-	-	-	506	-	-	-	-	5
Road	-	-	-	40,751	-	-	-	-	40,7
National navigation	-	-	-	1,073	-	-	-	-	1,0
Pipelines	-	-	-	-	-	-	-	-	
Other	**2,269**	**484**	**-**	**5,902**	**41,010**	**405**	**-**	**17,486**	**67,5**
Domestic	2,048	484	-	3,240	30,615	231	-	9,493	46,1
Public administration	206	-	-	1,224	4,416	95	-	1,951	7,8
Commercial	-	-	-	522	3,536	-	-	5,713	9,7
Agriculture	5	-	-	756	128	72	-	330	1,2
Miscellaneous	10	-	-	160	2,315	7	-	-	2,4
Non energy use	-	-	-	12,022	1,147	-	-	-	13,1

(1) Includes all manufactured solid fuels, benzole, tars, coke oven gas and blast furnace gas.
(2) Includes colliery methane.
(3) Includes geothermal and solar heat.
(4) Stock fall (+), stock rise (-).
(5) Primary supply minus primary demand.

1.2 Aggregate energy balance 1998

Thousand tonnes of oil equivalent

	Coal	Manufactured fuel (1)	Primary oils	Petroleum products	Natural gas (2)	Renewable & waste (3)	Primary electricity	Electricity	Total
Supply									
Indigenous production	25,858r	-	144,801r	-	90,178	2,146r	24,115	-	287,097r
Imports	14,806r	590	52,293	12,367r	910	-	-	1,086	82,051r
Exports	-707r	-225r	-92,415	-26,390r	-2,717	-	-	-14	-122,468r
Marine bunkers	-	-	-	-3,253	-	-	-	-	-3,253
Stock change (4)	+861	-98	-645	-96r	-32r	-	-	-	-10r
Primary supply	**40,818r**	**267r**	**104,033r**	**-17,372r**	**88,339r**	**2,146r**	**24,115**	**1,072**	**243,417r**
Statistical difference (5)	-306	-46r	-1,300r	+1,276r	+1,442r	-	-	+103r	+1,168r
Primary demand	**41,124**	**313r**	**105,334**	**-18,648r**	**86,897r**	**2,146r**	**24,115**	**969r**	**242,249r**
Transfers	-	-129	-2,732	2,708r	-52	-	-526	526	-206r
Transformation	**-37,343r**	**3,436r**	**-102,176**	**100,235r**	**-22,930r**	**-1,212r**	**-23,589**	**30,461r**	**-53,118r**
Electricity generation	-30,155r	-849r	-	-1,756r	-22,930r	-1,212r	-23,589	30,461r	-50,031r
Major power producers	-28,892r	-	-	-783	-20,377r	-147r	-23,589	28,302r	-45,486r
Autogenerators	-1,264r	-849r	-	-973r	-2,554r	-1,065r	-	2,159r	-4,545r
Petroleum refineries	-	-	-102,176	102,268	-	-	-	-	+93
Coke manufacture	-6,244	5,737r	-	-	-	-	-	-	-506r
Blast furnaces	-427	-1,904r	-	-277r	-	-	-	-	-2,608r
Patent fuel manufacture	-517	452	-	-	-	-	-	-	-66
Other	-	-	-	-	-	-	-	-	-
Energy industry use	**7**	**1,185r**	**426**	**6,713r**	**6,531r**	**-**	**-**	**2,412r**	**17,273r**
Electricity generation	-	-	-	-	-	-	-	1,521r	1,521r
Oil and gas extraction	-	-	426	-	5,629	-	-	46	6,101
Petroleum refineries	-	-	-	6,517r	324r	-	-	446r	7,288r
Coal extraction	7	-	-	-	26r	-	-	115	149r
Coke manufacture	-	583r	-	-	1	-	-	-	583r
Blast furnaces	-	573r	-	140r	45	-	-	79	838r
Patent fuel manufacture	-	29r	-	-	-	-	-	-	29r
Pumped storage	-	-	-	-	-	-	-	83	83
Other	-	-	-	55	505	-	-	121	681
Losses	-	156	-	-	689r	-	-	2,405	3,250r
Final consumption	**3,773r**	**2,280r**	**-**	**77,582r**	**56,695r**	**934r**	**-**	**27,139r**	**168,402r**
Industry	**1,690r**	**1,758r**	**-**	**6,153r**	**14,791r**	**530**	**-**	**9,075r**	**33,997r**
Unclassified	-	267r	-	1,930r	13r	530	-	-	2,739r
Iron and steel	7	1,396r	-	90r	1,807r	-	-	834r	4,133r
Non-ferrous metals	107r	95r	-	41r	432r	-	-	490r	1,166r
Mineral products	531r	-	-	241	1,218r	-	-	606r	2,596r
Chemicals	440r	-	-	487r	3,875r	-	-	1,676r	6,478r
Mechanical engineering etc	20r	-	-	213r	845r	-	-	735r	1,814
Electrical engineering etc	2	-	-	92	293r	-	-	516r	903r
Vehicles	61r	-	-	134r	841r	-	-	481r	1,517r
Food, beverages etc	182r	-	-	414r	2,365r	-	-	1,012r	3,973r
Textiles, leather, etc	50	-	-	100	582	-	-	315r	1,047r
Paper, printing etc	84r	-	-	121r	1,234r	-	-	916r	2,354r
Other industries	208r	-	-	1,740r	1,140r	-	-	1,362r	4,450r
Construction	-	-	-	550	146	-	-	132	828
Transport	-	-	-	**52,924r**	-	-	-	**720r**	**53,643r**
Air	-	-	-	10,238	-	-	-	-	10,238
Rail	-	-	-	523r	-	-	-	-	523r
Road	-	-	-	40,987r	-	-	-	-	40,987r
National navigation	-	-	-	1,176r	-	-	-	-	1,176r
Pipelines	-	-	-	-	-	-	-	-	-
Other	**2,083r**	**522r**	**-**	**6,675r**	**40,757r**	**404r**	**-**	**17,344r**	**67,785r**
Domestic	1,817r	522r	-	3,544r	30,600	230r	-	9,424	46,137r
Public administration	218r	-	-	1,483r	4,501r	96r	-	1,882r	8,180r
Commercial	-	-	-	601r	3,501r	-	-	5,705	9,807r
Agriculture	6	-	-	850	116	72	-	333	1,377
Miscellaneous	43r	-	-	196r	2,039	6r	-	-	2,284r
Non energy use	-	-	-	**11,831**	**1,147r**	-	-	-	**12,977r**

(1) Includes all manufactured solid fuels, benzole, tars, coke oven gas and blast furnace gas.
(2) Includes colliery methane.
(3) Includes geothermal and solar heat.
(4) Stock fall (+), stock rise (-).
(5) Primary supply minus primary demand.

1.3 Aggregate energy balance 1997

Thousand tonnes of oil equival[e]

	Coal	Manufactured fuel (1)	Primary oils	Petroleum products	Natural gas (2)	Renewable & waste (3)	Primary electricity	Electricity	Tot[al]
Supply									
Indigenous production	30,518r	-	140,580	-	85,883	1,913r	23,409r	-	282,30
Imports	13,779	617	54,617	9,325r	1,209	-	-	1,429	80,9
Exports	-836r	-225r	-86,887r	-29,050r	-1,863	-	-	-4	-118,8
Marine bunkers	-	-	-	-3,121	-	-	-	-	-3,1
Stock change (4)	-2,565r	-17	-226	+548	-354	-	-	-	-2,6
Primary supply	**40,895r**	**376r**	**108,084r**	**-22,297r**	**84,875**	**1,913r**	**23,409r**	**1,425**	**238,6**
Statistical difference (5)	+80r	+44r	-1,482r	-205r	+1,109r	-	-	-148r	-6
Primary demand	**40,815r**	**332r**	**109,566r**	**-22,092r**	**83,766r**	**1,913r**	**23,409r**	**1,573r**	**239,2**
Transfers	-	-60	-3,191r	3,174r	-43	-	-416r	416r	-1
Transformation	**-36,418r**	**3,385r**	**-105,953**	**103,176r**	**-21,516r**	**-982**	**-22,993**	**29,416r**	**-51,88**
Electricity generation	-29,031r	-849r	-	-2,299r	-21,516r	-982	-22,993	29,416r	-48,25
Major power producers	-27,713r	-	-	-1,377r	-19,233r	-138	-22,993	27,440r	-44,01
Autogenerators	-1,318r	-849r	-	-923r	-2,283r	-844r	-	1,976r	-4,24
Petroleum refineries	-	-	-105,953	105,945	-	-	-	-	
Coke manufacture	-6,224	5,756	-	-	-	-	-	-	-46
Blast furnaces	-464	-2,059	-	-470r	-	-	-	-	-2,99
Patent fuel manufacture	-700	537r	-	-	-	-	-	-	-16
Other	-	-	-	-	-	-	-	-	
Energy industry use	**5**	**1,148**	**423r**	**6,844r**	**5,818r**	**-**	**-**	**2,324r**	**16,5**
Electricity generation	-	-	-	-	-	-	-	1,419r	1,4
Oil and gas extraction	-	-	423r	-	5,011	-	-	58	5,4
Petroleum refineries	-	-	-	6,661r	268r	-	-	454r	7,3
Coal extraction	5	-	-	-	37r	-	-	131	1
Coke manufacture	-	520	-	-	1	-	-	-	5
Blast furnaces	-	594	-	128r	29	-	-	78	82
Patent fuel manufacture	-	34	-	-	-	-	-	-	
Pumped storage	-	-	-	-	-	-	-	85	
Other	-	-	-	54	471	-	-	99	62
Losses	-	**155**	-	-	**1,166**	-	-	**2,465**	**3,78**
Final consumption	**4,392r**	**2,354r**	**-**	**77,415r**	**55,223r**	**931r**	**-**	**26,616r**	**166,93**
Industry	**1,939r**	**1,877r**	**-**	**6,313r**	**14,685r**	**532**	**-**	**9,020r**	**34,36**
Unclassified	-	222r	-	1,914	13r	532	-	-	2,68
Iron and steel	1	1,526r	-	113r	1,769	-	-	830r	4,23
Non-ferrous metals	89r	129	-	53r	397r	-	-	452r	1,12
Mineral products	597r	-	-	259r	1,257r	-	-	610r	2,72
Chemicals	424r	-	-	558r	3,971r	-	-	1,667r	6,61
Mechanical engineering etc	16r	-	-	248	827r	-	-	716r	1,80
Electrical engineering etc	-	-	-	111r	251r	-	-	526	88
Vehicles	72r	-	-	156r	794r	-	-	485r	1,50
Food, beverages etc	252r	-	-	480r	2,295r	-	-	995r	4,02
Textiles, leather, etc	49r	-	-	101	580	-	-	321	1,05
Paper, printing etc	152r	-	-	126r	1,189r	-	-	927r	2,39
Other industries	286r	-	-	1,626r	1,194r	-	-	1,359r	4,46
Construction	-	-	-	568	148	-	-	133	84
Transport	-	-	-	**52,315r**	-	-	-	**723r**	**53,03**
Air	-	-	-	9,323	-	-	-	-	9,32
Rail	-	-	-	515r	-	-	-	-	51
Road	-	-	-	41,221	-	-	-	-	41,22
National navigation	-	-	-	1,255r	-	-	-	-	1,25
Pipelines	-	-	-	-	-	-	-	-	
Other	**2,453r**	**478r**	**-**	**6,748r**	**39,482r**	**399r**	**-**	**16,872r**	**66,43**
Domestic	1,991	478r	-	3,389r	29,709	225	-	8,981	44,77
Public administration	319	-	-	1,707r	4,574r	100r	-	1,865r	8,56
Commercial	-	-	-	633r	3,127r	-	-	5,699	9,46
Agriculture	5	-	-	816r	124	72	-	327	1,34
Miscellaneous	137	-	-	203r	1,947	2r	-	-	2,28
Non energy use	-	-	-	**12,039**	**1,056r**	-	-	-	**13,09**

(1) Includes all manufactured solid fuels, benzole, tars, coke oven gas and blast furnace gas.
(2) Includes colliery methane.
(3) Includes geothermal and solar heat.
(4) Stock fall (+), stock rise (-).
(5) Primary supply minus primary demand.

1.4 Value balance of traded energy in 1999[1]

£ million

	Coal	Crude oil	Natural gas	Electricity	Manufactured solid fuels	Petroleum products	Other fuels	Total
Supply								
Indigenous production	1,140	10,910	4,830	7,620	210	7,835	90	32,630
Imports	580	3,280	25	395	20	1,960	-	6,265
Exports	-40	-7,155	-225	-	-20	-2,855	-	-10,295
Marine bunkers	-	-	-	-	-	-190	-	-190
Stock change	-35	-30	-5	-	-	60	-	-
Basic value of inland consumption	1,645	7,005	4,630	8,015	210	6,810	90	28,405
Tax and margins								
Distribution costs and margins	**325**	-	**4,250**	**7,935**	**30**	**1,730**	-	**14,270**
Electricity generation	30	-	-	-	-	5	-	30
Solid fuel manufacture	10	-	-	-	-	-	-	10
of which iron & steel sector	5	-	-	-	-	-	-	5
Iron & steel final use	-	-	-	-	5	-	-	5
Other industry	25	-	-	-	15	30	-	70
Air transport	-	-	-	-	-	35	-	35
Rail and national navigation	-	-	-	-	-	-	-	-
Road transport	-	-	-	-	-	1,285	-	1,285
Domestic	260	-	-	-	15	40	-	315
Agriculture	-	-	-	-	-	10	-	10
Commercial and other services	5	-	-	-	-	30	-	35
Non energy use	-	-	75	-	-	290	-	365
VAT and duties	**20**	-	**245**	**355**	**5**	**29,680**	-	**30,310**
Electricity generation	-	-	-	-	-	30	-	30
Iron & steel final use	-	-	-	-	-	15	-	15
Other industry	-	-	-	-	-	120	-	120
Air transport	-	-	-	-	-	15	-	15
Rail and national navigation	-	-	-	-	-	50	-	50
Road transport	-	-	-	-	-	29,345	-	29,345
Domestic	20	-	245	355	5	25	-	655
Agriculture	-	-	-	-	-	25	-	25
Commercial and other services	-	-	-	-	-	60	-	60
Total tax and margins	**345**	-	**4,495**	**8,290**	**35**	**31,405**	-	**44,575**
Market value of inland consumption	**1,995**	**7,005**	**9,125**	**16,305**	**250**	**38,220**	**90**	**72,985**
Energy end use								
Total energy sector	**1,430**	**7,005**	**1,990**	**140**	-	**135**	**15**	**10,715**
Transformation	1,430	7,005	1,920	-	-	130	15	10,500
Electricity generation	1,155	-	1,920	-	-	130	15	3,215
of which from stocks	30	-	-	-	-	-	-	30
Petroleum refineries	-	7,005	-	-	-	-	-	7,005
Solid fuel manufacture	275	-	-	-	-	-	-	275
of which iron & steel sector	225	-	-	-	-	-	-	225
Other energy sector use	-	-	65	140	-	5	-	215
Oil & gas extraction	-	-	-	15	-	-	-	15
Petroleum refineries	-	-	20	80	-	-	-	100
Coal extraction	-	-	-	45	-	-	-	50
Other energy sector	-	-	45	-	-	5	-	50
Total non energy sector use	**565**	-	**7,070**	**16,165**	**250**	**36,550**	**75**	**60,675**
Industry	**115**	-	**1,015**	**3,705**	**115**	**755**	**45**	**5,750**
Iron & steel final use	15	-	115	230	75	50	-	490
Other industry	100	-	895	3,475	40	705	45	5,260
Transport	-	-	-	305	-	35,015	-	35,320
Air	-	-	-	-	-	1,175	-	1,175
Rail and national navigation	-	-	-	305	-	195	-	500
Road	-	-	-	-	-	33,650	-	33,650
Other final users	**450**	-	**6,060**	**12,155**	**135**	**780**	**30**	**19,605**
Domestic	435	-	5,175	7,450	135	425	30	13,650
Agriculture	-	-	15	220	-	100	-	335
Commercial and other services	15	-	870	4,480	-	250	-	5,620
Total value of energy end use	**1,995**	**7,005**	**9,060**	**16,305**	**250**	**36,685**	**90**	**71,385**
Value of non energy end use	-	-	**65**	-	-	**1,530**	-	**1,595**
Market value of inland consumption	**1,995**	**7,005**	**9,125**	**16,305**	**250**	**38,220**	**90**	**72,985**

(1) For further information see paragraphs 1.56 to 1.61.

1.5 Value balance of traded energy in 1998[1]

£ million

	Coal	Crude oil	Natural gas	Electricity	Manufactured solid fuels	Petroleum products	Other fuels	Total
Supply								
Indigenous production	1,335r	8,085	5,245r	7,640r	155r	8,890r	80	31,425r
Imports	640	2,275	45	375	45	1,410	-	4,785
Exports	-45	-5,090r	-80	-	-25	-2,300	-	-7,535
Marine bunkers	-	-	-	-	-	-230	-	-230
Stock change	20	-35	-	-	5	-5	-	-20
Basic value of inland consumption	1,950r	5,230r	5,205	8,015r	185r	7,765r	80	28,430r
Tax and margins								
Distribution costs and margins	325r	-	4,255r	8,310r	80	1,880	-	14,845r
Electricity generation	35	-	-	-	-	-	-	40r
Solid fuel manufacture	35	-	-	-	-	-	-	35
of which iron & steel sector	35	-	-	-	-	-	-	35
Iron & steel final use	-	-	-	-	10	-	-	10
Other industry	15r	-	-	-	20	60	-	95
Air transport	-	-	-	-	-	85	-	85
Rail and national navigation	-	-	-	-	-	15	-	15
Road transport	-	-	-	-	-	1,355	-	1,355
Domestic	225r	-	-	-	50	120	-	395r
Agriculture	-	-	-	-	-	25	-	30
Commercial and other services	10	-	-	-	-	70	-	80
Non energy use	-	-	75r	-	-	150	-	220r
VAT and duties	20r	-	285	365r	5	24,465r	-	25,145
Electricity generation	-	-	-	-	-	30r	-	30r
Iron & steel final use	-	-	-	-	-	10	-	10
Other industry	-	-	-	-	-	115	-	115
Air transport	-	-	-	-	-	10	-	10
Rail and national navigation	-	-	-	-	-	50	-	50
Road transport	-	-	-	-	-	24,130	-	24,130
Domestic	20r	-	285	365r	5	30	-	710
Agriculture	-	-	-	-	-	25	-	25
Commercial and other services	-	-	-	-	-	65	-	65
Total tax and margins	345r	-	4,540r	8,675r	85	26,345r	-	39,990r
Market value of inland consumption	2,295r	5,230	9,745r	16,690r	270r	34,105r	85	68,420r
Energy end use								
Total energy sector	1,780r	5,230r	1,815r	180	-	150r	10	9,160r
Transformation	1,780r	5,230r	1,750r	-	-	140r	10	8,910r
Electricity generation	1,440r	-	1,750r	-	-	140r	10	3,340r
of which from stocks	35	-	-	-	-	-	-	35
Petroleum refineries	-	5,230r	-	-	-	-	-	5,230r
Solid fuel manufacture	340	-	-	-	-	-	-	340
of which iron & steel sector	305	-	-	-	-	-	-	305
Other energy sector use	-	-	65r	180	-	5	-	250r
Oil & gas extraction	-	-	-	20	-	-	-	20
Petroleum refineries	-	-	25r	115	-	-	-	135
Coal extraction	-	-	-	45	-	-	-	50
Other energy sector	-	-	40	-	-	5	-	45
Total non energy sector use	515r	-	7,870r	16,515r	270r	32,540r	75	57,780r
Industry	110r	-	965r	3,535	130	700r	45	5,490r
Iron & steel final use	20	-	115r	225r	80	45	-	485r
Other industry	90r	-	850r	3,315r	50	655r	45	5,005r
Transport	-	-	-	300	-	30,965	-	31,265
Air	-	-	-	-	-	965	-	965
Rail and national navigation	-	-	-	300	-	195	-	495
Road	-	-	-	-	-	29,810	-	29,810
Other final users	405r	-	6,905r	12,675r	140r	875r	30	21,025r
Domestic	385r	-	6,015	7,700r	140r	465	30	14,730r
Agriculture	-	-	10	220r	-	115	-	345r
Commercial and other services	20	-	875	4,755r	-	295r	-	5,950r
Total value of energy end use	2,295r	5,230r	9,680r	16,690r	270r	32,690r	85	66,940r
Value of non energy end use	-	-	65	-	-	1,420	-	1,485
Market value of inland consumption	2,295r	5,230r	9,745r	16,690r	270r	34,105r	85	68,420r

(1) For further information see paragraphs 1.56 to 1.61.

1.6 Value balance of traded energy in 1997[1]

£ million

	Coal	Crude oil	Natural gas	Electricity	Manufactured solid fuels	Petroleum products	Other fuels	Total
Supply								
Indigenous production	1,665r	11,685	5,280r	7,745	140r	11,465r	80	38,070
Imports	660	3,655	105	405	55	1,655	-	6,530
Exports	-55	-7,045	-80	-	-25r	-3,140	-	-10,345r
Marine bunkers	-	-	-	-	-	-280	-	-280
Stock change	-125r	-30	-	-	-	60	-	-95
Basic value of inland consumption	2,140r	8,270	5,300r	8,150	175r	9,760r	80	33,880r
Tax and margins								
Distribution costs and margins	**390**	-	**3,890r**	**8,520r**	**85r**	**1,930r**	-	**14,815r**
Electricity generation	60	-	-	-	-	-	-	60
Solid fuel manufacture	40	-	-	-	-	-	-	40
of which iron & steel sector	35	-	-	-	-	-	-	35
Iron & steel final use	5	-	-	-	20	-	-	20
Other industry	10r	-	-	-	20	55	-	90r
Air transport	-	-	-	-	-	90	-	90
Rail and national navigation	-	-	-	-	-	20	-	20
Road transport	-	-	-	-	-	1,470	-	1,470
Domestic	260	-	-	-	45r	105	-	410r
Agriculture	-	-	-	-	-	20	-	20
Commercial and other services	15	-	-	-	-	60	-	80
Non energy use	-	-	65r	-	-	110	-	175r
VAT and duties	**30**	-	**400**	**520**	**10**	**22,210r**	-	**23,170r**
Electricity generation	-	-	-	-	-	40r	-	40r
Iron & steel final use	-	-	-	-	-	15	-	15
Other industry	-	-	-	-	-	105r	-	105r
Air transport	-	-	-	-	-	10	-	10
Rail and national navigation	-	-	-	-	-	45	-	45
Road transport	-	-	-	-	-	21,860	-	21,860
Domestic	30	-	400-	520	10	45	-	1,005
Agriculture	-	-	-	-	-	20	-	20
Commercial and other services	-	-	-	-	-	65	-	65
Total tax and margins	**420**	-	**4,290r**	**9,040r**	**95r**	**24,140**	-	**37,985r**
Market value of inland consumption	**2,560r**	**8,270**	**9,590r**	**17,190r**	**270r**	**33,905r**	**85**	**71,865r**
Energy end use					-			
Total energy sector	**1,965r**	**8,270**	**1,675r**	**180**	-	**220r**	**10**	**12,320r**
Transformation	1,965r	8,270	1,620r	-	-	215r	10	12,075r
Electricity generation	1,595r	-	1,620r	-	-	215r	10	3,435r
of which from stocks	-115	-	-	-	-	-	-	-115
Petroleum refineries	-	8,270	-	-	-	-	-	8,270
Solid fuel manufacture	370	-	-	-	-	-	-	370
of which iron & steel sector	300	-	-	-	-	-	-	300
Other energy sector use	-	-	**55r**	**180**	-	**10**	-	**245**
Oil & gas extraction	-	-	-	25	-	-	-	25
Petroleum refineries	-	-	20	100	-	-	-	120
Coal extraction	-	-	-	55	-	-	-	55
Other energy sector	-	-	35	-	-	10	-	45
Total non energy sector use	**595**	-	**7,850r**	**17,010r**	**270r**	**32,095r**	**75**	**57,895r**
Industry	**125r**	-	**870r**	**3,625**	**140**	**890r**	**45**	**5,700r**
Iron & steel final use	25	-	100	220r	90	70r	-	510r
Other industry	100r	-	770r	3,405r	50	820r	45	5,190r
Transport	-	-	-	**295**	-	**30,145**	-	**30,440**
Air	-	-	-	-	-	1,210	-	1,210
Rail and national navigation	-	-	-	295	-	250	-	545
Road	-	-	-	-	-	28,685	-	28,685
Other final users	**470**	-	**6,980r**	**13,090r**	**130r**	**1,060r**	**30**	**21,755r**
Domestic	430	-	6,125r	7,965r	130r	560	30	15,240r
Agriculture	-	-	15	245	-	125	-	385
Commercial and other services	35	-	840r	4,880	-	375r	-	6,130r
Total value of energy end use	**2,560r**	**8,270**	**9,525r**	**17,190r**	**270r**	**32,315r**	**85**	**70,210r**
Value of non energy end use	-	-	**65**	-	-	**1,590**	-	**1,655**
Market value of inland consumption	**2,560r**	**8,270**	**9,590r**	**17,190r**	**270r**	**33,905r**	**85**	**71,865r**

(1) For further information see paragraphs 1.56 to 1.61.

1.7 Final energy consumption by main industrial groups[1]

	1995	1996	1997	1998	1999
				Thousand tonnes of oil equivalent	
Iron and steel and non-ferrous metals					
Coal	143	82r	90r	114r	153
Manufactured solid fuels *(2)*	928	936	833	764r	778
Blast furnace gas	672	702	371r	340r	260
Coke oven gas	563	536	451r	388r	375
Natural gas	2,062r	2,236r	2,166r	2,239r	2,363
Petroleum	162r	141r	166r	131r	127
Electricity	1,328r	1,358r	1,282r	1,324r	1,358
Total iron and steel and non-ferrous metals	**5,858r**	**5,991r**	**5,359r**	**5,300r**	**5,414**
Chemicals					
Coal	546r	411r	424r	440r	555
Natural gas	2871r	3,135r	3,971r	3,875r	4,401
Petroleum	827r	704r	558r	487r	386
Electricity	1,642r	1,662r	1,667r	1,676r	1,707
Total chemicals	**5,886r**	**5,912r**	**6,620r**	**6,478r**	**7,049**
Metal products, machinery and equipment					
Coal	168	104	88r	83r	95
Natural gas	1,664r	1,920r	1,872r	1,979r	2,123
Petroleum	619r	616r	515	439r	406
Electricity	1,688r	1,713r	1,727r	1,732r	1,759
Total metal products, machinery and equipment	**4,139r**	**4,353r**	**4,202r**	**4,233r**	**4,383**
Food, beverages and tobacco					
Coal	314r	259r	252r	182r	180
Natural gas	2,032r	2,333r	2,295r	2,365r	2,494
Petroleum	645r	610r	480r	414r	355
Electricity	890r	970r	995r	1,012r	1,032
Total food, beverages and tobacco	**3,881r**	**4,172r**	**4,022r**	**3,973r**	**4,061**

(1) Industrial categories used are described in Table 1G.
(2) Includes tars, benzole, coke and breeze and other manufactured solid fuel.

1.7 Final energy consumption by main industrial groups[1] (continued)

					Thousand tonnes of oil equivalent
	1995	1996	1997	1998	1999
Paper, printing and publishing					
Coal	320r	177r	152r	84r	113
Natural gas	1,224r	1,289r	1,189r	1,234r	1,321
Petroleum	225r	179r	126r	121r	90
Electricity	819r	818r	927r	916r	938
Total paper, printing and publishing	**2,588r**	**2,463r**	**2,394r**	**2,355r**	2,463
Other industries					
Coal	1,190r	997r	932r	789	823
Natural gas	2,809r	3,069r	3,179r	3,086r	3,203
Petroleum	2,833r	2,733r	2,555	2,631r	2,663
Electricity	2,287r	2,345r	2,423r	2,415r	2,537
Total other industries	**9,619r**	**9,144r**	**9,089r**	**8,949r**	9,226
Unclassified					
Manufactured solid fuels *(2)*	125	153	202	256	220
Coke oven gas	14	18	19	10	11
Natural gas	17	17	13r	13r	13
Petroleum	1,755r	1,993	1,914	1,930r	2,080
Renewables & waste	531	533	532	530	536
Total unclassified	**2,291r**	**2,713**	**2,680**	**2,739**	2,860
Total					
Coal	2,840r	2,193r	1,939r	1,690r	1,919
Manufactured solid fuels *(2)*	1,053	1,088	1,035r	1,020r	997
Blast furnace gas	672	702	371r	340r	260
Coke oven gas	576	554	471r	398r	386
Natural gas	12,696	13,999r	14,685r	14,791r	15,916
Petroleum	7,066r	6,976r	6,313r	6,153r	6,109
Renewables & waste	531	533	532	530	536
Electricity	8,654r	8,866r	9,020r	9,075r	9,330
Total	**34,909r**	**35,817r**	**34,367r**	**33,997r**	35,453

1.8 Fuels consumed for electricity generation (autogeneration) by main industrial groups[1]

Thousand tonnes of oil equivalent
(except where shown otherwise)

	1995	1996	1997	1998	1999
Iron and steel and non-ferrous metals					
Coal	789	789	788	852	712
Blast furnace gas	695	694	693	693	687
Coke oven gas	138	140	136	111	147
Natural gas	25	31	32	32	38
Petroleum	64	62	59	58	59
Other (including renewables) (2)	75	62	64	87	91
Total fuel input (3)	**1,785**	**1,778**	**1,772**	**1,833**	**1,734**
Electricity generated by iron and steel and non-ferrous metals (4)	**542** 6,298 GWh	**532** 6,187 GWh	**553** 6,428 GWh	**547** 6,359GWh	**556** 6,466 GWh
Electricity consumed by iron and steel and non-ferrous metals from own generation (5)	**378** 4,392 GWh	**366** 4,257 GWh	**382** 4,444 GWh	**397** 4,625 GWh	**411** 4,777 GWh
Chemicals					
Coal	355	311	279	195	120
Natural gas	763	813	926	971	843
Petroleum	162	146	131	185	144
Other (including renewables) (2)	62	78	78	73	85
Total fuel input (3)	**1,342**	**1,348**	**1,414**	**1,424**	**1,192**
Electricity generated by chemicals (4)	**447** 5,200 GWh	**456** 5,303 GWh	**476** 5,538 GWh	**519** 6,033 GWh	**528** 6,143 GWh
Electricity consumed by chemicals from own generation (5)	**371** 4,315 GWh	**379** 4,408 GWh	**397** 4,613 GWh	**428** 4,977 GWh	**428** 4,979 GWh
Metal products, machinery and equipment					
Coal	7	4	2	1	-
Natural gas	25	24	29	28	31
Petroleum	-	-	-	7	7
Other (including renewables) (2)	-	-	-	-	-
Total fuel input (3)	**32**	**28**	**31**	**36**	**38**
Electricity generated by metal products, machinery and equipment (4)	**15** 176 GWh	**13** 145 GWh	**14** 166 GWh	**17** 192 GWh	**17** 200 GWh
Electricity consumed by metal products, machinery and equipment from own generation (5)	**14** 168 GWh	**12** 139 GWh	**14** 159 GWh	**16** 183 GWh	**16** 191 GWh
Food, beverages and tobacco					
Coal	30	30	33	30	27
Natural gas	125	126	150	216	359
Petroleum	29	26	18	14	13
Other (including renewables) (2)	-	-	-	-	-
Total fuel input (3)	**184**	**182**	**201**	**260**	**399**
Electricity generated by food, beverages and tobacco (4)	**76** 886 GWh	**77** 896 GWh	**85** 991 GWh	**112** 1,304 GWh	**189** 2,194 GWh
Electricity consumed by food, beverages and tobacco from own generation (5)	**64** 741 GWh	**64** 749 GWh	**72** 838 GWh	**96** 1,113 GWh	**171** 1,984 GWh

(1) Industrial categories used are described in Table 1G.
(2) Includes hydro electricity, solid and gaseous renewables and waste.
(3) Total fuels used for generation of electricity. Consistent with figures for fuels used by other generators in Table 5.4.

1.8 Fuels consumed for electricity generation (autogeneration) by main industrial groups[1] (continued)

Thousand tonnes of oil equivalent
(except where shown otherwise)

	1995	1996	1997	1998	1999
Paper, printing and publishing					
Coal	76	76	56	54	40
Natural gas	274	294	417	417	504
Petroleum	11	11	12	11	14
Other (including renewables) (2)	-	-	-	9	-
Total fuel input (3)	**361**	**381**	**485**	**491**	**558**
Electricity generated by paper, printing and publishing (4)	**156** 1,812 GWh	**168** 1,953 GWh	**222** 2,580 GWh	**225** 2,620 GWh	**264** 3,068 GWh
Electricity consumed by paper, printing and publishing from own generation (5)	**144** 1,672 GWh	**156** 1,808 GWh	**207** 2,407 GWh	**208** 2,421 GWh	**236** 2,744 GWh
Other industries					
Coal	-	-	-	-	-
Coke oven gas	5	20	20	45	20
Natural gas	20	35	30	65	73
Petroleum	9	6	8	11	14
Other (including renewables) (2)	319	367	435	516	693
Total fuel input (3)	**353**	**428**	**493**	**637**	**800**
Electricity generated by other industries (4)	**74** 865 GWh	**88** 1,023 GWh	**86** 1,001 GWh	**102** 1,190 GWh	**108** 1,256 GWh
Electricity consumed by other industries from own generation (5)	**20** 233 GWh	**27** 311 GWh	**20** 238 GWh	**51** 591 GWh	**41** 477 GWh
Total					
Coal	1,257	1,210	1,158	1,132	899
Blast furnace gas	695	694	693	693	687
Coke oven gas	143	160	156	156	167
Natural gas	1,232	1,323	1,584	1,729	1,848
Petroleum	275	251	228	286	251
Other (including renewables) (2)	455	507	577	685	869
Total fuel input (3)	**4,057**	**4,145**	**4,396**	**4,681**	**4,721**
Electricity generated (4)	**1,310** 15,237 GWh	**1,333** 15,507 GWh	**1,436** 16,704 GWh	**1,522** 17,698 GWh	**1,662** 19,327 GWh
Electricity consumed from own generation (5)	**991** 11,521 GWh	**1,004** 11,672 GWh	**1,092** 12,699 GWh	**1,196** 13,912 GWh	**1,303** 15,150 GWh

(4) Combined heat and power (CHP) generation (i.e. electrical output from Table 6.8) plus non-chp generation, so that the total electricity generated is consistent with the "other generators" figures in Table 5.5.

(5) This is the electricity consumed by the industrial sector from its own generation and is consistent with the other generators final users figures used within the electricity balances (Tables 5.1 and 5.2). These figures are less than the total generated because some of the electricity is sold to the public distribution system and other users.

(6) The figures presented here are consistent with other figures presented elsewhere in this publication as detailed at (3), (4), and (5) above but are further dissaggregated. Overall totals covering all autogenerators can be derived by adding in figures for transport, services and the fuel industries. This can be summarised as follows:

Fuel input	1995	1996	1997	1998	1999
			Thousand tonnes of oil equivalent		
All industry	4,057	4,145	4,396	4,681	4,721
Fuel industries	1,027	1,087	1,239	1,273	1,279
Transport	352	373	385	396	402
Services	396	412	498	683	684
Total fuel input	**5,832**	**6,017**	**6,518**	**7,033**	**7,086**
Electricity generated	**1,884**	**1,952**	**2,105**	**2,310**	**2,550**
Electricity consumed	**1,403**	**1,446**	**1,543**	**1,709**	**1,886**
					GWh
Electricity generated	**21,914**	**22,704**	**24,478**	**26,870**	**29,656**
Electricity consumed	**16,322**	**16,823**	**17,945**	**19,872**	**21,935**

Inland consumption of primary fuels and equivalents for energy use, 1960 to 1999

		1960	1965	1966	1967	1968	1969	1970	1971	1972
In original units of measurement	Unit									
Coal *(1)*	M.tonnes	198.6	187.5	176.8	165.8	167.3	164.1	156.9	139.3	122.4
Petroleum *(2)*	"	40.1	62.4	67.5	72.1	76.1	82.1	88.2	89.0	95.4
Natural gas *(3)*	GWh	820	9,554	9,320	15,679	35,403	68,989	131,472	212,037	300,808
Nuclear electricity *(4)*	"	2,374	16,151	21,573	24,839	27,948	29,425	26,039	27,418	29,275
Hydro electricity *(4)(5)*	"	3,130	4,624	4,513	4,888	3,564	3,256	4,539	3,397	3,429
Million tonnes of oil equivalent										
Coal *(1)*		124.2	121.0	115.2	107.6	107.7	105.0	99.0	87.7	76.8
Petroleum *(2)*		42.9	65.6	71.2	76.5	81.9	88.4	94.8	95.7	102.7
Natural gas *(3)*		0.1	0.8	0.8	1.3	3.0	5.9	11.3	18.2	25.9
Nuclear electricity		0.6	4.4	5.8	6.7	7.5	7.9	7.0	7.4	7.9
Hydro electricity *(5)*		0.3	0.4	0.4	0.4	0.3	0.3	0.4	0.3	0.3
Total		168.1	192.2	193.4	192.5	200.5	207.5	212.4	209.3	213.6
Percentage shares (energy supplied basis)										
Coal		73.9	63.0	59.6	55.9	53.7	50.6	46.6	41.9	36.0
Petroleum		25.5	34.1	36.8	39.7	40.9	42.6	44.6	45.7	48.1
Natural gas		-	0.4	0.4	0.7	1.5	2.9	5.3	8.7	12.1
Nuclear electricity		0.4	2.3	3.0	3.5	3.7	3.8	3.3	3.5	3.7
Hydro electricity		0.2	0.2	0.2	0.2	0.2	0.1	0.2	0.1	0.1

		1973	1974	1975	1976	1977	1978	1979	1980	1981
In original units of measurement	Unit									
Coal *(1)*	M.tonnes	133.0	117.9	120.0	122.0	122.7	119.9	129.6	120.8	118.2
Petroleum *(2)*	"	96.6	89.7	80.3	78.9	80.3	82.0	81.8	71.4	65.2
Natural gas *(3)*	GWh	325,455	389,286	407,750	432,661	459,858	477,002	521,197	521,051	528,114
Nuclear electricity *(4)*	"	27,757	33,377	30,215	35,570	39,575	37,065	38,062	36,870	37,897
Hydro electricity *(4)(5)*	"	3,874	4,095	3,789	4,552	3,919	4,038	4,289	3,934	4,383
Million tonnes of oil equivalent										
Coal *(1)*		83.2	73.3	73.7	75.0	75.3	73.3	78.8	73.3	72.9
Petroleum *(2)*		104.2	96.2	86.3	85.0	86.5	88.3	88.2	77.4	70.8
Natural gas *(3)*		28.0	33.5	35.0	37.2	39.5	41.0	44.9	44.8	45.4
Nuclear electricity		7.5	9.0	8.1	9.6	10.6	10.0	10.2	9.9	10.2
Hydro electricity *(5)*		0.3	0.4	0.3	0.4	0.3	0.3	0.4	0.3	0.4
Total		223.2	212.3	203.5	207.2	212.3	213.0	222.5	205.7	199.6
Percentage shares (energy supplied basis)										
Coal		37.3	34.5	36.2	36.2	35.5	34.4	35.4	35.6	36.5
Petroleum		46.7	45.3	42.4	41.0	40.8	41.5	39.6	37.6	35.5
Natural gas		12.5	15.8	17.2	17.9	18.6	19.3	20.2	21.8	22.7
Nuclear electricity		3.3	4.2	4.0	4.6	5.0	4.7	4.6	4.8	5.1
Hydro electricity		0.1	0.2	0.2	0.2	0.2	0.2	0.2	0.2	0.2

(1) Includes other solid fuels.
(2) Excludes petroleum for non-energy use and marine bunkers.
(3) Includes colliery methane, non-energy use of natural gas up to 1988. Following the introduction of the presentation of energy data in the format of an energy balance it has been possible to separately identify the losses from the statistical difference for gas, bringing gas onto the same basis as other fuels. This has resulted in downwards revisions of the consumption figures for gas from 1994 onwards.
(4) Electricity generated i.e. including own use.

1.9 Inland consumption of primary fuels and equivalents for energy use, 1960 to 1999 (continued)

		1982	1983	1984	1985	1986	1987	1988	1989	1990
In original units of measurement	Unit									
Coal *(1)*	M.tonnes	110.7	111.5	79.0	105.3	113.5	116.2	112.0	108.1	108.4
Petroleum *(2)*	"	65.4	62.4	79.5	67.7	66.2	64.3	68.3	70.8	71.6
Natural gas *(3)*	GWh	525,476	547,750	560,410	602,701	612,724	629,311	597,220	571,187	595,131
Nuclear electricity *(4)*	"	44,212	50,138	53,957	61,391	59,079	55,238	63,456	71,734	65,749
Hydro electricity *(4)(5)*	"	4,558	4,563	4,005	4,093	4,780	4,198	4,919	4,758	5,216
Net electricity imports	"	-	-	-	-	4,255	11,635	12,830	12,631	11,943
Million tonnes of oil equivalent										
Coal *(1)*		68.0	68.6	48.7	64.8	70.0	71.7	70.0	67.0	66.9
Petroleum *(2)*		71.0	68.1	85.5	73.5	72.3	70.4	74.7	76.3	78.3
Natural gas *(3)*		45.2	47.1	48.2	51.8	52.7	54.1	51.3	49.5	51.2
Nuclear electricity		11.9	13.5	14.5	16.5	15.4	14.4	16.6	17.7	16.3
Hydro electricity *(5)*		0.4	0.4	0.3	0.4	0.4	0.4	0.4	0.4	0.4
Net electricity imports		-	-	-	-	0.4	1.0	1.1	1.1	1.0
Total *(6)*		196.4	197.7	197.2	207.0	211.2	212.0	214.4	212.8	214.9
Percentage shares (energy supplied basis)										
Coal		34.6	34.7	24.7	31.3	33.1	33.8	32.6	31.5	31.1
Petroleum		36.2	34.5	43.3	35.5	34.2	33.2	34.8	35.9	36.4
Natural gas		23.0	23.8	24.4	25.0	24.9	25.5	23.9	23.3	23.8
Nuclear electricity		6.0	6.8	7.4	8.0	7.2	6.8	7.7	8.3	7.6
Hydro electricity		0.2	0.2	0.2	0.2	0.2	0.2	0.2	0.2	0.2
Net electricity imports		-	-	-	-	0.2	0.5	0.5	0.5	0.5

		1991	1992	1993	1994	1995	1996	1997	1998	1999
In original units of measurement	Unit									
Coal *(1)*	M.tonnes	107.6	101.1	87.4	82.1	77.2	75.5	63.4r	63.4r	56.1
Petroleum *(2)*	"	71.3	71.5	72.1	71.2	69.4	71.3	68.7r	68.3	66.7
Natural gas *(3)*	GWh	643,863	640,459	732,090	754,284	805,058	940,372	961,968	997,329r	1,057,066
Nuclear electricity *(4)*	"	70,543	76,807	76,807	89,353	88,282	94,671	98,146	100,140	96,281
Hydro electricity *(4)(5)*	"	4,635	5,465	5,465	4,521	5,438	3,847	4,836r	6,114r	6,250
Net electricity imports	"	16,408	16,694	16,716	16,887	16,313	16,677	16,754r	12,468	14,244
Million tonnes of oil equivalent										
Coal *(1)*		67.1	63.0	55.0	51.3	48.9	46.2	41.1	41.4	36.7
Petroleum *(2)*		77.8	78.3	78.5	77.6	75.7	77.9	75.4	74.9	73.4
Natural gas *(3)*		55.3	55.1	63.0	64.8	69.2	80.9	82.7	85.8	90.9
Nuclear electricity		17.4	18.5	21.6	21.2	21.2	22.1	23.0	23.6	22.7
Hydro electricity *(5)*		0.4	0.5	0.4	0.5	0.5	0.3	0.4	0.5	0.5
Net electricity imports		1.4	1.4	1.4	1.5	1.4	1.4	1.4	1.1	1.2
Total *(6)*		220.0	217.7	221.0	218.5	218.7	230.3	226.2	229.4	227.8
Percentage shares (energy supplied basis)										
Coal		30.5	28.9	24.9	23.5r	22.4	19.8	18.2	18.1	16.1
Petroleum		35.3	36.0	35.5	35.5	34.6	33.8	33.3	32.6	32.2
Natural gas		25.1	25.3	28.5	29.7	31.6	35.5	36.6	37.4	39.9
Nuclear electricity		7.9	8.5	9.8	9.7	9.6	9.6	10.2	10.3	10.0
Hydro electricity		0.2	0.2	0.2	0.2	0.2	0.1	0.2	0.2	0.2
Net electricity imports		0.6	0.6	0.6	0.7	0.6	0.6	0.6	0.5	0.5

(5) Excludes pumped storage. Includes generation at wind stations from 1988.
(6) From 1988 includes renewable sources (wood, waste, landfill gas, sewage gas etc). Following the introduction of the presentation of

data in the format of an energy balance it has been possible to separately identify the losses from the statistical difference for electricity, bringing electricity onto the same basis as other fuels. This has been accounted for in the total from 1994 onwards.

1.10 Availability and consumption of primary fuels and equivalents (energy supplied basis) 1965 to 1999

Thousand tonnes of oil equivalent

	Available supply												
	Production					Imports					Exports (1)		
	Coal	Petroleum (2)	Natural gas (3)	Primary electricity (4)	Total (5)	Coal (6)	Petroleum (7)	Natural gas	Elec-tricity	Total	Coal (6)	Petroleum (7)	Total (8)
1965	123,672	499	166	4,739	129,076	-	90,525	655	8	91,188	2,464	17,049	19,512
1966	116,134	461	161	6,186	122,942	-	100,065	640	30	100,735	1,895	18,704	20,597
1967	115,324	96	579	7,095	123,094	-	104,702	768	13	105,483	1,746	13,391	15,137
1968	109,352	88	2,045	7,817	119,302	-	113,771	998	63	114,831	2,312	15,111	17,424
1969	99,574	113	4,882	8,188	112,757	-	122,338	1,048	50	123,436	2,874	15,612	18,487
1970	92,792	166	10,461	7,388	110,807	81	131,142	839	48	132,109	2,620	19,762	22,381
1971	94,178	227	17,384	7,661	119,450	2,887	136,359	836	10	140,092	2,048	20,024	22,071
1972	76,484	358	25,084	8,163	110,089	3,408	138,253	771	40	142,472	1,433	21,160	22,593
1973	82,636	400	27,235	7,793	118,064	1,214	144,117	738	5	146,074	2,131	22,026	24,157
1974	68,630	438	32,847	9,322	111,237	2,317	136,472	612	5	139,407	2,149	17,283	19,432
1975	79,172	1,675	34,203	8,446	123,496	3,209	111,703	844	8	115,763	1,975	16,517	18,492
1976	75,988	13,114	36,221	9,951	135,274	2,010	108,818	967	-	111,796	1,506	21,671	23,177
1977	74,769	41,186	37,845	10,973	164,773	1,761	90,004	1,680	-	93,445	1,753	33,112	34,865
1978	75,479	58,184	36,241	10,308	180,212	1,736	85,815	4,758	-	92,309	2,164	41,289	43,460
1979	74,028	83,966	36,596	10,598	205,188	3,169	77,903	8,323	-	89,394	2,025	57,607	59,632
1980	78,502	86,911	34,790	10,247	210,450	5,030	60,385	9,995	-	75,411	3,320	58,385	61,705
1981	78,008	96,941	34,712	10,562	220,223	3,192	50,040	10,681	-	63,912	6,884	69,615	76,500
1982	76,069	112,519	35,281	12,274	236,143	3,360	49,944	9,885	-	63,189	5,693	80,595	86,288
1983	72,696	125,482	36,379	13,866	248,423	3,713	43,543	10,701	-	57,957	4,844	90,608	95,452
1984	30,719	137,646	35,563	14,845	218,773	7,980	59,146	12,606	-	79,731	1,668	101,289	102,957
1985	56,572	139,404	39,679	16,851	252,506	9,482	52,577	12,645	-	74,703	2,441	106,602	109,043
1986	65,592	139,084	41,717	15,839	262,232	7,794	57,610	11,784	366	77,553	2,615	112,166	114,796
1987	63,189	135,071	43,674	14,797	256,731	7,363	54,305	11,079	1,000	73,746	1,872	107,108	108,980
1988	63,303	125,469	42,059	16,990	248,469	9,270	58,254	9,922	1,103	78,550	1,595	97,266	98,861
1989	60,882	100,373	41,188	18,150	221,320	8,840	64,153	9,784	1,163	83,941	1,738	74,434	76,249
1990	56,443	100,104	45,480	16,706	219,446	10,271	69,217	6,866	1,031	87,385	1,880	80,408	82,293
1991	57,555	99,890	50,638	17,830	226,669	13,493	72,942	6,193	1,412	94,040	1,526	81,105	82,632
1992	51,514	103,734	51,494	18,924	226,547	13,955	74,025	5,268	1,438	94,686	854	85,245	86,155
1993	41,588	109,613	60,542	21,969	234,882	13,103	77,612	4,173	1,438	96,326	954	95,312	96,854
1994	29,704	138,937	64,636	21,670	256,559	10,840	68,680	2,843	1,452	83,815	1,098	114,083	116,003
1995	32,751	142,746	70,807	21,735	269,738	11,615	63,341	1,673	1,405	78,034	889	116,001	117,859
1996	31,683	142,226	84,176	22,390	282,247	13,127	64,385	1,703	1,437	80,651	899	114,984	117,194
1997	30,517	140,580	85,883	23,409	282,303	14,396	63,942	1,209	1,429	80,977	836	115,937	118,864
1998	25,858	144,801	90,178	24,115	287,097	15,396	64,660	910	1,086	82,051	707	118,805	122,468
1999	23,543	150,308	98,930	23,217	298,373	14,710	62,434	1,106	1,247	79,498	580	123,984	132,040

(1) Includes marine bunkers prior to 1967.
(2) Crude oil plus all condensates and petroleum gases extracted at gas separation plants. For 1965 and 1966 includes products derived from coal and marketed by the petroleum industry.
(3) Includes colliery methane.
(4) Nuclear and natural flow hydro electricty excluding generation of pumped storage stations. From 1988 includes generation at wind stations.
(5) Includes solar and geothermal heat, solid renewable sources (wood, waste, etc), and gaseous renewable sources (landfill gas, sewage gas) from 1988.
(6) Includes other solid fuels.
(7) Crude and process oils and petroleum products.
(8) Includes exports of natural gas and electricity.

1.10 Availability and consumption of primary fuels and equivalents (energy supplied basis) 1965 to 1999 (continued)

Thousand tonnes of oil equivalent

	Marine Bunkers	Stock changes (9)			Statistical Difference (10)			Gross inland consumption	Non-energy use	Inland consumption for energy use				
	Petroleum	Coal (6)	Petroleum (7)	Natural gas	Coal (6)	Petroleum (7)	Total (14)	(15)	(11)	Coal (6)	Petroleum (7)	Natural gas (3)(12)	Primary Electricity (4)(13)	Total (5)
65	-	-262	-622	-	+81	-892	-811	**199,057**	6,889	121,028	65,572	821	4,746	**192,167**
66	-	+718	-1,197	-	+229	-2,093	-1,864	**200,736**	7,318	115,186	71,215	801	6,216	**193,418**
67	5,224	-5,406	+60	-	-574	-995	-1,569	**201,302**	8,794	107,599	76,454	1,348	7,107	**192,508**
68	5,804	+559	-1,040	-	+80	-252	-172	**210,253**	9,733	107,680	81,918	3,043	7,879	**200,520**
69	5,791	+8,128	-2,040	-	+131	+76	+207	**218,209**	10,658	104,959	88,424	5,930	8,238	**207,552**
70	5,721	+8,542	-680	-	+199	+466	+665	**223,341**	10,859	98,994	94,752	11,300	7,435	**212,482**
71	5,874	-7,046	-3,489	-	-239	-652	-891	**220,170**	10,839	87,732	95,707	18,220	7,672	**209,331**
72	5,265	-1,370	+2,904	-	-242	-887	-1,129	**225,109**	11,474	76,847	102,730	25,855	8,203	**213,635**
73	5,769	+1,456	+458	-	+60	-340	-280	**235,847**	12,635	83,235	104,206	27,974	7,797	**223,212**
74	4,922	+4,839	-5,139	-	-360	-514	-874	**225,116**	12,865	73,278	96,188	33,460	9,326	**212,252**
75	3,572	-6,489	+3,660	-	-202	-395	-597	**213,769**	10,255	73,716	86,298	35,047	8,453	**203,514**
76	3,698	-1,597	-348	-	+121	-254	-133	**218,116**	10,925	75,016	85,036	37,188	9,951	**207,191**
77	2,942	+600	+2,466	-	-113	-557	-670	**222,806**	10,517	75,263	86,528	39,526	10,973	**212,289**
78	2,733	-1,368	-814	-	-363	-569	-932	**223,214**	10,245	73,321	88,349	40,999	10,301	**212,970**
79	2,789	+3,600	-2,229	-	+43	-806	-763	**232,768**	10,232	78,814	88,205	44,919	10,597	**222,536**
80	2,562	-6,789	+40	-	-171	-1,567	-1,738	**213,118**	7,464	73,263	77,358	44,785	10,247	**205,654**
81	2,156	-2,013	+3,882	-	+562	-154	+408	**207,756**	8,111	72,865	70,827	45,392	10,564	**199,645**
82	2,715	-5,660	+2,305	-	-118	-2,315	-2,433	**204,540**	8,134	67,958	71,008	45,166	12,274	**196,406**
83	2,118	-3,209	+1,010	-	+234	-544	-310	**206,290**	8,625	68,590	68,129	47,080	13,866	**197,665**
84	2,370	+11,842	+922	-	-136	+247	+111	**206,052**	8,847	48,738	85,455	48,168	14,845	**197,205**
85	2,239	+1,461	+297	-521	-249	-731	-980	**216,184**	9,230	64,824	73,477	51,803	16,851	**206,955**
86	2,212	-1,889	+338	-836	+1,126	-83	+1,043	**221,432**	10,247	70,008	72,323	52,665	16,189	**211,185**
87	1,756	+3,396	+338	-662	-355	-146	-501	**222,311**	10,290	71,721	70,414	54,090	15,796	**212,021**
88	1,932	-1,547	+1,272	-637	+189	-111	+78	**225,392**	10,970	69,621	74,716	51,344	18,083	**214,421**
89	2,525	-1,787	-628	-281	+817	+159	+976	**224,767**	12,039	67,014	76,281	49,470	19,236	**212,727**
90	2,666	+891	+1,049	+108	+1,229	+990	+2,219	**226,139**	11,252	66,954	78,301	51,187	17,733	**214,887**
91	2,618	-3,402	-851	-273	+947	+448	+1,395	**232,330**	12,184	67,067	77,794	55,286	19,240	**220,145**
92	2,688	-2,439	+709	-348	+884	-647	+237	**230,549**	12,890	63,060	78,279	55,080	20,359	**217,659**
93	2,618	+766	-631	+84	+411	+1,597	+2,008	**233,964**	13,012	54,913	78,501	62,961	23,406	**220,951**
94	2,451	+11,055	+454	+233	+772	-1,668	-87	**231,956**	13,521	51,272	77,596	64,868	23,087	**218,435**
95	2,602	+5,088	+1,122	+820	+820	-426	+1,752	**232,458**	13,735	48,924	75,748	69,236	23,116	**218,723**
96	2,812	+2,531	-315	-236	+284	-1,750	+933	**243,938**	13,665	46,157	77,863	80,853	23,824	**230,273**
97	3,121	-2,582r	+322	-354	+124	-1,687r	-601r	**239,282r**	13,095r	41,147r	75,435r	82,710r	24,833r	**226,187r**
98	3,253	+763	-741r	-32r	-352r	-24r	+1,168r	**242,249r**	12,977r	41,437r	74,855r	85,750r	25,187r	**229,374r**
99	2,469	-327	+422	-524	-427	+1,327	+2,109	**240,822**	13,169	36,725	73,361	90,886	24,442	**227,789**

(9) Stock fall (+), stock rise (-).

(10) Recorded demand minus supply.

(11) Petroleum products for feedstock for petrochemical plants, industrial and white spirits, lubricants bitumen and wax. Also includes from 1968 miscellaneous petroleum products mainly for inland consumption but excludes small quantities derived from coal. From 1989 also includes estimated quantities of natural gas used for non-energy purposes. Data for non-energy use of natural gas from 1994 can be found in Tables 1.1-1.3 and 4.1 and 4.2.

(12) Includes non-energy use of natural gas up to 1988. (See footnote 11).

(13) Includes net imports of electricity.

(14) As of 1994 this total includes the statistical differences for electricity and natural gas.

(15) Equivalent to primary demand as in Tables 1.1 , 1.2 and 1.3.

1.11 Comparison of net imports of fuel with total consumption of primary fuels and equivalents 1950 to 1999

	Gross inland consumption of primary fuels (1) plus marine bunkers (A)	Net imports (+) /net exports (-) of fuels (B)	Import dependency (2) (C)	Export ratio (3) (D)
	Million tonnes of oil equivalent		Per cent	
1950	151.6	+12.4	8.0	-
1955	170.1	+32.4	19.1	-
1960	180.2	+49.7	27.6	-
1965	199.1	+71.7	36.0	-
1970	229.1	+109.7	47.9	
1971	226.0	+118.0	52.2	
1972	230.4	+119.9	52.0	
1973	241.6	+121.9	50.5	
1974	230.0	+120.0	52.2	
1975	217.3	+97.3	44.8	
1976	221.8	+88.6	40.0	
1977	225.7	+58.6	25.9	
1978	225.9	+48.8	21.6	
1979	235.6	+29.8	12.6	-
1980	215.7	+13.7	6.4	-
1981	209.9	-12.6	-	6.0
1982	207.3	-23.1	-	11.1
1983	208.4	-37.5	-	18.0
1984	208.4	-23.2	-	11.1
1985	218.4	-34.3	-	15.7
1986	223.6	-37.2	-	16.7
1987	224.1	-35.2	-	15.7
1988	227.3	-20.3	-	8.9
1989	227.3	+7.7	3.4	-
1990	228.8	+ 5.1	2.2	-
1991	234.9	+11.4	4.9	-
1992	233.2	+8.5	3.7	-
1993	236.6	-0.6	-	0.2
1994	234.4	-32.2	-	13.7
1995	235.1	-39.8	-	16.9
1996	246.8	-36.5	-	14.8
1997	242.4	-37.9	-	15.7
1998	245.5	-40.4	-	16.5
1999	243.3	-52.5	-	21.5

(1) Includes non-energy use. Equivalent to primary demand plus marine bunkers.

(2) Import dependency (C) = $\dfrac{\text{Net imports (B)}}{\text{(A)}} \times 100$

(3) Export ratio (D) = $\dfrac{\text{Net exports (B)}}{\text{(A)}} \times 100$

1.12 Primary energy consumption, gross domestic product and the energy ratio[1], 1950 to 1999

	Total inland consumption of primary energy (temperature corrected) *(2)*	Gross domestic product at market prices (1995 prices)*(3)*	Energy ratio *(4)*	
	Million tonnes of oil equivalent *(A)*	£ billion *(B)*	Tonnes of oil equivalent per £1 million GDP *(C)*	Index 1995 = 100 *(D)*
1950	143.5	239.5	599	190.9
1951	148.4	246.1	603	192.1
1952	147.4	247.2	596	190.0
1953	152.0	256.7	592	188.6
1954	156.6	267.7	585	186.3
1955	159.4	276.0	578	184.0
1956	160.7	278.7	577	183.7
1957	159.8	283.6	563	179.5
1958	159.2	284.6	559	178.2
1959	160.5	297.0	540	172.2
1960	170.7	312.9	546	173.8
1961	171.5	320.6	535	170.4
1962	172.4	324.6	531	169.2
1963	177.8	339.9	523	166.6
1964	183.5	358.5	512	163.0
1965	189.5	367.5	516	164.2
1966	190.8	374.6	509	162.2
1967	191.1	383.2	499	158.9
1968	197.2	398.9	494	157.5
1969	203.5	407.1	500	159.2
1970	211.9	416.8	508	161.9
1971	209.7	425.2	493	157.1
1972	212.6	440.4	483	153.8
1973	223.1	472.7	472	150.3
1974	212.4	464.8	457	145.5
1975	206.0	461.6	446	142.1
1976	208.9	474.5	440	140.2
1977	213.1	485.7	439	139.8
1978	213.7	502.2	426	135.5
1979	220.0	516.1	426	135.8
1980	206.2	504.8	409	130.1
1981	198.7	498.3	399	127.0
1982	196.3	507.3	387	123.3
1983	197.5	526.3	375	119.5
1984	196.7	539.0	365	116.2
1985	203.1	559.5	363	115.6
1986	206.8	583.2	355	113.0
1987	210.0	609.0	345	109.8
1988	217.7	640.6	340	108.3
1989	217.8	654.3	333	106.0
1990	221.6	658.5	337	107.2
1991	221.4	648.6	341	108.7
1992	220.6	649.0	340	108.3
1993	222.5	664.0	335	106.7
1994	221.5	693.2	320	101.8
1995	223.7	712.5	314	100.0
1996	229.2	730.8	314	99.9
1997	231.7	756.7	306	97.5
1998	235.2	772.8r	304r	96.9r
1999	235.9	788.4	299	95.2

(1) See paragraphs 1.31 to 34
(2) The methodology used to temperature correct gas consumption has been modified from 1990 onwards. See paragraph 1.50.
(3) GDP revised to be on ESA95 basis.
(4) Energy ratio (C) = $\dfrac{(A)}{(B)}$

1.13 Energy consumption by final user (energy supplied basis)[1] 1960 to 1999

Thousand tonnes of oil equivalent

Industry (2)

	Coal	Coke and breeze	Other solid fuels(3)	Coke oven gas	Town gas	Natural gas (4)	Electricity	Petroleum	Creosote/ pitch	Total(5)
1960	22,784	12,116	181	1,388	2,171	-	3,826	10,381	995	**53,843**
1961	21,278	11,489	176	1,302	2,139	-	3,957	12,184	851	**53,378**
1962	20,066	10,293	169	1,108	2,134	-	4,091	14,127	965	**52,952**
1963	18,983	10,134	166	1,126	2,169	-	4,287	16,202	965	**54,032**
1964	18,225	11,074	172	1,317	2,257	-	4,738	18,207	786	**56,776**
1965	17,781	10,973	149	1,335	2,315	-	5,033	20,419	756	**58,761**
1966	16,809	9,749	184	1,217	2,348	-	5,169	21,981	801	**58,257**
1967	15,250	9,146	186	1,161	2,272	-	5,234	23,507	698	**57,455**
1968	14,834	9,625	189	1,217	2,252	128	5,700	24,895	557	**59,398**
1969	14,036	9,711	199	1,139	2,141	496	6,046	26,734	446	**60,948**
1970	12,681	9,655	209	1,164	1,778	1,788	6,275	28,397	385	**62,333**
1971	10,232	8,298	176	1,118	1,038	5,194	6,313	28,130	247	**60,746**
1972	7,675	7,832	252	1,111	1,154	8,136	6,292	28,674	181	**61,307**
1973	7,950	8,340	226	1,290	788	10,791	6,884	28,691	186	**65,149**
1974	7,290	7,167	201	975	494	12,320	6,517	24,968	126	**60,058**
1975	6,373	6,338	199	1,038	222	12,555	6,479	22,145	96	**55,444**
1976	5,902	7,129	131	1,091	68	14,237	6,950	21,966	111	**57,584**
1977	5,947	6,368	158	1,010	30	14,940	7,053	21,978	88	**57,574**
1978	5,627	5,932	179	899	15	15,149	7,222	21,570	78	**56,673**
1979	6,081	6,512	148	977	18	15,663	7,527	21,590	48	**58,564**
1980	5,083	3,335	133	642	13	15,258	6,854	16,938	35	**48,291**
1981	4,534	4,564	116	665	13	14,489	6,622	14,761	10	**45,776**
1982	4,668	4,083	144	605	8	14,588	6,353	13,530	30	**44,007**
1983	4,708	4,307	126	635	5	14,021	6,376	11,988	25	**42,191**
1984	3,796	4,408	68	537	5	14,686	6,758	10,859	20	**41,138**
1985	4,708	4,655	151	768	3	14,865	6,837	9,701	15	**41,702**
1986(11)	5,242	4,144	98	778	3	13,542	6,884	10,240	-	**40,931**
1987	4,048	4,660	80	821	3	14,137	8,005	8,456	-	**40,211**
1988	4,166	5,041	55	771	-	12,883	8,350	9,441	-	**40,833**
1989	4,489	4,286	30	613	-	12,515	8,550	8,820	-	**39,401**
1990	4,172	3,951	42	602	-	12,889	8,655	8,242	-	**38,658**
1991	4,270	3,691	14	570	-	12,311	8,563	8,729	-	**38,257**
1992	4,375	3,601	14	534	-	11,380	8,194	8,334	-	**36,535**
1993	3,553	3,613	7	560	-	11,521	8,328	8,592	-	**36,440**
1994	3,402	1,091	194	804	-	12,885	8,082	8,253	-	**35,167**
1995	2,840r	868	184r	576	-	12,696	8,654r	7,066r	-	**34,909r**
1996	2,193r	855	233	554	-	13,999r	8,866r	6,976r	-	35,817r
1997	1,939r	787r	249r	471	-	14,685r	9,020r	6,313r	-	34,367r
1998	1,690r	777r	243r	398	-	14,791r	9,075r	6,153r	-	33,997r
1999	1,919	780	217	386	-	15,916r	9,330	6,109	-	**35,453**

(1) Excluding non-energy use of fuels.
(2) Includes the iron and steel industry, but from 1994 onwards excludes Iron and Steel use of fuels for transformation and energy industry own use purposes.
(3) Includes, from 1994, manufactured liquid fuels.
(4) Includes colliery methane. Up to 1988 also includes non-energy use of natural gas.
(5) Includes renewable fuels from 1988.

1.13 Energy consumption by final user (energy supplied basis)[1] 1960 to 1999 (continued)

Thousand tonnes of oil equivalent

| | Transport | | | | | | | | | | | |
| | Rail | | | | | Road | | | Water | | Air | |
	Coal	Coke and breeze	Other solid fuels	Electricity (6)	Petroleum	Electricity	Petroleum	Coal derived fuel	Coal	Petroleum	Petroleum	Total (7)
1960	6,310	83	385	156	264	38	11,381	134	315	1,121	2,010	**22,197**
1961	5,444	81	395	171	562	25	12,230	156	280	1,151	2,459	**22,953**
1962	4,368	78	375	189	703	18	12,945	141	254	1,237	2,401	**22,709**
1963	3,476	76	438	191	826	15	13,799	136	224	1,224	2,547	**22,953**
1964	2,715	63	254	191	930	13	15,258	144	186	1,164	2,617	**23,535**
1965	1,962	58	-	191	1,013	10	16,348	108	161	1,209	2,710	**23,772**
1966	1,285	48	-	207	1,033	8	17,222	116	144	1,199	2,932	**24,192**
1967	602	43	-	217	1,058	5	18,379	108	121	1,179	3,297	**25,009**
1968	166	43	-	224	1,139	3	19,563	71	98	1,083	3,587	**25,976**
1969	106	38	-	232	1,184	3	20,321	30	96	1,131	3,728	**26,868**
1970	88	35	-	234	1,254	3	21,406	15	88	1,184	3,869	**28,174**
1971	68	13	-	237	1,186	-	22,412	-	63	1,081	4,247	**29,306**
1972	53	5	-	229	1,121	-	23,535	-	23	962	4,514	**30,442**
1973	58	-	-	224	1,123	-	25,125	-	10	1,088	4,806	**32,435**
1974	50	-	-	234	1,048	-	24,465	-	10	1,239	4,219	**31,266**
1975	40	-	-	249	1,000	-	23,948	-	8	1,300	4,340	**30,885**
1976	43	3	-	247	945	-	24,994	-	8	1,317	4,476	**32,032**
1977	40	3	-	252	950	-	25,633	-	8	1,312	4,678	**32,875**
1978	45	3	-	254	967	-	26,946	-	5	1,300	5,051	**34,571**
1979	43	3	-	254	947	-	27,520	-	5	1,363	5,224	**35,359**
1980	38	3	-	262	919	-	27,815	-	5	1,257	5,242	**35,541**
1981	38	-	-	259	877	-	27,009	-	-	1,101	5,020	**34,304**
1982	35	-	-	229	793	-	27,797	-	3	1,186	4,993	**35,037**
1983	15	-	-	247	849	-	28,646	-	3	1,207	5,093	**36,059**
1984	3	-	-	247	816	-	30,006	-	-	1,328	5,383	**37,782**
1985	3	-	-	254	821	-	30,586	-	-	1,254	5,582	**38,500**
1986 (11)	3	-	-	259	809	-	32,606	-	-	1,151	6,126	**40,954**
1987	3	-	-	264	761	-	34,062	-	-	1,103	6,479	**42,672**
1988	-	-	-	282	766	-	36,233	-	-	1,159	6,905	**45,345**
1989	3	-	-	272	702	-	37,801	-	-	1,355	7,308	**47,442**
1990	2	-	-	455	668	-	38,816	-	-	1,363	7,332	**48,635**
1991	-	-	-	454	685	-	38,535	-	-	1,424	6,872	**47,973**
1992	-	-	-	461	715	-	39,363	-	-	1,377	7,435	**49,355**
1993	-	-	-	641	665	-	39,502	-	-	1,341	7,871	**50,024**
1994	-	-	-	599	651	-	39,690	-	-	1,239	8,070	**50,253**
1995	-	-	-	636	654	-	39,268	-	-	1,193	8,485	**50,238**
1996	-	-	-	638	628	-	40,734	-	-	1,293	8,918	**52,212**
1997	-	-	-	723r	515r	-	41,221	-	-	1,255	9,323	**53,037r**
1998	-	-	-	720r	523r	-	40,987r	-	-	1,176r	10,238	**53,643r**
1999	-	-	-	727	506	-	40,751	-	-	1,073	10,708	**53,765**

(6) Includes, from 1990, electricity used at transport premises (see footnote 8). See Chapter 5, paragraph 5.15.
(7) Includes small amounts of natural gas for road transport.

1.13 Energy consumption by final user (energy supplied basis)[1] 1960 to 1999 (continued)

Thousand tonnes of oil equivalent

| | Domestic | | | | | | | Other final users (8) | | | | | |
	Coal	Coke and breeze	Other solid fuels	Natural gas (9)	Electricity	Petroleum	Total (4)	Coal	Coke and breeze	Natural gas (9)	Electricity	Petroleum	Total (4)
1960	25,190	2,378	907	3,270	2,894	1,690	**36,329**	4,423	3,353	1,199	1,579	4,350	**14,905**
1961	23,676	2,423	967	3,275	3,287	1,841	**35,470**	4,051	3,008	1,181	1,718	3,867	**13,828**
1962	24,049	2,776	995	3,529	3,927	2,189	**37,465**	4,116	3,290	1,267	1,967	4,519	**15,159**
1963	23,532	2,998	1,068	3,872	4,496	2,499	**38,465**	4,310	3,471	1,320	2,174	5,219	**16,495**
1964	20,568	2,942	1,189	4,066	4,501	2,295	**35,561**	3,768	2,831	1,300	2,277	5,272	**15,449**
1965	20,207	3,008	1,292	4,708	4,920	2,441	**36,576**	3,741	2,721	1,385	2,496	5,786	**16,129**
1966	18,865	2,980	1,385	5,484	5,141	2,436	**36,291**	3,509	2,418	1,451	2,630	6,182	**16,190**
1967	17,180	2,857	1,494	6,230	5,360	2,534	**35,654**	3,126	2,151	1,534	2,763	6,703	**16,278**
1968	16,585	2,605	1,743	7,127	5,731	2,809	**36,599**	2,607	2,184	1,647	3,015	7,572	**17,026**
1969	15,469	2,280	1,957	8,088	6,207	3,078	**37,080**	2,746	1,960	1,773	3,244	8,381	**18,104**
1970	14,242	1,761	1,975	8,922	6,622	3,363	**36,884**	2,723	1,499	1,919	3,408	9,038	**18,586**
1971	12,164	1,136	2,156	9,900	6,937	3,328	**35,621**	2,328	688	2,181	3,534	9,184	**17,915**
1972	10,602	849	2,144	11,359	7,471	3,836	**36,261**	2,013	537	2,509	3,650	9,487	**18,195**
1973	10,565	778	2,053	12,129	7,849	4,202	**37,576**	1,731	602	2,728	3,940	9,585	**18,586**
1974	9,968	821	1,955	13,562	7,963	3,733	**38,002**	1,685	567	3,197	3,642	8,401	**17,492**
1975	8,517	645	1,778	14,840	7,670	3,612	**37,062**	1,234	408	3,393	3,894	8,431	**17,360**
1976	7,910	549	1,640	15,602	7,318	3,615	**36,634**	1,300	335	3,831	4,023	8,668	**18,157**
1977	8,136	534	1,589	16,600	7,386	3,653	**37,898**	1,370	315	3,998	4,257	9,157	**19,097**
1978	7,476	471	1,464	18,291	7,378	3,610	**38,689**	1,300	275	4,393	4,481	8,764	**19,213**
1979	7,688	479	1,431	20,718	7,711	3,539	**41,566**	1,307	285	4,955	4,731	8,754	**20,031**
1980	6,575	401	1,370	21,258	7,403	2,834	**39,841**	1,154	237	5,194	4,733	7,403	**18,721**
1981	6,214	368	1,202	22,076	7,260	2,554	**39,674**	1,174	204	5,315	4,804	7,096	**18,592**
1982	6,242	365	1,146	21,963	7,116	2,385	**39,218**	1,222	212	5,486	4,867	6,678	**18,464**
1983	5,796	335	1,141	22,346	7,129	2,267	**39,014**	1,166	257	5,915	5,106	6,403	**18,847r**
1984	4,733	335	728	22,502	7,212	2,385	**37,896**	1,141	252	6,101	5,063	6,381	**18,938**
1985	6,290	385	957	24,394	7,582	2,454	**42,062**	1,123	297	6,718	5,446	6,018	**19,603**
1986(11)	6,121	335	965	25,797	7,892	2,590	**43,700**	982	390	7,308	5,731	5,723	**20,135**
1987	5,189	315	1,018	26,450	8,015	2,474	**43,460**	935	368	7,534	5,965	4,988	**19,790**
1988	4,741	300	907	25,833	7,940	2,441	**42,346**	831	264	7,569	6,240	5,008	**20,038**
1989	3,719	239	815	24,988	7,935	2,355	**40,236**	698	119	7,278	6,497	4,345	**19,089**
1990	3,153	254	762	25,835	8,066	2,480	**40,758**	795	127	7,329	6,426	4,402	**19,217**
1991	3,582	210	785	28,721	8,436	2,825	**44,769**	753	105	8,640	6,717	4,456	**20,819**
1992	3,105	176	709	28,389	8,555	2,889	**44,067**	622	88	8,585	6,996	4,518	**20,959**
1993	3,498	147	751	29,254	8,639	3,019	**45,549**	566	74	8,504	6,999	4,446	**20,733**
1994	2,957	67	601	28,355	8,721	3,004	**43,947**	496	34	8,695	6,951	4,289	**20,638**
1995	2,077	78	470	28,037	8,790	2,997	**42,689**	362	39	9,374	7,199	4,016	**20,990**
1996	2,085	129	587	32,315	9,244	3,518	**48,118**	418	-	10,260	7,487	3,906	**20,071**
1997	1,991	59r	419r	29,709	8,981	3,389r	**44,773r**	461	-	9,772r	7,891r	3,359r	**21,659r**
1998	1,817	82r	440r	30,600	9,424	3,544r	**46,137r**	267r	-	10,157r	7,920r	3,130r	**21,648r**
1999	2,048	83	401	30,615	9,493	3,240	**46,111**	221	-	10,395	7,994	2,662	**21,446**

(8) Mainly agriculture, public administration and commerce. Prior to 1990, including electricity used at transport premises (see footnote 6).
(9) Includes town gas prior to 1989. (Separate figures maybe found in previous editions of this Digest).

1.13 Energy consumption by final user (energy supplied basis)[1] 1960 to 1999 (continued)

Thousand tonnes of oil equivalent

					All final users					
	Coal	Coke and breeze	Other solid fuels (2)	Coke oven gas	Town gas	Natural gas (4)	Electricity	Petroleum	Creosote/ pitch	Total (4)(10)
1960	59,023	17,930	1,474	1,388	6,640	-	8,494	31,198	995	**127,275**
1961	54,728	17,001	1,539	1,302	6,595	-	9,159	34,294	851	**125,629**
1962	52,854	16,436	1,539	1,108	6,930	-	10,192	38,120	965	**128,285**
1963	50,526	16,678	1,673	1,126	7,361	-	11,164	42,317	965	**131,945**
1964	45,463	16,910	1,615	1,317	7,622	-	11,721	45,743	786	**131,321**
1965	43,853	16,759	1,441	1,335	8,408	-	12,650	49,927	756	**135,238**
1966	40,611	15,195	1,569	1,217	9,283	-	13,154	52,985	801	**134,930**
1967	36,279	14,197	1,680	1,161	10,033	3	13,580	56,657	698	**134,396**
1968	34,291	14,457	1,932	1,217	10,947	207	14,673	60,648	557	**138,999**
1969	32,452	13,988	2,156	1,139	11,426	1,073	15,731	64,557	446	**142,999**
1970	29,822	12,950	2,184	1,164	10,746	3,662	16,542	68,511	385	**145,977**
1971	24,855	10,134	2,333	1,118	8,882	9,431	17,021	69,568	247 .	**143,589**
1972	20,366	9,222	2,396	1,111	8,094	15,063	17,643	72,129	181	**146,205**
1973	20,313	9,721	2,280	1,290	5,852	20,584	18,898	74,620	186	**153,744**
1974	19,003	8,555	2,156	975	3,836	25,736	18,356	68,072	126	**146,818**
1975	16,172	7,391	1,977	1,038	1,796	29,212	18,293	64,776	96	**140,751**
1976	15,162	8,016	1,771	1,091	534	33,204	18,537	65,981	111	**144,407**
1977	15,502	7,220	1,748	1,010	174	35,393	18,948	67,361	88	**147,444**
1978	14,454	6,681	1,642	899	81	37,766	19,336	68,208	78	**149,146**
1979	15,124	7,279	1,579	977	91	41,262	20,223	68,937	48	**155,521**
1980	12,854	3,975	1,504	642	76	41,647	19,252	62,408	35	**142,394**
1981	11,960	5,136	1,317	665	65	41,828	18,945	58,420	10	**138,346**
1982	12,169	4,660	1,290	605	55	41,990	18,567	57,360	30	**136,726**
1983	11,688	4,899	1,267	635	45	42,242	18,856	56,453	25	**136,111**
1984	9,673	4,995	796	537	43	43,251	19,280	57,158	20	**135,753**
1985	12,124	5,338	1,108	768	40	45,940	20,118	56,416	15	**141,867**
1986(11)	12,348	4,869	1,063	778	28	46,622	20,763	59,245	-	**145,719**
1987	10,174	5,343	1,098	821	28	48,096	22,252	58,325	-	**146,132**
1988	9,738	5,605	949	771	8	46,262	22,811	61,952	-	**148,565**
1989	8,909	4,645	833	613	-	44,780	23,254	62,685	-	**146,168**
1990	8,122	4,333	805	602	-	46,052	23,601	63,302	-	**147,268**
1991	8,605	4,006	799	570	-	49,676	24,170	63,525	-	**151,818**
1992	8,101	3,866	723	534	-	48,357	24,206	64,632	-	**151,916**
1993	7,617	3,833	758	560	-	49,282	24,607	65,437	-	**152,746**
1994	6,855	1,191	794	804	-	49,908	24,353	65,196	-	**150,005**
1995	5,520	1,163	655	728	-	51,768	25,182	63,821	-	**148,783**
1996	4,896	984	820	1,256	-	57,354	26,177	66,054	-	**158,496**
1997	4,392r	846	668	842r	-	54,166r	26,615r	65,376r	-	**153,836r**
1998	3,773r	859	683r	738r	-	55,548r	27,139r	65,751r	-	**155,425r**
1999	4,188	863	618	646	-	56,926	27,544	65,049	-	**156,775**

(10) Before 1971 includes the use for transport of liquid fuel made from coal.
(11) See paragraph 1.36 about changed treatment of electricity produced, and fuel used by, companies other than major power producers.

1.14 Expenditure on energy by final user, 1970 to 1999[1]

United Kingdom

£ million

	Industry					Domestic				
	Coal and solid fuels [3]	Natural gas [4]	Electricity	Petroleum products [5]	Total [6]	Coal and solid fuels [3]	Natural gas [4]	Electricity	Petroleum products [5]	Total [6]
1970	285	70	475	300	**1,130**	395	385	645	85	**1,510**
1971	285	85	530	350	**1,250**	385	430	730	90	**1,635**
1972	280	120	540	345	**1,285**	360	505	830	110	**1,805**
1973	320	150	595	390	**1,455**	370	535	885	140	**1,930**
1974	410	195	775	880	**2,260**	405	605	1,070	200	**2,280**
1975	545	240	1,015	920	**2,720**	440	760	1,495	235	**2,930**
1976	720	380	1,260	1,065	**3,425**	500	1,000	1,825	295	**3,620**
1977	780	535	1,470	1,305	**4,090**	595	1,205	2,135	360	**4,295**
1978	800	695	1,670	1,255	**4,420**	620	1,365	2,380	370	**4,735**
1979	1,010	820	1,925	1,570	**5,325**	770	1,575	2,675	475	**5,495**
1980	675	1,060	2,185	1,815	**5,735**	920	1,875	3,310	510	**6,615**
1981	850	1,215	2,420	1,890	**6,375**	960	2,460	3,905	560	**7,885**
1982	860	1,335	2,560	1,870	**6,625**	995	3,070	4,200	610	**8,875**
1983	900	1,375	2,655	1,800	**6,730**	1,015	3,520	4,300	645	**9,480**
1984	845	1,555	2,695	1,810	**6,905**	830	3,655	4,495	640	**9,620**
1985	990	1,735	2,750	1,740	**7,215**	1,120	4,090	4,840	665	**10,715**
1986	1,000	1,350	2,765	1,065	**6,180**	1,135	4,385	5,105	460	**11,085**
1987	865	1,375	3,285	865	**6,390**	990	4,465	5,140	410	**11,005**
1988	880	1,225	3,590	785	**6,480**	830	4,385	5,340	365	**10,920**
1989	905	1,210	3,965	845	**6,925**	730	4,455	5,800	390	**11,375**
1990	930	1,260	3,985	900	**7,075**	700	4,865	6,255	485	**12,305**
1991	910	1,115	4,120	905	**7,050**	795	5,775	7,105	460	**14,135**
1992	775	970	4,180	790	**6,715**	710	5,685	7,460	460	**14,315**
1993	740	915	3,940	895	**6,490**	780	5,705	7,590	465	**14,540**
1994	650	1,010	3,855	865	**6,380**	685	6,020	7,870	455	**15,030**
1995	605r	1,015r	3,970r	830r	**6,420r**	615	6,010	8,060	470	**15,155**
1996	590r	755r	3,900r	965r	**6,210r**	640	6,510	8,380	630	**16,165**
1997	565r	870r	3,625	890r	**5,950r**	560r	6,125r	7,965r	560	**15,210r**
1998	545r	965r	3,535	700r	**5,745r**	525r	6,015	7,700r	465	**14,705r**
1999	455	1,015	3,705	755	**5,930**	570	5,175	7,450	425	**13,620**

(1) All data is to the nearest £5 million. VAT is only included where not refundable. Methodology used to calculate the series has changed over the years, as such the data provides a guide to changing patterns of expenditure on energy, but not too much significance should be drawn from small changes.

(2) Includes commercial, public administration, agriculture and all fuels used for transport purposes.

(3) Includes coal, coke, breeze and other manufactured solid fuel. Prior to 1996 an estimate of the value of coke produced in coke ovens owned by the iron and steel industry was included, this has now been replaced by an estimate of the value of coal purchased for such ovens, which is the actual monetary trade.

(4) Includes town gas.

(5) Includes heating oils, LPG etc. Excludes motor transport fuels.

(6) Excludes other fuels not listed e.g. crude oil, coke oven gas etc.

1.14 Expenditure on energy by final user, 1970 to 1999[1](continued)

United Kingdom

£ million

	Other final users (2)						All final users				
	Coal and solid fuels (3)	Natural gas (4)	Electricity	Petroleum products	Of which road transport	Total (6)	Coal and solid fuels (3)	Natural gas (4)	Electricity	Petroleum	Total (6)
1970	60	70	390	1,910	1,720	**2,430**	740	525	1,510	2,295	**5,070**
1971	45	80	435	2,105	1,885	**2,665**	715	595	1,695	2,545	**5,550**
1972	45	80	480	2,305	2,070	**2,910**	685	705	1,850	2760	**6,000**
1973	45	90	515	2,580	2,305	**3,230**	735	775	1,995	3,110	**6,615**
1974	60	105	590	3,885	3,150	**4,640**	875	905	2,435	4,965	**9,180**
1975	70	140	835	4,685	3,845	**5,730**	1,05	1,140	3,345	5,840	**11,380**
1976	90	200	1,030	5,305	4,325	**6,625**	1,310	1,580	4,115	6,665	**13,670**
1977	115	255	1,200	6,030	4,835	**7,600**	1,490	1,995	4,805	7,695	**15,985**
1978	115	310	1,375	6,075	4,890	**7,875**	1,535	2,370	5,425	7,700	**17,030**
1979	130	385	1,655	8,265	6,660	**10,435**	1,910	2,780	6,255	10,310	**21,255**
1980	115	520	1,985	10,735	8,650	**13,355**	1,710	3,455	7,480	13,060	**25,705**
1981	110	585	2,460	12,345	10,060	**15,500**	1,920	4,260	8,785	14,795	**29,760**
1982	135	655	2,690	13,470	10,950	**16,950**	1,990	5,060	9,450	15,950	**32,450**
1983	135	745	2,855	14,965	12,240	**18,700**	2,050	5,640	9,810	17,410	**34,910**
1984	135	795	2,980	16,140	13,250	**20,050**	1,810	6,005	10,170	18,590	**36,575**
1985	155	920	3,265	17,640	14,615	**21,980**	2,265	6,745	10,855	20,045	**39,910**
1986	140	1,045	3,485	15,845	13,745	**20,515**	2,275	6,780	11,355	17,370	**37,780**
1987	125	1,035	3,490	16,630	14,525	**21,280**	1,980	6,870	11,915	17,905	**38,670**
1988	95	1,025	3,810	16,855	14,960	**21,785**	1,805	6,635	12,740	18,005	**39,185**
1989	95	1,015	4,185	18,755	16,690	**24,050**	1,730	6,680	13,950	19,980	**42,340**
1990	105	1,085	4,465	21,120	19,020	**26,775**	1,735	7,210	14,705	22,505	**46,155**
1991	85	1,310	4,960	21,900	19,995	**28,255**	1,790	8,200	16,185	23,265	**49,440**
1992	95	1,245	5,495	22,455	20,825	**29,290**	1,580	7,900	17,135	23,705	**50,320**
1993	70	1,155	5,555	24,365	22,540	**31,145**	1,590	7,775	17,115	25,725	**52,205**
1994	50	1,125	5,380	25,190	23,515	**31,745**	1,385	8,155	17,140	26,510	**53,190**
1995	35	1,110	5,300r	25,895	24,140	**32,340r**	1,255r	8,135r	17,330r	27,195r	**53,915r**
1996	30	975r	5,405r	28,240	26,145	**34,650r**	1,260r	8,240r	17,685r	29,835r	**57,020r**
1997	35	855r	5,420	30,645r	28,685	**36,955r**	1,165r	7,850r	17,010r	32,095r	**58,120r**
1998	20	895r	5,275r	31,375r	29,810	**37,565r**	1,090r	7,865r	16,510r	32,540r	**58,005r**
1999	15	885	5,010	35,370	33,650	**41,280**	1,040	7,070	16,165	36,550	**60,825**

1.15 Mean air temperatures[1][2]

Great Britain

Degrees Celsius

	Average 1961-90	Deviations from normal (average 1961-90)					
		1995	1996	1997	1998	1999	2000
Calendar year	9.5	+1.1	-0.1	+1.2	+1.0	+1.1	
First half year	7.7	+0.9	-0.2	+1.1	+1.7	+1.4	
Second half year	11.1	+1.4	+0.2	+1.5	+0.5	+0.9	
First quarter	4.5	+1.2	-0.3	+1.6	+2.6	+1.8	+2.0
Second quarter	10.9	+0.6	-0.1	+0.7	+0.8	+1.1	
Third quarter	15.0	+2.1	+0.6	+1.7	+0.4	+1.5	
Fourth quarter	7.3	+0.7	-0.1	+1.2	+0.6	-2.3	
Summer (3)	12.9	+1.4	+0.3	+1.2	+0.6	+1.3	
Winter (3)	5.9	+0.3	+0.8	+1.6	+1.6	+1.1	
January	3.9	+1.0	+0.9	-1.0	+1.6	+1.9	+1.6
February	3.9	+2.8	-0.8	+3.0	+3.8	+1.7	+2.5
March	5.7	-0.1	-1.1	+2.7	+2.3	+1.7	+1.8
April	7.8	+1.1	+0.9	+1.3	-	+1.6	+0.1
May	10.9	+0.7	-1.6	+0.6	+2.0	+1.9	+1.2
June	13.9	+0.1	+0.5	+0.1	+0.2	-0.2	
July	15.8	+2.6	+0.6	+1.1	-0.3	+1.7	
August	15.6	+3.3	+1.1	+3.0	+0.3	+0.7	
September	13.5	+0.3	+0.2	+1.0	+1.3	+2.2	
October	10.6	+2.6	+1.2	-0.1	-	+0.4	
November	6.6	+1.5	-0.4	+2.3	+0.7	+1.5	
December	4.7	-1.9	-1.2	+1.4	+1.2	-0.9	

(1) See Energy Trends Table 24 for the latest monthly figures.
(2) Based on data provided by the Meteorological Office. The figures are averages of the monthly mean temperatures as recorded at 16 meteorological stations selected as representative of fuel consumption in Great Britain - 2 in Scotland, 2 in Wales and 12 in England, four of which are counted twice.(Prior to September 1990, recordings were from 15 stations - 2 in Scotland, 2 in Wales and 11 in England, five of which were counted twice.)
(3) The summer period is from April to September inclusive, and the winter period is the six months beginning in October and ending with March of the following year.

1.16 Mean air temperatures[1][2], 1960 to 1999

Great Britain

Degrees Celsius

	January	February	March	April	May	June	July	August	September	October	November	December
1960	4.4	4.1	5.5	9.2	12.8	15.0	15.3	15.2	13.3	10.5	7.5	4.2
1961	3.6	6.9	8.4	9.9	11.1	14.4	15.2	15.6	12.6	11.0	6.3	2.6
1962	4.4	4.5	2.9	7.8	10.3	13.8	14.8	14.5	12.7	10.7	5.7	2.3
1963	-1.4	-0.3	6.0	8.5	10.5	14.7	15.1	14.4	13.1	11.1	8.2	3.0
1964	3.6	4.8	4.5	8.9	13.2	13.9	16.1	15.7	14.2	9.2	7.5	3.8
1965	3.7	3.7	5.4	8.2	11.5	14.3	14.0	15.0	12.6	11.2	4.9	4.6
1966	3.1	5.7	6.8	7.3	11.1	15.3	15.0	14.8	14.1	10.3	5.7	5.5
1967	4.6	5.5	7.4	7.9	10.2	14.2	16.8	15.7	13.8	10.9	5.7	4.7
1968	4.6	2.3	6.6	8.2	9.8	14.9	14.9	15.5	14.0	12.7	6.7	3.3
1969	5.7	1.2	3.6	7.5	11.0	13.8	16.8	16.3	14.1	12.9	5.5	3.4
1970	4.0	3.2	4.0	6.8	12.7	16.1	15.4	16.1	14.5	10.9	7.9	4.5
1971	4.7	5.0	5.4	7.8	11.5	12.5	16.9	15.6	14.3	11.6	6.4	7.1
1972	4.2	4.6	6.5	8.6	10.6	11.9	15.5	15.2	11.9	10.7	6.4	5.8
1973	4.7	4.7	6.5	7.2	11.3	14.9	15.7	16.5	14.3	9.4	6.2	5.1
1974	6.1	5.8	5.8	8.0	10.9	13.7	15.1	15.2	12.1	7.9	6.7	8.0
1975	6.7	4.7	5.0	8.3	9.7	14.5	17.2	18.2	13.4	10.2	6.3	5.3
1976	5.9	4.8	5.0	8.0	11.8	16.7	18.3	17.3	13.4	10.7	6.2	2.2
1977	3.0	5.1	7.0	7.3	10.4	12.4	15.9	15.3	13.1	11.7	6.4	6.2
1978	3.4	3.6	6.8	6.4	11.3	13.6	14.7	14.9	14.0	11.9	8.6	4.3
1979	0.5	1.4	4.8	7.6	9.7	14.1	16.2	14.9	13.2	11.2	7.0	5.5
1980	2.4	6.0	4.9	8.7	11.0	13.8	14.5	15.7	14.6	9.0	6.6	5.8
1981	4.8	3.3	6.6	7.8	10.5	13.3	15.6	16.2	14.6	7.6	7.7	0.8
1982	2.8	4.8	5.8	8.2	11.1	11.2	16.2	15.4	13.8	9.8	7.4	4.1
1983	6.2	1.9	6.1	6.3	9.6	13.6	18.4	16.8	13.2	10.0	7.3	5.5
1984	3.3	3.5	4.5	7.7	9.5	13.9	16.2	17.0	13.2	10.7	7.7	5.0
1985	1.0	2.5	4.4	8.0	10.4	12.2	15.6	14.2	14.1	10.7	4.0	6.1
1986	3.2	-0.5	4.9	5.4	10.6	14.1	15.4	13.2	11.0	10.6	7.3	5.8
1987	1.1	3.7	4.1	9.4	9.7	12.2	15.5	15.2	13.3	9.3	6.4	4.7
1988	4.9	4.5	5.8	7.8	11.2	14.0	14.4	14.9	13.2	9.4	5.3	7.1
1989	6.1	5.8	7.0	6.1	12.5	14.0	17.4	16.1	14.1	11.5	6.4	4.5
1990	6.3	7.0	8.0	7.7	12.1	13.3	16.3	17.6	13.1	12.0	7.2	5.1
1991	3.7	2.4	7.8	8.0	11.0	12.2	17.1	17.0	14.7	10.3	7.0	5.0
1992	4.0	5.9	7.4	8.6	13.1	15.5	16.1	15.3	13.2	7.8	7.5	4.1
1993	6.0	5.4	6.6	9.3	11.2	14.4	15.1	14.4	12.5	8.5	5.0	5.3
1994	5.2	3.5	7.6	8.1	10.4	14.3	17.6	15.9	12.7	10.2	10.1	6.4
1995	4.9	6.7	5.6	8.9	11.6	14.0	18.4	18.9	13.8	13.2	8.1	2.8
1996	4.8	3.1	4.6	8.7	9.3	14.4	16.4	16.7	13.7	11.8	6.2	3.5
1997	2.9	6.9	8.4	9.1	11.5	14.0	16.9	18.6	14.5	10.5	8.9	6.1
1998	5.5	7.7	8.0	7.8	12.9	14.1	15.5	15.9	14.8	10.6	7.3	5.9
1999	5.8	5.6	7.4	9.4	12.8	13.7	17.5	16.3	15.7	11.0	8.1	3.8

(1) See Energy Trends Table 24 for the latest monthly figures.

(2) Average mean air temperatures calculated from the maximum and minimum daily temperature as recorded at 16 meteorological stations (17 up to 1976, 15 between 1977 and August 1990), selected as representative of fuel consumption in Great Britain - 2 in Scotland, 2 in Wales and 12 in England, 4 of which are counted twice (13 in England up to 1976, 7 of which were counted twice, and between 1977 and 1990 11 in England, 5 of which were counted twice). Data on temperatures recorded are provided by the Meteorological Office.

Chapter 2
Solid fuels and derived gases

Introduction

2.1   This chapter presents figures on the supply and demand for coal and solid fuels derived from coal, and on the production and consumption of gases derived from the processing of solid fuels.

2.2   The structure of this chapter is unchanged this year. Balances for coal and for manufactured fuels covering each of the last three years form the first six tables (Tables 2.1 to 2.6). These are followed by a 5 year table showing the supply and consumption of coal as a time series (Table 2.7). Comparable 5 year tables bring together data for coke oven coke, coke breeze and manufactured solid fuels (Table 2.8) and coke oven gas, blast furnace gas, benzoles and tars (Table 2.9). The long term trends tables on coal production and stocks (Table 2.10) and on coal consumption (Table 2.11) complete the chapter.

2.3   Imports and exports of solid fuels are given in Chapter 8, and solid renewables are covered in Chapter 7.

2.4   Figures for actual consumption of coal are available for all fuel producers and for final use by the iron and steel industry. For the remaining final users consumption figures are based on information on disposals by producers and on imports. For further details see the technical notes and definitions section which begins at paragraph 2.36 of this chapter.

Structure of the coal industry

2.5   Following the privatisation of the coal industry at the end of 1994 the four main coal producers were RJB Mining, Mining Scotland, Celtic Energy and Coal Investments. The last mentioned went into receivership in June 1996 and only two of their mines remained as going concerns under the company name of Midlands Mining. One of these two mines closed at the end of 1998 and the other at the beginning of 2000. The other four collieries reverted to the Coal Authority and were subsequently closed.

2.6   There are also a number of independent deep mines of which the main ones are Goitre Tower Anthracite, Hatfield Coal Company and Betws Anthracite, and there are several independent opencast coal producers, particularly in Scotland and the north east of England. Some further coal and slurry are supplied from recovery operations.

Commodity balances for coal (Tables 2.1, 2.2 and 2.3).

2.7   These balance tables separately identify the three main types of coal, steam coal, coking coal, and anthracite and show the variation both in the sources of supply and where the various types of coal are mainly used.

2.8   In 1999, 80½ per cent of coal demand was for steam coal, 15½ per cent was for coking coal and 4 per cent was for anthracite. Electricity generation accounted for 89 per cent of demand for steam coal and 28 per cent of demand for anthracite. Coking coal was nearly all used in coke ovens (94 per cent), but 6 per cent was directly injected into blast furnaces.

2.9   Only 11 per cent of the total demand for coal was for final consumption, where it would be used for steam raising, space or hot water heating or heat for processing. Steam coal accounted for 82 per cent of this final consumption with nearly half of final consumption being steam coal for industry where mineral products (eg cement, glass and bricks) and chemicals were the largest users. The domestic sector accounted for nearly half of the final demand for coal with two thirds of this demand being for steam coal and one third for anthracite.

2.10   Chart 2.1, below, compares the sources of coal supplies in the UK in 1999, along with a breakdown of consumption by user and serves to illustrate some of the features brought out below.

2.11   In 1999, 37 per cent of the total supply figures in Table 2.1 was met from deep-mined production, 27 per cent from opencast operations, 36 per cent from net imports and 1½ per cent from other sources such as slurry. Supply from all these sources was higher than demand resulting in 1.0 million tonnes (equivalent to 1½ per cent of total supply) being added to stocks.

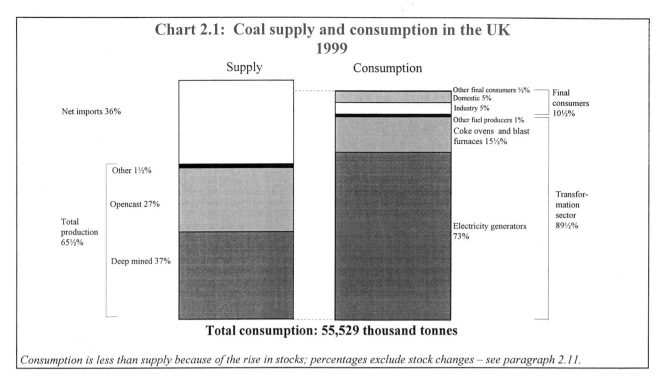

Chart 2.1: Coal supply and consumption in the UK
1999

Supply   Consumption

Net imports 36%

Other 1½%

Opencast 27%

Total
production
65½%

Deep mined 37%

Other final consumers ½% — Final
Domestic 5% — consumers
Industry 5% — 10½%
Other fuel producers 1%
Coke ovens and blast
furnaces 15½%

Transfor-
mation
sector
89½%

Electricity generators
73%

Total consumption: 55,529 thousand tonnes

Consumption is less than supply because of the rise in stocks; percentages exclude stock changes – see paragraph 2.11.

2.12   Recent trends in coal production and consumption are described in paragraphs 2.20 to 2.26.

Commodity balances for manufactured fuels (Table 2.4, 2.5 and 2.6).

2.13   The balance tables for manufactured fuels cover fuels manufactured from coal and gases produced when coal is used in coke ovens and blast furnaces. Definitions of terms associated with coke, breeze and other manufactured solid fuels are to be found in paragraphs 2.48 to 2.51.

2.14   The majority of **coke oven coke**, is home produced with imports amounting to less than 7 per cent of the home produced volume. About 1½ per cent of home production was exported in 1999. The amount screened out by producers as breeze and fines, amounted to about 15 per cent of production plus imports in 1999, and this 1.0 million tonnes appears in the coke breeze column of the balance. Transfers out are not equal to transfers in because of differences in the timing and location of measurements. Over 90 per cent of the demand for coke is at blast furnaces (part of the transformation sector) with the remainder going into final consumption in either the non-ferrous metals sector (eg foundry coke) or the domestic sector in 1999.

2.15   Most of the supply of **coke breeze** is from re-screened coke oven coke, with exports more than matching the small quantities that are produced manufacture or in blast furnaces, but the majority is boiler fuel.

2.16   Patent fuels are manufactured smokeless fuels produced mainly for the domestic market, as the balances show. A small amount of these fuels (only 1 per cent of total supply in 1999) is imported, but exports generally exceed imports. Imports and exports of manufactured fuels can contain small quantities of non-smokeless fuels.

2.17   Chart 2.3 shows the sources of coke, breeze and other manufactured solid fuels and a breakdown of their consumption.

2.18   The carbonisation and gasification of solid fuels at coke ovens produces **coke oven gas** as a by-product. Some of this (40 per cent in 1999) is used to fuel the coke ovens themselves while at steel works some is piped to blast furnaces and used in the production of steel (9 per cent in 1999). Elsewhere at steel works the gas is used for electricity generation (15 per cent) or for heat production for other iron and steel making processes (33 per cent).

2.19   **Blast furnace gas** is a by-product of iron making in a blast furnace. A similar product is obtained when steel is made in basic oxygen steel converters, and "BOS" gas is included in this category. Most of this gas is used in other parts of integrated steel works, with 42 per cent being used for electricity generation in 1999, 33 per cent being used in coke ovens and blast furnaces themselves, and 16 per cent being used for general heat production. The remaining 9 per cent is lost or burned as waste.

Supply and consumption of coal (Table 2.7).

2.20 **Production** - Figures for 1999 show that coal production fell by 9½ per cent compared to production in 1998. Deep-mined production fell by 17½ per cent compared to 1998 but opencast production rose by 3½ per cent. Overall demand for coal was 12 per cent down on 1998 resulting in 1 million tonnes being added to stocks and a 1½ per cent fall in the volume of net imports of coal. One major deep mine (Calverton) ceased production at the end of July 1999, and Annesley-Bentinck closed in January 2000. Longer term trends in production are illustrated in Chart 2.4.

2.21 Table 2A shows how production of coal is divided between England, Wales and Scotland. In 1999/2000 70 per cent of the coal output was in England, 24 per cent in Scotland, and 6 per cent in Wales.

Table 2A: Output from UK coal mines [1]

	Million tonnes		
	April 1997 to March 1998	April 1998 to March 1999	April 1999 to March 2000
Deep-mined			
England	25.4	21.4	17.9
Scotland	1.9	1.6	1.0
Wales	0.8	0.7	0.6
Total	28.1	23.7	19.5
Opencast			
England	8.1	7.0	6.2
Scotland	6.3	6.4	7.2
Wales	1.8	1.5	1.5
Total	16.2	14.9	14.9
Total			
England	33.6	28.4	24.1
Scotland	8.2	8.0	8.2
Wales	2.5	2.2	2.1
Total	44.3	38.6	34.4

Source: The Coal Authority

(1) Output is the tonnage declared by operators to the Coal Authority, including estimated tonnages. It excludes estimates of slurry recovered from dumps, ponds, rivers, etc.

2.22 Table 2B shows how numbers employed in the production of coal have changed over the last three years. During 1999/2000 employment including contractors fell by 21 per cent. At 31 March 2000, 75 per cent of the 11,200 people employed in UK coal mining worked in England while 17 per cent were in Scotland and 8 per cent in Wales.

2.23 **Foreign trade** - Net imports of coal were 1½ per cent lower in 1999 than the record level set in 1998. Within this imports of steam coal rose by ½ per cent. The trends in coal imports and exports are discussed in greater detail in Chapter 8, paragraphs 8.19 to 8.22.

Table 2B: Employment in UK coal mines [1]

	end March 1998	end March 1999	end March 2000
Deep-mined			
England	10,906	9,016	7,064
Scotland	876	802	643
Wales	841	664	536
Total	12,623	10,482	8,243
Opencast			
England	2,265	1,882	1,300
Scotland	1,600	1,301	1,252
Wales	694	550	433
Total	4,559	3,733	2,985
Total			
England	13,171	10,898	8,364
Scotland	2,476	2,103	1,895
Wales	1,535	1,214	969
Total	17,182	14,215	11,228

Source: The Coal Authority

(1) Employment includes contractors and is as declared by licensees to the Coal Authority at 31 March each year.

2.24 **Transformation** - The 12 per cent fall in coal consumption during 1999 mainly results from a 16½ per cent fall in consumption by major power producers. However, consumption for generation in 1998 was higher than in 1997 mainly because of reduced imports of electricity from France, and the rate of decline over the last two years taken together has been a more modest 7½ per cent per year. Even so, the volume of coal used for electricity generation in 1999 was a third lower than in 1995.

2.25 **Consumption** - Consumption by final consumers showed a 11 per cent rise compared with 1998, but a 6 per cent fall compared with two years earlier. Some of this rise could be due to consumers replenishing stocks after a substantial fall during 1998 coupled with a 15 per cent fall in consumption by final consumers. Domestic sector consumption was 14 per cent up, but in each of the last five years the volume consumed has been around 2.5 million tonnes. Industrial consumption was 13 per cent up, but 4½ per cent down on two years earlier.

2.26 Long term trends in the consumption of coal in the UK since 1960 onwards are presented in Table 2.11.

2.27 **Stocks** – Production and net imports together in 1999 were 7 per cent lower than in 1998, while demand for coal fell more substantially (by 12 per cent). This led to stocks of coal rising by 1.0 million tonnes (5½ per cent). This contrasts with 1.2 million tonnes being withdrawn from stocks in 1998. Stocks held at collieries and opencast sites at the end of 1999 were 0.4 million tonnes higher than a year earlier while stocks held by the major power producers were 0.8 million tonnes higher. The recent changes in coal stocks are illustrated in Chart 2.2.

Supply and consumption of coke oven coke, coke breeze and other manufactured fuels (Table 2.8)

2.28   This table presents figures for the most recent five years on the same basis as the balance tables. Figures for stocks are also included. Coal used to produce these manufactured fuels is given in Table 2.7. For **coke oven coke,** 1999 saw an increase in demand of 6 per cent. However, production was 4 per cent lower and imports 50 per cent lower, although already small in volume. As a result stocks of coke fell.

2.29   In 1999, the demand for **coke breeze** fell by 4½ per cent, but imports and rescreenings were also down so there was a rise in stocks. There was a 9 per cent decline in the demand for **other manufactured solid fuels,** mainly because of a 7½ per cent fall in domestic sector demand. UK production was up 3 per cent on 1998 levels.

Supply and consumption of coke oven gas, blast furnace gas, benzole and tars (Table 2.9)

2.30   This table presents figures for the most recent five years on the same basis as the balance tables.

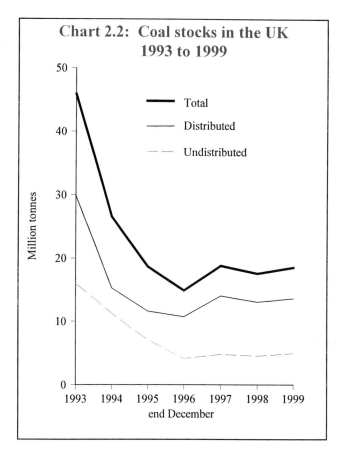

Demand for **coke oven gas** fell by 3½ per cent in 1999, but demand has remained close to 13,000 GWh in each of the last 5 years. Use in electricity generation and in blast furnaces rose but use in coke ovens and for general heating purposes in the iron and steel industry both fell. Both production and demand for **blast furnace gas** were lower in 1999 than in 1998 by around 5½ per cent. This decline cut across all areas of use.

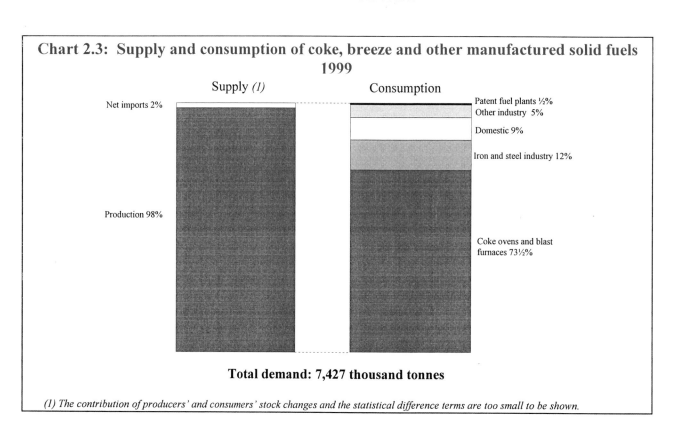

52

Long term trends

Coal production and stocks (Table 2.10)

2.31   Figures for coal production, imports, overseas shipments and stocks are given in Table 2.10 which is in turn based on Table 2.7 with the series extended back to 1960.

2.32   Table 2.10 shows a decline in deep-mined production of more than 89 per cent since the highest level shown in this table in 1963. Opencast production in 1999 was double the average level recorded in the 1960s. Table 2.10 also shows that imports, initially of coal types in short supply in this country, started in 1970 and in 1999 accounted for over a third of total coal supplies.   Both these trends are illustrated in Chart 2.4.   Stock levels in the early '90s were relatively high reaching a peak of 53 per cent of annual inland coal consumption in 1993. After this electricity generators began to run down their stocks sharply so that at the end of 1996 stocks were 21 per cent of annual consumption, but in 1997 they rose again to 30 per cent of consumption, and in 1999 they rose further in proportionate terms to 33½ per cent of annual consumption (see paragraph 2.27, above, and Chart 2.2).

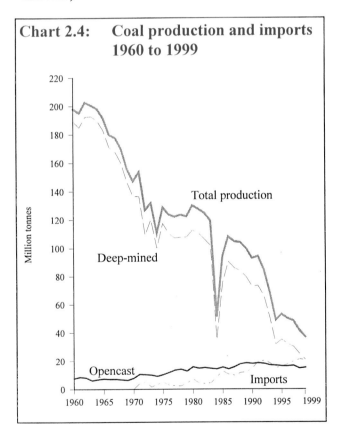

Chart 2.4:   Coal production and imports 1960 to 1999

Inland consumption of solid fuels (Table 2.11)

2.33   Figures for inland consumption of coal by fuel producers and final users are given in Table 2.11, which is in turn based on Table 2.7. The table also shows final consumption figures for coke and breeze and other solid fuels based on Table 2.8   These products are mainly supplied from the conversion of coal, supplemented by a small amount of foreign trade. Where possible the series have been extended back to 1960.

2.34   Trends in inland consumption of coal, in total and by power stations, coke ovens, and final consumers, are illustrated in Chart 2.5 below.

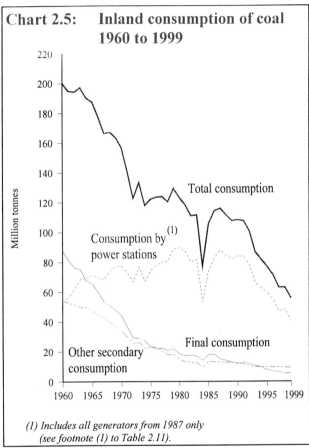

Chart 2.5:   Inland consumption of coal 1960 to 1999

(1) Includes all generators from 1987 only (see footnote (1) to Table 2.11).

2.35   Total inland consumption of coal fell by 72 per cent from 200 million tonnes in 1960 to 56 million tonnes in 1999.   Consumption by the electricity generators increased from 53 million tonnes in 1960 to a peak of 90 million tonnes in 1980 and continued in the 80-90 million tonnes range until 1991 with the exception of the miners strike years.   With the increased use of nuclear power and natural gas, the consumption of coal by the generators has fallen steadily since 1991. However, there was a pause in this trend in 1998 when coal fired generation was called upon to make up for the temporary reduction in imported electricity from France. The proportion of electricity supplied from coal in the early 1990s was around 70 per cent, falling to 28 per cent in 1999 (see Chapter 5). However, at 40 million tonnes in 1999, use of coal at power stations represents 73 per cent of total coal consumption, compared with only 26 per cent in 1960.

Technical notes and definitions

2.36 These notes and definitions are in addition to the technical notes and definitions covering all fuels and energy as a whole in Chapter 1, paragraphs 1.45 to 1.79. For notes on the commodity balances and definitions of the terms used in the row headings see the Annex A, paragraphs A.7 to A.41.

Steam coal, coking coal, and anthracite

2.37 **Steam coal** is coal classified as such by UK coal producers and by importers of coal. It tends to be coal having lower calorific values.

2.38 **Coking coal** is coal sold by producers for use in coke ovens and similar carbonising processes. The definition is not therefore determined by the calorific value or caking qualities of each batch of coal sold, although calorific values tend to be higher than for steam coal.

2.39 **Anthracite** is coal classified as such by UK coal producers and importers of coal. Typically it has a high heat content making it particularly suitable for certain industrial processes and for use as a domestic fuel.

Coal production

2.40 **Deep-mined** - The statistics cover saleable output from deep mines including coal obtained from working on both revenue and capital accounts. All licensed collieries (and British Coal collieries prior to 1995) are included, even where coal is only a subsidiary product.

2.41 **Opencast** - The figures cover saleable output and include the output of sites worked by private operators under agency agreements and licences as well as the output of sites licensed for the production of coal as a subsidiary to the production of other minerals.

2.42 **Other** - Estimates of slurry etc. recovered and disposed of from dumps, ponds, rivers, etc.

Imports and exports of coal and other solid fuels

2.43 Figures are derived from returns made to HM Customs and Excise and are broken down in greater detail in Chapter 8, Table 8.5.

2.44 However, in Tables 2.4, 2.5, 2.6 and 2.8, the export figures used for hard coke, coke breeze and other manufactured solid fuels for the years before 1998 are quantities of fuel exported as reported to DTI by the companies concerned, rather than quantities recorded by HM Customs and Excise in their Trade Statistics.

Allocation of imported coal

2.45 Although data are available on consumption of home produced coal, and also on consumption of imported coal by secondary fuel producers there is only very limited direct information on consumption of imported coal by final users. Following surveys conducted in 1992 and 1995, 75 per cent of steam coal imports, excluding those used by electricity generators, was allocated to industry each year and 25 per cent to the domestic sector. A further survey in 1998 showed that it was more appropriate to allocate 15 per cent of steam coal imports to the public administration sector in 1998, 1997 and 1996, 10 per cent in 1995 and 5 per cent in 1994 because of coal consumption in schools and hospitals. Correspondingly the allocation to industry has been reduced by the same percentages (ie to 60 per cent in 1996 to 1998). The 1998 proportions have been retained for 1999. In addition, 10 per cent of anthracite imports, excluding cleaned smalls, are allocated to industry, with 90 per cent to the domestic sector in all years shown in the tables. All imports of coking coal and cleaned anthracite smalls are allocated to coke and other solid fuel producers.

Stocks of coal

2.46 Undistributed stocks are those held at collieries and opencast sites. It is not possible to distinguish these two locations separately in the stock figures. Distributed stocks are those held at power stations and stocking grounds of the major power producing companies (as defined in Chapter 5, paragraph 5.58), coke ovens and low temperature carbonisation plants, and patent fuel plants.

Transformation, energy industry use and consumption of solid fuels

2.47 The introductory section on balances defines the activities that fall within these parts of the balances, but the following additional notes relevant to solid fuels are given below:

Transformation: Blast furnaces - Coking coal injected into blast furnaces is shown separately within the balance tables.

Transformation: Low temperature carbonisation plants and patent fuel plants - Coal used at these plants for the manufacture of domestic coke such as Coalite and of briquetted fuels such as Phurnacite and Homefire.

Consumption: Industry - The statistics comprise sales of coal by the eight main coal producers to the iron and steel industry (excluding that used at coke ovens and blast furnaces) and to other industrial sectors and estimated proportions of anthracite and steam coal imports. The figures exclude coal used for industries' own generation of electricity which appears separately under transformation.

Consumption: Domestic - Coal supplied free of charge or at reduced prices to current and retired miners, officials, etc in the coalfields. The concessionary fuel provided to miners in 1999 is estimated at 307 thousand tonnes. This estimate is included in the domestic steam coal and domestic anthracite figures.

Consumption of coke and other manufactured solid fuels - These are disposals from coke ovens to merchants. The figures also include estimated proportions of coke imports.

Coke oven coke (hard coke) and hard coke breeze

2.48   The statistics cover coke produced at coke ovens owned by the British Steel/Corus plc, Coal Products Ltd and other producers. Low temperature carbonisation plants are not included (see paragraph 2.51, below). Breeze (as defined in paragraph 2.49) is excluded from the figures for coke oven coke.

2.49   Breeze can generally be described as coke screened below 19 mm (¾ inch) with no fines removed, but the screen size may vary in different areas and to meet the requirements of particular markets. Coke that has been transported from one location to another is usually re-screened before use to remove smaller sizes, giving rise to further breeze.

2.50   In 1998, an assessment using industry data showed that on average over the last five years 91 per cent of imports have been coke and 9 per cent breeze and it is these proportions that have been used for 1998 and 1999 in Tables 2.4, 2.5, 2.6 and 2.8.

2.51   Other manufactured solid fuels are mainly solid smokeless fuels for the domestic market for use in both open fires and in boilers. A smaller quantity is exported (although exports are largely offset by similar quantities of imports in most years). Manufacture takes place in patented fuel plants and low temperature carbonisation plants. The brand names used for these fuels include Homefire, Phurnacite, Ancit and Coalite.

Blast furnace gas, coke oven gas, benzole and tars

2.52   The following definitions are used in the tables that include these fuels:

Blast furnace gas - includes basic oxygen steel furnace (BOS) gas. Blast furnace gas is the gas produced during iron ore smelting when hot air passes over coke within the blast ovens. It contains carbon monoxide, carbon dioxide, hydrogen and nitrogen. In a basic oxygen steel furnace the aim is not to introduce nitrogen or hydrogen into the steel making process, so pure oxygen gas and suitable fluxes are used to remove the carbon and phosphorous from the molten pig iron and steel scrap. A similar fuel gas is thus produced.

Coke oven gas - is a gas produced during the carbonisation of coal to form coke at coke ovens.

Synthetic coke oven gas - is mainly natural gas which is mixed with smaller amounts of blast furnace and BOS gas to produce a gas with almost the same quantities as coke oven gas. The transfers row of Tables 2.4, 2.5 and 2.6 shows the quantities of blast furnace gas used for this purpose and the total input of gases to the synthetic coke oven gas process. There is a corresponding outward transfer from natural gas in Chapter 4,Table 4.1.

Benzole - a colourless liquid, flammable, aromatic hydrocarbon by-product of the iron and steel making process. It is used as a solvent in the manufacture of styrenes and phenols but is also used as a motor fuel.

Tars - viscous materials usually derived from the destructive distillation of coal which are a by-product of the coke and iron making processes.

Periods covered

2.53   Figures in this chapter generally relate to periods of 52 weeks or 53 weeks as follows:

Year	52 weeks ended
1995	23 December 1995
	53 weeks ended
1996	28 December 1996
	52 weeks ended
1997	27 December 1997
1998	26 December 1998
1999	25 December 1999

The 53 week data for 1996 have been adjusted to 52 week equivalents by taking 13/14ths of the 14 week first quarter of 1996.

2.54 Data for coal used for electricity generation by major power producers follow the electricity industry calendar (see Chapter 5, paragraph 5.66) and coal use by other generators is for the 12 months ended 31 December each year. HM Customs and Excise data on imports and exports are also for the 12 months ended 31 December each year. Data for coal and coke use in the iron and steel industry, and for gases, benzoles and tars produced by the iron and steel industry follow the iron and steel industry calendar (see Chapter 5, paragraph 5.67).

Data collection

2.55 Since 1995 aggregate data on coal production have been obtained from the Coal Authority. In addition eight main producers (Betws Anthracite, Celtic Energy, Goitre Tower Anthracite, Hatfield Coal Company, Midlands Mining, Mining Scotland, Monktonhall Colliery and RJB Mining)[1] have provided data in response to an annual DTI inquiry covering production (deep-mined and opencast), trade, stocks, and disposals. The Iron and Steel Statistics Bureau (ISSB) provides DTI with an annual statement of coke and breeze production and use of coal, coke and breeze within that industry. The ISSB is also the source of data on gases produced by the iron and steel industry (coke oven gas, blast furnace gas and basic oxygen steel furnace gas). DTI directly surveys producers of manufactured fuels other than coke or breeze.

2.56 Trade in solid fuels is also covered by using data from HM Customs and Excise as described in Chapter 8 paragraphs 8.23 to 8.28. Consumption of coal for electricity generation is covered by data collected by DTI from electricity generators described in Chapter 5, paragraphs 5.71 to 5.73.

Monthly and quarterly data

2.57 Monthly data on coal production, foreign trade, consumption and stocks are provided in DTI's monthly statistical bulletin *Energy Trends* in Tables 4, 5 and 6. Quarterly data on coke, breeze and other solid fuels production, foreign trade and consumption is given in Table 7 of that bulletin. See Annex F for more information about *Energy Trends*.

Statistical differences

2.58 Tables 2.1 to 2.9 each contain a statistical difference term covering the difference between recorded supply and recorded demand. These statistical differences arise for a number of reasons. Firstly the data within each table are taken from varied sources, as described above, such as producers, intermediate consumers (such as electricity generators), final consumers (such as the iron and steel industry), and Customs and Excise. Secondly some of these industries work to different statistical calendars (see paragraphs 2.53 and 2.54, above), and thirdly some of the figures are estimated either because data in the required detail are not readily available within the industry or because the methods of collecting the data do not cover the smallest members of the industry.

Contact: *Mike Janes (Statistician)*
mike.janes@dti.gsi.gov.uk
020-7215 5186

James Achur
james.achur@dti.gsi.gov.uk
020-7215 2717

---

[1] *Midlands Mining succeeded Coal Investments during 1996 and Monktonhall colliery ceased production in early 1997.*

2.1 Commodity balances 1999

Coal

Thousand tonnes

	Steam coal	Coking coal	Anthracite	Total
Supply				
Production	..	..	..	36,163
Other sources	..	-	..	914
Imports	12,139	8,020	599	20,758
Exports	-434	-	-328	-762
Marine bunkers	-	-	-	-
Stock change (1)	..	258	..	-983
Transfers	-	-	-	-
Total supply	..	8,546	..	56,090
Statistical difference (2)	..	-12	..	+561
Total demand	44,633	8,558	2,338	55,529
Transformation	39,807	8,558	1,307	49,672
Electricity generation	39,807	-	664	40,471
Major power producers	38,321	-	664	38,985
Autogenerators	1,486	-	-	1,486
Petroleum refineries	-	-	-	-
Coke manufacture	-	8,064	-	8,064
Blast furnaces	-	494	-	494
Patent fuel manufacture and low temperature carbonisation	-	-	643	643
Energy industry use	..	-	..	10
Electricity generation	-	-	-	-
Oil and gas extraction	-	-	-	-
Petroleum refineries	-	-	-	-
Coal extraction	..	-	..	10
Coke manufacture	-	-	-	-
Blast furnaces	-	-	-	-
Patent fuel manufacture	-	-	-	-
Pumped storage	-	-	-	-
Other	-	-	-	-
Losses	-	-	-	-
Final consumption	4,817	-	1,030	5,847
Industry	2,783	-	53	2,836
Unclassified	-	-	-	-
Iron and steel	10	-	-	10
Non-ferrous metals	..	-	..	244
Mineral products	..	-	..	890
Chemicals	..	-	..	782
Mechanical engineering etc	..	-	..	30
Electrical engineering etc	..	-	..	8
Vehicles	..	-	..	97
Food, beverages etc	..	-	..	257
Textiles, leather, etc	..	-	..	63
Paper, printing etc	..	-	..	164
Other industries	..	-	..	291
Construction	-	-	-	-
Transport	-	-	-	-
Air	-	-	-	-
Rail	-	-	-	-
Road	-	-	-	-
National navigation	-	-	-	-
Pipelines	-	-	-	-
Other	..	-	..	3,011
Domestic	1,718	-	975	2,693
Public administration	..	-	..	293
Commercial	-	-	-	4
Agriculture	..	-	..	7
Miscellaneous	..	-	..	14
Non energy use	-	-	-	-

(1) Stock fall (+), stock rise (-).
(2) Total supply minus total demand.

2.2 Commodity balances 1998

Coal

Thousand tonnes

	Steam coal	Coking coal	Anthracite	Total
Supply				
Production	..	588r	..	40,046r
Other sources	..	-	..	1,131r
Imports	12,079r	8,646	519r	21,244r
Exports	-689r	-	-282r	-971r
Marine bunkers	-	-	-	-
Stock change (1)	..	-184	..	+1,224
Transfers	-	-	-	-
Total supply	..	9,050r	..	62,674r
Statistical difference (2)	..	+322r	..	-465r
Total demand	51,940r	8,728	2,471r	63,139r
Transformation	47,708r	8,728	1,437r	57,873r
Electricity generation	47,708r	-	802r	48,510r
Major power producers	45,825r	-	802r	46,627
Autogenerators	1,883r	-	-	1,883r
Petroleum refineries	-	-	-	-
Coke manufacture	-	8,169	-	8,169
Blast furnaces	-	559	-	559
Patent fuel manufacture and low temperature carbonisation	-	-	635	635
Energy industry use	..	-	..	10
Electricity generation	-	-	-	-
Oil and gas extraction	-	-	-	-
Petroleum refineries	-	-	-	-
Coal extraction	..	-	..	10
Coke manufacture	-	-	-	-
Blast furnaces	-	-	-	-
Patent fuel manufacture	-	-	-	-
Pumped storage	-	-	-	-
Other	-	-	-	-
Losses	-	-	-	-
Final consumption	4,223r	-	1,033r	5,256r
Industry	2,453r	-	61r	2,514r
Unclassified	-	-	-	-
Iron and steel	9	-	-	9
Non-ferrous metals	..	-	..	183r
Mineral products	..	-	..	835r
Chemicals	..	-	..	637r
Mechanical engineering etc	..	-	..	28
Electrical engineering etc	..	-	..	3
Vehicles	..	-	..	87r
Food, beverages etc	..	-	..	257r
Textiles, leather, etc	..	-	..	69
Paper, printing etc	..	-	..	121r
Other industries	..	-	..	285r
Construction	-	-	-	-
Transport	-	-	-	-
Air	-	-	-	-
Rail	-	-	-	-
Road	-	-	-	-
National navigation	-	-	-	-
Pipelines	-	-	-	-
Other	..	-	..	2,742r
Domestic	1,402r	-	960r	2,362r
Public administration	..	-	..	308
Commercial	-	-	-	4
Agriculture	..	-	..	9
Miscellaneous	..	-	..	59
Non energy use	-	-	-	-

(1) Stock fall (+), stock rise (-).
(2) Total supply minus total demand.

2.3 Commodity balances 1997

Coal

Thousand tonnes

	Steam coal	Coking coal	Anthracite	Total
Supply				
Production	..	1,076	..	46,981
Other sources	..	-	..	1,514
Imports	10,756r	8,072	931r	19,759r
Exports	-731	-1	-414	-1,146r
Marine bunkers	-	-	.. -	-
Stock change *(1)*	..	102r	..	-3,973r
Transfers	-	-	-	-
Total supply	..	9,249r	..	63,135r
Statistical difference *(2)*	..	+499r	..	+55r
Total demand	52,043r	8,750	2,287r	63,080r
Transformation	47,250r	8,750	864	56,864r
Electricity generation	47,250r	-	-	47,250r
Major power producers	45,323r	-	-	45,323r
Autogenerators	1,927r	-	-	1,927r
Petroleum refineries	-	-	-	-
Coke manufacture	-	8,143	-	8,143
Blast furnaces	-	607	-	607
Patent fuel manufacture and low temperature carbonisation	-	-	864	864
Energy industry use	..	-	..	8
Electricity generation	-	-	-	-
Oil and gas extraction	-	-	-	-
Petroleum refineries	-	-	-	-
Coal extraction	..	-	..	8
Coke manufacture	-	-	-	-
Blast furnaces	-	-	-	-
Patent fuel manufacture	-	-	-	-
Pumped storage	-	-	-	-
Other	-	-	-	-
Losses	-	-	-	-
Final consumption	4,786r	-	1,422r	6,208r
Industry	2,879r	-	91	2,970r
Unclassified	-	-	-	-
Iron and steel	1	-	-	1
Non-ferrous metals	..	-	..	149r
Mineral products	..	-	..	926r
Chemicals	..	-	..	650r
Mechanical engineering etc	..	-	..	22r
Electrical engineering etc	-	-	-	-
Vehicles	..	-	..	102r
Food, beverages etc	..	-	..	368r
Textiles, leather, etc	..	-	..	68
Paper, printing etc	..	-	..	233r
Other industries	..	-	..	451r
Construction	-	-	-	-
Transport	-	-	-	-
Air	-	-	-	-
Rail	-	-	-	-
Road	-	-	-	-
National navigation	-	-	-	-
Pipelines	-	-	-	-
Other	..	-	..	3,238r
Domestic	1,309	-	1,278	2,587
Public administration	..	-	..	455
Commercial	-	-	-	-
Agriculture	..	-	..	7
Miscellaneous	..	-	..	189
Non energy use	-	-	-	-

(1) Stock fall (+), stock rise (-).
(2) Total supply minus total demand.

2.4 Commodity balances 1999

Manufactured fuels

	Thousand tonnes				GWh		
	Coke oven coke	Coke breeze	Other manuf. solid fuel	Total manuf. solid fuel	Benzole and tars (4)	Coke oven gas	Blast furnace gas
Supply							
Production	5,837	33	635	6,504	2,369	12,619	19,037
Other sources	-	-	-	-	-	-	-
Imports	389	40	6	436	-	-	-
Exports	-79	-165	-54	-297	-	-	-
Marine bunkers	-	-	-	-	-	-	-
Stock change (1)	+353	+6	-7	+352	-	-	-
Transfers (3)	-951	+1,035	-	+83	-	+528	-22
Total supply	**5,549**	**948**	**580**	**7,077**	**2,369**	**13,147**	**19,015**
Statistical difference (2)	-73	-152	-5	-230	-	+1	-29
Total demand	**5,622**	**1,100**	**585**	**7,307**	**2,369**	**13,146**	**19,044**
Transformation	**5,119**	**225**	**-**	**5,344**	**-**	**1,946**	**7,992**
Electricity generation	-	-	-	-	-	1,946	7,992
Major power producers	-	-	-	-	-	-	-
Autogenerators	-	-	-	-	-	1,946	7,992
Petroleum refineries	-	-	-	-	-	-	-
Coke manufacture	-	25	-	25	-	-	-
Blast furnaces	5,119	200	-	5,319	-	-	-
Patent fuel manufacture	-	-	-	-	-	-	-
Low temperature carbonisation	-	-	-	-	-	-	-
Energy industry use	**20**	**-**	**13**	**33**	**-**	**6,522**	**6,282**
Electricity generation	-	-	-	-	-	-	-
Oil and gas extraction	-	-	-	-	-	-	-
Petroleum refineries	-	-	-	-	-	-	-
Coal extraction	-	-	-	-	-	-	-
Coke manufacture	-	-	-	-	-	5,283	1,094
Blast furnaces	-	-	-	-	-	1,239	5,188
Patent fuel manufacture	20	-	13	33	-	-	-
Pumped storage	-	-	-	-	-	-	-
Other	-	-	-	-	-	-	-
Losses	**-**	**-**	**-**	**-**	**-**	**194**	**1,741**
Final consumption	**484**	**875**	**572**	**1,931**	**2,369**	**4,484**	**3,029**
Industry	**368**	**875**	**18**	**1,261**	**2,369**	**4,484**	**3,029**
Unclassified	206	14	18	239	606	124	-
Iron and steel	19	861	-	880	1,763	4,360	3,029
Non-ferrous metals	143	-	-	143	-	-	-
Mineral products	-	-	-	-	-	-	-
Chemicals	-	-	-	-	-	-	-
Mechanical engineering etc	-	-	-	-	-	-	-
Electrical engineering etc	-	-	-	-	-	-	-
Vehicles	-	-	-	-	-	-	-
Food, beverages etc	-	-	-	-	-	-	-
Textiles, leather, etc	-	-	-	-	-	-	-
Paper, printing etc	-	-	-	-	-	-	-
Other industries	-	-	-	-	-	-	-
Construction	-	-	-	-	-	-	-
Transport	**-**	**-**	**-**	**-**	**-**	**-**	**-**
Air	-	-	-	-	-	-	-
Rail	-	-	-	-	-	-	-
Road	-	-	-	-	-	-	-
National navigation	-	-	-	-	-	-	-
Pipelines	-	-	-	-	-	-	-
Other	**116**	**-**	**554**	**670**	**-**	**-**	**-**
Domestic	116	-	554	670	-	-	-
Public administration	-	-	-	-	-	-	-
Commercial	-	-	-	-	-	-	-
Agriculture	-	-	-	-	-	-	-
Miscellaneous	-	-	-	-	-	-	-
Non energy use	**-**	**-**	**-**	**-**	**-**	**-**	**-**

(1) Stock fall (+), stock rise (-).
(2) Total supply minus total demand.
(3) Coke oven gas and blast furnace gas transfers are for synthetic coke oven gas, see paragraph 2.52

(4) Because of the small number of benzole suppliers, figures for benzole and tars cannot be given separately.

2.5 Commodity balances 1998

Manufactured fuels

	Thousand tonnes				GWh		
	Coke oven coke	Coke breeze	Other manuf. solid fuel	Total manuf. solid fuel	Benzole and tars (4)	Coke oven gas	Blast furnace gas
Supply							
Production	6,178	37	616	6,831	2,542r	13,126r	20,116r
Other sources	-	-	-	-	-	-	-
Imports	753	78	10	841	-	-	-
Exports	-93r	-196r	-56	-345r	-	-	-
Marine bunkers	-	-	-	-	-	-	-
Stock change (1)	-264	+42	+88r	-134r	-	-	-
Transfers (3)	-1,223	+1,163	-	-60	-	+630	-22
Total supply	**5,351r**	**1,124r**	**658r**	**7,133r**	**2,542r**	**13,756r**	**20,094r**
Statistical difference (2)	-57r	-46r	+13r	-90r	-	+126r	+34r
Total demand	**5,408r**	**1,170r**	**645r**	**7,223r**	**2,542r**	**13,630r**	**20,060r**
Transformation	**4,908**	**287**	**-**	**5,195r**	**-**	**1,814r**	**8,057r**
Electricity generation	-	-	-	-	-	1,814r	8,057r
Major power producers	-	-	-	-	-	-	-
Autogenerators	-	-	-	-	-	1,814r	8,057r
Petroleum refineries	-	-	-	-	-	-	-
Coke manufacture	-	50	-	50	-	-	-
Blast furnaces	4,908	237	-	5,145	-	-	-
Patent fuel manufacture	-	-	-	-	-	-	-
Low temperature carbonisation	-	-	-	-	-	-	-
Energy industry use	**27**	**-**	**14**	**41**	**-**	**6,855r**	**6,579r**
Electricity generation	-	-	-	-	-	-	-
Oil and gas extraction	-	-	-	-	-	-	-
Petroleum refineries	-	-	-	-	-	-	-
Coal extraction	-	-	-	-	-	-	-
Coke manufacture	-	-	-	-	-	5,690r	1,085r
Blast furnaces	-	-	-	-	-	1,165r	5,494r
Patent fuel manufacture	27	-	14	41	-	-	-
Pumped storage	-	-	-	-	-	-	-
Other	-	-	-	-	-	-	-
Losses	**-**	**-**	**-**	**-**	**-**	**335r**	**1,474**
Final consumption	**473r**	**883r**	**631r**	**1,987r**	**2,542r**	**4,626r**	**3,950r**
Industry	**358r**	**883r**	**32**	**1,273r**	**2,542r**	**4,626r**	**3,950r**
Unclassified	201r	61r	32	294r	617r	116	-
Iron and steel	23	822	-	845	1,925r	4,510r	3,950r
Non-ferrous metals	134r	-	-	134r	-	-	-
Mineral products	-	-	-	-	-	-	-
Chemicals	-	-	-	-	-	-	-
Mechanical engineering etc	-	-	-	-	-	-	-
Electrical engineering etc	-	-	-	-	-	-	-
Vehicles	-	-	-	-	-	-	-
Food, beverages etc	-	-	-	-	-	-	-
Textiles, leather, etc	-	-	-	-	-	-	-
Paper, printing etc	-	-	-	-	-	-	-
Other industries	-	-	-	-	-	-	-
Construction	-	-	-	-	-	-	-
Transport	**-**	**-**	**-**	**-**	**-**	**-**	**-**
Air	-	-	-	-	-	-	-
Rail	-	-	-	-	-	-	-
Road	-	-	-	-	-	-	-
National navigation	-	-	-	-	-	-	-
Pipelines	-	-	-	-	-	-	-
Other	**115r**	**-**	**599r**	**714r**	**-**	**-**	**-**
Domestic	115r	-	599r	714r	-	-	-
Public administration	-	-	-	-	-	-	-
Commercial	-	-	-	-	-	-	-
Agriculture	-	-	-	-	-	-	-
Miscellaneous	-	-	-	-	-	-	-
Non energy use	**-**	**-**	**-**	**-**	**-**	**-**	**-**

(1) Stock fall (+), stock rise (-).
(2) Total supply minus total demand.
(3) Coke oven gas and blast furnace gas transfers are for synthetic coke oven gas, see paragraph 2.52.

(4) Because of the small number of benzole suppliers, figures for benzole and tars cannot be given separately.

2.6 Commodity balances 1997

Manufactured fuels

	Thousand tonnes				GWh		
	Coke oven coke	Coke breeze	Other manuf. solid fuel	Total manuf. solid fuel	Benzole and tars *(4)*	Coke oven gas	Blast furnace gas
Supply							
Production	6,192	41	741r	6,974r	2,567	13,282	20,762
Other sources	-	-	-	-	-	-	-
Imports	750	112	24	886	-	-	-
Exports	-61r	-201r	-83	-345r	-	-	-
Marine bunkers	-	-	-	-	-	-	-
Stock change *(1)*	-72	+64	-6	-14	-	-	-
Transfers *(3)*	-1,257	+1,336	-	+79	-	+559	-63
Total supply	**5,552r**	**1,352r**	**676r**	**7,580r**	**2,567**	**13,841**	**20,699r**
Statistical difference *(2)*	-93r	+131r	+31r	+69r	-	+346	-206
Total demand	**5,645r**	**1,221r**	**645r**	**7,511r**	**2,567**	**13,495**	**20,905r**
Transformation	**5,196**	**310**	**-**	**5,506**	**-**	**1,814r**	**8,057r**
Electricity generation	-	-	-	-	-	1,814r	8,057r
Major power producers	-	-	-	-	-	-	-
Autogenerators	-	-	-	-	-	1,814r	8,057r
Petroleum refineries	-	-	-	-	-	-	-
Coke manufacture	-	64	-	64	-	-	-
Blast furnaces	5,196	246	-	5,442	-	-	-
Patent fuel manufacture	-	-	-	-	-	-	-
Low temperature carbonisation	-	-	-	-	-	-	-
Energy industry use	**18**	**-**	**29**	**47**	**-**	**6,013**	**6,939**
Electricity generation	-	-	-	-	-	-	-
Oil and gas extraction	-	-	-	-	-	-	-
Petroleum refineries	-	-	-	-	-	-	-
Coal extraction	-	-	-	-	-	-	-
Coke manufacture	-	-	-	-	-	4,909	1,137
Blast furnaces	-	-	-	-	-	1,104	5,802
Patent fuel manufacture	18	-	29	47	-	-	-
Pumped storage	-	-	-	-	-	-	-
Other	-	-	-	-	-	-	-
Losses	**-**	**-**	**-**	**-**	**-**	**202**	**1,597**
Final consumption	**431r**	**911r**	**616r**	**1,958r**	**2,567**	**5,466r**	**4,312r**
Industry	**348**	**911r**	**38**	**1,297r**	**2,567**	**5,466r**	**4,312r**
Unclassified	127	55r	38	220r	606	220	-
Iron and steel	40	856	-	896	1,961	5,246r	4,312r
Non-ferrous metals	181	-	-	181	-	-	-
Mineral products	-	-	-	-	-	-	-
Chemicals	-	-	-	-	-	-	-
Mechanical engineering etc	-	-	-	-	-	-	-
Electrical engineering etc	-	-	-	-	-	-	-
Vehicles	-	-	-	-	-	-	-
Food, beverages etc	-	-	-	-	-	-	-
Textiles, leather, etc	-	-	-	-	-	-	-
Paper, printing etc	-	-	-	-	-	-	-
Other industries	-	-	-	-	-	-	-
Construction	-	-	-	-	-	-	-
Transport	-	-	-	-	-	-	-
Air	-	-	-	-	-	-	-
Rail	-	-	-	-	-	-	-
Road	-	-	-	-	-	-	-
National navigation	-	-	-	-	-	-	-
Pipelines	-	-	-	-	-	-	-
Other	**83r**	**-**	**578r**	**661r**	**-**	**-**	**-**
Domestic	83r	-	578r	661r	-	-	-
Public administration	-	-	-	-	-	-	-
Commercial	-	-	-	-	-	-	-
Agriculture	-	-	-	-	-	-	-
Miscellaneous	-	-	-	-	-	-	-
Non energy use	**-**	**-**	**-**	**-**	**-**	**-**	**-**

(1) Stock fall (+), stock rise (-).
(2) Total supply minus total demand.
(3) Coke oven gas and blast furnace gas transfers are for synthetic coke oven gas, see paragraph 2.52.
(4) Because of the small number of benzole suppliers, figures for benzole and tars cannot be given separately.

2.7 Supply and consumption of coal

Thousand tonnes

	1995	1996	1997	1998	1999
Supply					
Production	51,519	48,538	46,981	40,046r	36,163
Deep-mined	35,150	32,223	30,281	25,279r	20,888
Opencast	16,369	16,315	16,700	14,767r	15,275
Other sources *(3)*	1,518	1,659	1,514	1,131r	914
Imports	15,896	17,799	19,759r	21,244r	20,758
Exports	-859	-988	-1,146r	-971r	-762
Stock change *(1)*	+8,542	+3,825	-3,973r	+1,224	-983
Total supply	**76,616**	**70,833**	**63,135r**	**62,674r**	**56,090**
Statistical difference *(2)*	-332	-567	+55r	-465r	+561
Total demand	**76,948**	**71,400r**	**63,080r**	**63,139r**	**55,529**
Transformation	**69,681r**	**65,017r**	**56,864r**	**57,873r**	**49,672**
Electricity generation	60,035r	55,439r	47,250r	48,510r	40,471
Major power producers	57,943	53,423	45,323r	46,627	38,985
Autogenerators	2,092r	2,016r	1,927r	1,883r	1,486
Coke manufacture	8,079	8,049	8,143	8,169	8,064
Blast furnaces	585	583	607	559	494
Patent fuel manufacture and low temperature carbonisation	982	946	864	635	643
Energy industry use	**8**	**8**	**8**	**10**	**10**
Coal extraction	8	8	8	10	10
Final consumption	**7,259r**	**6,375r**	**6,208r**	**5,256r**	**5,847**
Industry	**4,046r**	**3,093r**	**2,970r**	**2,514r**	**2,836**
Unclassified	-	-	-	-	-
Iron and steel	8	3	1	9	10
Non-ferrous metals	230	132r	149r	183r	244
Mineral products	950	925r	926r	835r	890
Chemicals	841r	635r	650r	637r	782
Mechanical engineering etc	29	29	22r	28	30
Electrical engineering etc	-	-	-	3	8
Vehicles	222	119	102r	87r	97
Food, beverages etc	463r	384	368r	257r	257
Textiles, clothing, leather, etc	229	126	68	69	63
Pulp, paper, printing etc	472r	267r	233r	121r	164
Other industries	602r	473r	451r	285r	291
Construction	-	-	-	-	-
Transport	-	-	-	-	-
Other	**3,213**	**3,282**	**3,238r**	**2,742r**	**3,011**
Domestic	2,690	2,705	2,587	2,362r	2,693
Public administration	348	388	455	308r	293
Commercial	-	-	-	4	4
Agriculture	13	13	7	9	7
Miscellaneous	162	176	189	59	14
Non energy use	-	-	-	-	-
Stocks at end of year *(4)*					
Distributed stocks	11,626	10,752	14,064	13,092	13,661
Of which:					
Major power producers	10,587	9,495	12,897	11,760	12,584
Coke ovens	961	1,228	1,128	1,312	1,054
Undistributed stocks	7,104	4,153	4,814r	4,562r	4,976
Total stocks	**18,730**	**14,905**	**18,878r**	**17,654r**	**18,637**

(1) Stock fall (+), stock rise (-).
(2) Total supply minus total demand.
(3) Estimates of slurry etc. recovered from ponds, dumps, rivers etc.
(4) Excludes distributed stocks held in merchants' yards, etc., mainly for the domestic market, and stocks held by the industrial sector.

2.8 Supply and consumption of coke oven coke, coke breeze and other manufactured solid fuels

Thousand tonnes

	1995	1996	1997	1998	1999
Coke oven coke					
Supply					
Production	6,187	6,178	6,192	6,178	5,837
Imports	553	668	750	753	389
Exports	-89	-88	-61r	-93r	-79
Stock change *(1)*	-124	+167	-72	-264	+353
Transfers	-1,080	-1,209	-1,257	-1,223	-951
Total supply	**5,447**	**5,716**	**5,552r**	**5,351r**	**5,549**
Statistical difference *(2)*	-78	-102	-93r	-57r	-73
Total demand	**5,525**	**5,818**	**5,645r**	**5,408r**	**5,622**
Transformation	**4,846**	**5,180**	**5,196**	**4,908**	**5,119**
Blast furnaces	4,846	5,180	5,196	4,908	5,119
Energy industry use	-	-	18	27	20
Final consumption	**679**	**638**	**431r**	**473r**	**484**
Industry	**505**	**457**	**348**	**358r**	**368**
Unclassified	131	78	127	201r	206
Iron and steel	14	25	40	23	19
Non-ferrous metals	360	354	181	134r	143
Other	**174**	**181**	**83r**	**115r**	**116**
Domestic	174	181	83r	115r	116
Stocks at end of year *(3)*	**404**	**237**	**308**	**573**	**220**
Coke breeze					
Supply					
Production	41	44	41	37	33
Imports	87	100	112	78	40
Exports	-175	-123	-201r	-196r	-165
Stock change *(1)*	-9	-110	+64	+42	+6
Transfers	+1,126	+1,281	+1,336	+1,163	+1,035
Total supply	**1,070**	**1,192**	**1,381**	**1,124r**	**948**
Statistical difference *(2)*	-20	+17	+131r	-46r	-152
Total demand	**1,090**	**1,175**	**1,221r**	**1,170r**	**1,100**
Transformation	**230**	**280**	**310**	**287**	**225**
Coke manufacture	74	68	64	50	25
Blast furnaces	156	212	246	237	200
Energy industry use	-	-	-	-	-
Final consumption	**860**	**895**	**911r**	**883r**	**875**
Industry	**860**	**895**	**911r**	**883r**	**875**
Unclassified	16	54	55r	61r	14
Iron and steel	844	841	856	822	861
Stocks at end of year *(3)*	**185**	**295**	**231**	**189**	**277**
Other manufactured solid fuels					
Supply					
Production	841	862	741r	616	635
Imports	46	50	24	10	6
Exports	-103	-90	-83	-56	-54
Stock change *(1)*	-51	+36	-6	+88r	-7
Total supply	**733**	**858**	**676r**	**658r**	**580**
Statistical difference *(2)*	-38	-10	+31r	+13r	-5
Total demand	**771**	**868**	**645r**	**645r**	**585**
Transformation	-	-	-	-	-
Energy industry use	**29**	**33**	**29**	**14**	**13**
Patent fuel manufacture	29	33	29	14	13
Final consumption	**742**	**835**	**616r**	**631r**	**572**
Industry	**34**	**20**	**38**	**32**	**18**
Unclassified	34	20	38	32	18
Other	**708**	**815**	**578r**	**599r**	**554**
Domestic	708	815	578r	599r	554
Stocks at end of year *(3)*	**251**	**216**	**222**	**134**	**140**

(1) Stock fall (+), stock rise (-).
(2) Total supply minus total demand.

(3) Producers stocks and distributed stocks.

2.9 Supply and consumption of coke oven gas, blast furnace gas, benzole and tars

GWh

	1995	1996	1997	1998	1999
Coke oven gas					
Supply					
Production	12,854	13,246	13,282	13,126r	12,619
Imports	-	-	-	-	-
Exports	-	-	-	-	-
Transfers *(1)*	..	+286	+559	+630	+528
Total supply	**12,854**	**13,532**	**13,841**	**13,756r**	**13,147**
Statistical difference *(2)*	+158	+239	+346	+126r	+1
Total demand	**12,696**	**13,293**	**13,495**	**13,630r**	**13,146**
Transformation	**1,600r**	**1,860r**	**1,814r**	**1,814r**	**1,946**
Electricity generation	1,600r	1,860r	1,814r	1,814r	1,946
Other	-	-	-	-	-
Energy industry use	**5,183**	**6,030**	**6,013**	**6,855r**	**6,522**
Coke manufacture	4,181	4,983	4,909	5,690r	5,283
Blast furnaces	1,002	1,047	1,104	1,165r	1,239
Other	-	-	-	-	-
Losses	**303**	**296**	**202**	**335r**	**194**
Final consumption	**5,610r**	**5,107r**	**5,466r**	**4,626r**	**4,484**
Industry	**5,610r**	**5,107r**	**5,466r**	**4,626r**	**4,484**
Unclassified	158	208	220	116	124
Iron and steel	5,452r	4,899r	5,246r	4,510r	4,360
Blast furnace gas					
Supply					
Production	19,226	20,109	20,762	20,116r	19,037
Imports	-	-	-	-	-
Exports	-	-	-	-	-
Transfers *(1)*	..	-16	-63	-22	-22
Total supply	**19,226**	**20,093r**	**20,699r**	**20,094r**	**19,015**
Statistical difference *(2)*	-74	-166r	-206	+34r	-29
Total demand	**19,300**	**20,259**	**20,905r**	**20,060r**	**19,044**
Transformation	**8,089r**	**8,072r**	**8,057r**	**8,057r**	**7,992**
Electricity generation	8,089r	8,072r	8,057r	8,057r	7,992
Other	-	-	-	-	-
Energy industry use	**6,424**	**6,754**	**6,939**	**6,579r**	**6,282**
Coke manufacture	1,068	1,122	1,137	1,085r	1,094
Blast furnaces	5,356	5,632	5,802	5,494r	5,188
Other	-	-	-	-	-
Losses	**1,615**	**1,850**	**1,597**	**1,474**	**1,741**
Final consumption	**3,172r**	**3,583r**	**4,312r**	**3,950r**	**3,029**
Industry	**3,172r**	**3,583r**	**4,312r**	**3,950r**	**3,029**
Unclassified	-	-	-	-	-
Iron and steel	3,172r	3,583r	4,312r	3,950r	3,029
Benzole and tars *(3)*					
Supply					
Production	2,502	2,534	2,567	2,542r	2,369
Final consumption	**2,502**	**2,534**	**2,567**	**2,542r**	**2,369**
Unclassified	621	582	606	617r	606
Iron and steel	1,881	1,952	1,961	1,925r	1,763

(1) To and from synthetic coke oven gas, see paragraph 2.52. Small quantities of synthetic coke oven gas were made in 1995 but the details are not available and the production figure is net of transfers.

(2) Total supply minus total demand.

(3) Because of the small number of benzole suppliers, figures for benzole and tars cannot be given separately.

2.10 Coal production and stocks, 1960 to 1999

Thousand tonnes

	Coal production					Coal stocks (at year end) (1)		
	Total (2)	Deep-mined	Opencast	Imports (3)	Exports (4)	Total	Distributed	Undistributed
1960	197,831	189,039	7,674	-	5,588	43,312	13,656	29,657
1961	195,047	184,858	8,665	-	5,738	37,740	15,979	21,761
1962	202,639	192,368	8,238	-	4,918	41,111	15,344	25,766
1963	200,462	192,702	6,237	-	7,649	36,660	16,740	19,920
1964	198,056	189,816	6,919	-	6,067	38,251	17,547	20,704
1965	191,627	183,059	7,450	-	3,817	38,661	16,800	21,860
1966	179,909	171,069	7,139	-	2,921	37,555	18,709	18,847
1967	177,651	167,709	7,199	-	1,921	47,003	18,865	28,138
1968	169,920	159,739	6,794	-	2,721	44,426	15,941	28,485
1969	155,694	146,557	6,421	-	3,511	33,066	14,289	18,778
1970	147,195	136,686	7,885	79	3,191	20,630	13,414	7,216
1971	153,683	136,478	10,666	4,241	2,667	28,664	18,271	10,393
1972	126,834	109,086	10,438	4,998	1,796	30,460	19,351	11,110
1973	131,984	120,030	10,123	1,675	2,693	27,886	17,035	10,850
1974	110,452	99,993	9,231	3,547	1,865	21,807	15,827	5,979
1975	128,683	117,412	10,414	5,083	2,182	31,159	20,541	10,618
1976	123,801	110,265	11,944	2,837	1,436	33,115	22,457	10,658
1977	122,150	107,123	13,551	2,439	1,835	31,444	21,704	9,740
1978	123,577	107,528	14,167	2,352	2,253	34,475	22,038	12,437
1979	122,369	107,775	12,862	4,375	2,175	27,908	18,339	9,569
1980	130,097	112,430	15,779	7,334	3,809	37,687	20,370	17,317
1981	127,469	110,473	14,828	4,290	9,113	42,253	20,136	22,117
1982	124,711	106,161	15,266	4,063	7,447	52,377	30,422	21,955
1983	119,254	101,742	14,706	4,456	6,561	57,960	33,964	23,996
1984	51,182	35,243	14,306	8,894	2,293	36,548	15,794	20,753
1985	94,111	75,289	15,569	12,732	2,432	34,979	25,752	9,228
1986	108,099	90,366	14,275	10,554	2,677	38,481	29,776	8,704
1987	104,533	85,957	15,786	9,781	2,353	33,246	27,104	6,142
1988	104,001	83,762	17,899	11,685	1,822	36,166	28,834	7,332
1989	99,820	79,628	18,657	12,137	2,049	39,244	29,191	10,053
1990	92,762	72,899	18,134	14,783	2,307	37,760	28,747	9,013
1991	94,202	73,357	18,636	19,611	1,824	43,321	32,343	10,977
1992	84,493	65,800	18,187	20,339	973	47,207	33,493	13,714
1993	68,199	50,457	17,006	18,400	1,114	45,860	29,872	15,989
1994	49,785r	31,854	16,804	15,088	1,236	27,272	16,001	11,271
1995	53,037	35,150	16,369	15,896	859	18,730	11,626	7,104
1996	50,197	32,223	16,315	17,799	988	14,905	10,752	4,153
1997	48,495	30,281	16,700	19,759r	1,146r	18,878r	14,064	4,814r
1998	41,177r	25,279r	14,767r	21,244r	971r	17,654r	13,092	4,562r
1999	37,077	20,888	15,275	20,758	762	18,637	13,661	4,976

(1) Excludes distributed stocks held in merchants' yards etc mainly for the domestic market and stocks held by the industrial sector.
(2) Includes estimates for slurry etc recovered from dumps, ponds, rivers etc.
(3) The 1993 import figure includes an additional estimate for unrecorded trade.
(4) From 1990 based on HM Custom and Excise data; before 1990 based on British Coal's shipments.

2.11 Inland consumption of solid fuels, 1960 to 1999

Thousand tonnes

| | | Coal consumption by fuel producers | | | | | | Final consumption | | | | | |
| | | Primary | Secondary | | | | | Coal (1) | | | | Coke and breeze | Other solid fuel |
	Total inland consumption of coal	Collieries	Power stations(1)	Coke ovens (2)	Other solid fuel plants (3)	Gas works	Total	Industry	Domestic	Other	Total	(3)	(4)
1960	199,858	5,080	52,733	29,262	2,337	22,962	107,295	45,212	36,068	6,202	87,482	19,112	1,402
1961	194,879	4,572	56,289	27,433	2,439	22,861	109,022	41,656	33,833	5,796	81,285	..	..
1962	194,269	4,267	62,081	24,080	2,642	24,760	113,563	36,307	34,341	5,791	76,439	..	..
1963	197,114	3,963	68,584	24,080	2,743	22,760	118,167	35,154	33,633	6,198	74,985	..	..
1964	190,205	3,759	69,092	26,316	2,743	20,829	118,980	32,817	29,362	5,287	67,466	18,280	1,887
1965	187,564	3,455	71,124	26,722	2,540	18,492	118,878	30,988	28,854	5,389	65,231	18,490	2,028
1966	177,505	3,150	69,701	25,198	2,743	17,171	114,814	27,737	26,927	4,877	59,541	17,010	2,288
1967	166,430	2,947	68,279	23,979	3,251	14,834	110,340	24,545	24,485	4,166	53,196	16,410	2,445
1968	167,148	2,459	74,366	25,334	3,530	10,817	114,045	22,946	23,584	4,114	50,644	17,610	2,814
1969	163,746	2,065	77,101	25,780	3,908	6,977	113,766	21,709	21,974	4,232	47,915	18,180	3,205
1970	156,886	1,916	77,237	25,340	4,150	4,280	111,006	19,613	20,190	4,159	43,962	18,090	3,203
1971	140,932	1,581	72,847	23,554	4,477	1,855	102,733	16,105	17,185	3,327	36,617	15,100	3,456
1972	122,884	1,405	66,664	20,476	4,547	575	92,261	11,663	14,554	2,999	29,216	14,090	3,514
1973	133,370	1,381	76,838	21,888	3,607	512	102,845	12,062	14,502	2,581	29,145	15,000	3,375
1974	117,888	1,256	67,026	18,461	3,788	107	89,382	11,077	13,667	2,505	27,249	13,220	3,184
1975	122,217	1,238	74,569	19,085	4,063	9	97,725	9,685	11,616	1,948	23,253	11,640	2,919
1976	123,604	1,132	77,819	19,402	3,405	8	100,632	8,970	10,823	2,045	21,838	12,460	2,647
1977	123,978	1,124	79,956	17,406	3,173	-	100,536	9,033	11,136	2,149	22,318	11,310	2,609
1978	120,477	1,010	80,643	14,946	3,070	-	98,659	8,550	10,217	2,041	20,808	10,484	2,453
1979	129,378	834	88,790	15,081	2,883	-	106,753	9,232	10,508	2,051	21,791	11,361	2,364
1980	123,460	663	89,569	11,610	3,022	-	104,201	7,898	8,946	1,752	18,596	6,221	2,252
1981	118,386	616	87,226	10,805	2,458	-	100,489	7,046	8,454	1,781	17,281	7,952	1,975
1982	110,998	534	80,228	10,406	2,326	-	92,960	7,175	8,474	1,855	17,504	7,248	1,921
1983	111,475	486	81,565	10,448	2,114	-	94,127	7,218	7,872	1,772	16,862	7,600	1,889
1984	77,309	209	53,411	8,246	1,300	-	62,957	7,006	5,406	1,731	14,143	7,653	1,186
1985	105,386	332	73,940	11,122	2,176	-	87,237	8,313	7,799	1,704	17,817	8,230	1,658
1986	114,234	306	82,652	11,122	1,959	-	95,732	9,278	7,421	1,496	18,196	7,558	1,601
1987	115,894	235	87,960	10,859	2,052	-	100,871	6,827	6,536	1,425	14,789	8,233	1,652
1988	111,498	196	84,258	10,902	2,006	-	97,166	7,131	5,741	1,265	14,135	8,591	1,443
1989	107,581	146	82,053	10,792	1,717	-	94,562	6,763	5,048	1,062	12,873	8,159	1,253
1990	108,256	117	84,014	10,852	1,544	-	96,409	6,280	4,239	1,211	11,730	7,637	1,214
1991	107,513	112	83,542	10,011	1,501	-	95,054	6,426	4,778	1,144	12,348	7,136	1,200
1992	100,580	79	78,469	9,031	1,319	-	88,819	6,581	4,156	945	11,682	6,887	1,089
1993	86,757	48	66,136	8,479	1,329	-	75,944	5,300	4,638	826	10,765	6,638	1,138
1994	81,783	22	62,406	8,595	1,190	-	72,191	4,948	3,901	721	9,570	6,578	949
1995	76,948	8	60,035r	8,664	982	-	69,681r	4,046r	2,690	523	7,259r	6,541	742
1996	71,400r	8	55,439r	8,632	946	-	65,017r	3,093r	2,705	577	6,375r	6,925	835
1997	63,080r	8	47,250r	8,750r	864	-	56,864r	2,970r	2,587	651	6,208r	6,784r	616
1998	63,139r	10	48,510r	8,728	635	-	57,873r	2,514r	2,362r	380	5,256r	6,501r	631r
1999	55,529	10	40,471	8,558	643	-	49,672	2,836	2,693	318	5,847	6,678	572

(1) Up to 1986 power stations include those in the public electricity supply, railways and transport industries. Consumption by other generators is included in final coal consumption. From 1987, coal consumption at power stations also includes other generators' consumption, which is therefore excluded from final coal consumption (see also Table 2.7).

(2) Includes blast furnaces.

(3) This series comprises final consumption and consumption at blast furnaces which can now be separated following production of energy balances in Tables 2.4 to 2.6. Pure final consumption figures for coke and breeze in 1997, 1998, and 1999 were 1,342, 1,356, and 1,376 thousand tonnes respectively.

(4) Low temperature carbonisation and patent fuel plants and their products.

Chapter 3
Petroleum

Introduction

3.1   This chapter contains commodity balances covering the supply and disposal of primary oils (crude oil and natural gas liquids), feedstocks (including partly processed oils) and petroleum products in the UK in the period 1997 to 1999. These balances are given in Tables 3.1 to 3.6. Additional data has been included in supplementary tables on areas not covered by the format of the balances. This extra information includes details on refinery capacities and aggregates for refinery operations, and extra detail on deliveries into consumption, including breakdowns by country, sector and industry. Long term series for selected headings covering the period 1960 to 1999 are presented in Tables 3.11 and 3.12.

3.2   Statistics of imports and exports of crude oil, other refinery feedstocks and petroleum products, refinery receipts, refinery throughput and output and deliveries of petroleum products are obtained from the United Kingdom Petroleum Industry Association, and the Department of Trade and Industry's Petroleum Production Reporting System.

3.3   The annual figures relate to calendar years or the ends of calendar years. In the majority of tables the data cover the United Kingdom.

3.4   Also included at the end of the Digest is an annex on the oil and gas resources of the UK (Annex C). This information was presented as a separate chapter in previous versions of the Digest, and is included to provide a more complete picture of the UK oil and gas production sector.

Commodity balances for primary oil (Tables 3.1, 3.2 and 3.3)

3.5   These tables show details of the production, supply and disposals of primary oils (crude oil and natural gas liquids (NGLs)) and feedstocks in 1999, 1998 and 1997. The upper half of the table (Supply) equates to the upstream oil industry, covering the supply chain from the production of oil and NGLs recorded by individual fields to the disposal of oil and NGLs to export or to UK refineries. The lower half of the table covers the uses made of these primary oils, including the amounts recorded as used as a fuel during the extraction process (i.e. burned to provide power for drilling and pumping operations) and as inputs into refineries, as recorded by refineries. The statistical difference in the tables thus represents the differences between data reported by these different sources and the sites of production and consumption.

3.6   Gross production of crude oil and NGLs in 1999 was 137 million tonnes, the highest annual production since oil was first produced from the North Sea in 1975, exceeding the highest previous production total of 132 million tonnes in 1998. In line with previous years, two-thirds of the United Kingdom's primary oil production in 1999 was exported. Feedstocks (including partly processed oils) made up about one-eighth of total imports of oil in 1999. Exports of primary oils and feedstocks exceeded imports by 59 per cent in 1997 and 78 per cent in 1998, and in 1999 exports were more than double the level of imports, making a significant contribution to the UK economy (see Chapter 8). Increased UK production combined with a reduction in refinery activity in the UK led to increased supplies being available for export in 1999, such that exports of primary oils and feedstocks in 1999 were 8½ per cent higher than in 1998. Chart 3.1 illustrates recent trends in production, imports and exports of crude oil, NGLs and feedstocks.

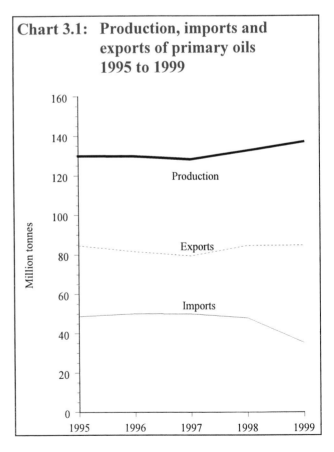

Chart 3.1: Production, imports and exports of primary oils 1995 to 1999

3.7   The UK produces more than enough crude oil to meet its own needs, but imports still take place. As it generally contains lower levels of contaminants such as sulphur (which can make the crude oil difficult to refine), UK crude oil can command a higher price than other crude oils on the international market. It also contains a higher proportion of the lighter hydrocarbon molecules, resulting in higher yields of products such as motor spirit and other transport fuels. These two factors together make it financially attractive to export the crude oil rather than use it in the UK, with imports being brought in to make up the difference. In addition, some crude oils are specifically imported for the heavier hydrocarbons they contain which are needed for the manufacture of various petroleum products, such as bitumen and lubricating oils.

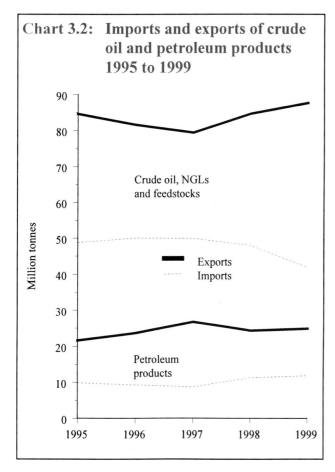

Chart 3.2: Imports and exports of crude oil and petroleum products 1995 to 1999

lower than in 1998. Additional analysis of the exports and imports of oil products is given in paragraphs 3.11 to 3.18 and 3.78 to 3.85, and additional explanations for the decline in exports of crude oil in 1997 are given in Annex C, paragraphs C.32 to C.34.

3.9   It will be seen from the balances in Tables 3.1 to 3.3 that while the overall statistical difference in the primary oil balance for 1999 is relatively small (+168 thousand tonnes), there are large statistical differences in the primary oil balances for 1998 and 1997. That is, there are large differences being seen between the total quantities of crude oil and NGLs reported as being produced by the individual production fields in the UK and the totals reported by UK oil companies as being received by refineries or going for export. This is discussed later in paragraphs 3.33 to 3.42.

Commodity balances - Petroleum products (Tables 3.4 to 3.6)

3.10   These tables show details of the production, supply and disposals of petroleum products into the UK market in 1999, 1998 and 1997. The upper half of the table (Supply) covers details of the overall availability of these products in the UK as calculated by observing production at refineries, and adding in the impact of trade (imports and exports), stock changes, product transfers and deliveries to international marine bunkers. The lower half of the table covers the uses made of these products, including the uses made within refineries as fuels in the refining process, and details of the amounts reported by oil companies within the UK as delivered for final consumption.

3.8   Chart 3.2 compares the level of imports and exports of crude oil, NGLs and feedstocks with those for petroleum products over the period 1995 to 1998. Imports of crude oil, which increased between 1992 and 1993, were lower between 1995 and 1999 due to the increased levels of output from the United Kingdom continental shelf during this period. Imports of petroleum products during the period continued to decline between 1995 and 1997 while exports of products increased fuelled by increased production by of UK refineries. In 1998 a reversal of these trends occurred due to the closure of the Gulf Oil refinery in Milford Haven in December 1997. The impact of this reduction in refinery processing capacity in the UK carried through into 1999, and there will continue to be a reduction in processing in the UK in 2000 due to the closure of Shell's Shellhaven refinery at the end of November 1999. The reduction in capacity has led to both a reduction in the level of exports of petroleum products along with a corresponding increase in the level of imports of products to ensure UK demand for the products continues to be met. At the same time there has been a reduction in imports of crude oil and feedstocks, with imports in 1999 being 6½ per cent

Supply of petroleum products

3.11   Looking at refinery operations first, total output from UK refineries in 1999, at 88 million tonnes of products, was 6 per cent lower than in 1998, primarily due to the loss of the output from the Gulf Oil refinery mentioned above. The Gulf refinery accounted for around 6 per cent of total UK refinery distillation capacity when it closed, leading to the reduction of output seen in 1999 with some effect also being seen in 1998. The loss of this refinery capacity has been somewhat compensated by other refineries increasing their processing capacity over the past two years, in particular those owned by BP and Conoco. However, the first half of 1999 saw depressed international markets for crude oil and oil products, resulting in decreased profit margins for refining companies, and thus led to a general slowdown in refinery activity during the early part of 1999. This was countered in the second half of the year by increased prices for crude oil and products due to increased demand and a reduction in world supplies. This is explained in more detail in Annex C, paragraphs C.30 to C.34.

3.12 In terms of output of individual products, as might be expected, the level of output of aviation turbine fuel, motor spirit, gas oil/diesel oil and fuel oil were all lower in 1999 than in 1998 (by 7½, 6½, 6, and 8 per cent respectively). These decreases are all in line with the impact of the closure of the Gulf Oil refinery, which was predominantly a producer of these categories of products. More information on refinery capacity in the UK and refinery capacity utilisation is given in paragraphs 3.43 and 3.44.

3.13 Every year since 1974, with the exception of 1984 due to the effects of the industrial action in the coal-mining sector, the UK has been a net exporter of oil products. In 1999, exports of petroleum products were 22 million tonnes, 11 per cent lower than exports in 1998 and 19 per cent lower than in 1997, which, at 27 million tonnes was the highest annual total ever recorded. The decrease in exports is mainly due to the closure of the Gulf Oil refinery mentioned above. Related to this closure, the 12 million tonnes of oil products imported into the UK in 1999 were 8 per cent higher than in 1998 and 41 per cent higher than in 1997. However, despite these increased imports the UK remained very much a net exporter of products in 1999, by some 9½ million tonnes, but at a lower level than in 1998.

3.14 The US remains one of the key markets for UK exports of oil products, with 3 million tonnes being exported there from the UK in 1999. These exports made up 14 per cent of total UK exports of oil products in 1999, with the main other countries receiving UK exports of petroleum products being Ireland, Italy, France, the Netherlands, Germany and Spain. The main sources of the UK's imports of petroleum products in 1999 were France, the Netherlands and Norway.

3.15 Chart 3.3 shows how the UK has penetrated selected overseas markets for its petroleum products. UK products supplied 6 per cent of the total volume of US imports of petroleum products in 1999 (mostly in the form of motor spirit) and 15, 3 and 6 per cent of total imports of petroleum products into France, Germany and the Netherlands (mostly as gas oil for heating, motor spirit and fuel oil). The UK regularly supplies the vast majority of total oil products imported into Ireland (mostly motor spirit and DERV fuel for transport and gas oil for heating).

3.16 It may be asked why the UK imports petroleum products at all if it has such a surplus of them available for export. This can be explained if you look at the difference in the product detail. Exports in 1999 were mainly made up of motor spirit (6 million tonnes), gas oil/diesel oil (7 million tonnes) and fuel oil (5 million tonnes). Imports were made up of aviation turbine fuel (3 million tonnes), motor spirit (2 million tonnes) and gas oil/diesel oil (5 million tonnes). The make up of refinery structure in the UK is such that it contains a surplus of motor spirit capacity, leading to a surplus

availability within the UK, which is exported. The imports of motor spirit are of grades of products which UK refineries are not currently able to manufacture. For example, this includes low sulphur versions of motor spirit. The imports also take place to cover specific periods of heavy demand within the UK, such as around the time of the Budget in March or during the summer.

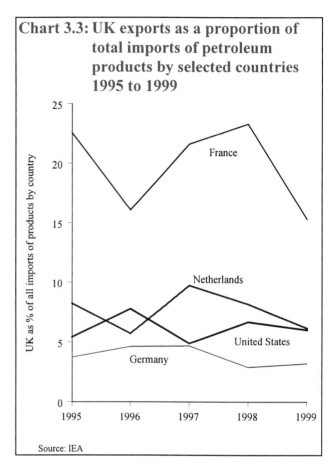

Chart 3.3: UK exports as a proportion of total imports of petroleum products by selected countries 1995 to 1999

Source: IEA

3.17 Similarly for gas oil/diesel oil, the exports from the UK tend to be of lower grades of gas oil/diesel oil for use as heating fuels, while the imports tend to be of higher grade gas oil/diesel oil with a low sulphur content. For example, the imports cover products such as Ultra Low Sulphur Diesel (ULSD). Aviation turbine fuel is imported simply because the UK cannot make enough of it to meet demand. It is derived from the same sort of hydrocarbons as gas oil/diesel oil, and as such there is a physical limit to how much can be made from the amount of oil processed in the UK.

3.18 More information on the structure of refineries in the UK and trends in imports and exports of crude oil and oil products is given in the section discussing Table 3.11 (paragraphs 3.78 to 3.85).

3.19 In 1999, 9½ per cent of UK production of fuel oil and 4 per cent of gas oil/diesel oil production went into international bunkers, totalling some 2 million tonnes of products, 2½ per cent of total UK refinery production in the year. These are sales of fuels that are destined for consumption on ocean going vessels. As such the products cannot be classified as being consumed within the UK, and these quantities are thus treated in a similar way to exports in the commodity balances. It should be noted that these quantities do not include deliveries of fuels for use in UK coastal waters. These deliveries are counted as UK consumption and the figures given in the transport section of the commodity balances.

3.20 Details are given in the balances of stocks of products held within the UK either at refineries or oil distribution centres such as coastal oil terminals (undistributed stocks). In addition, some information is available on stocks of oil products held by major electricity generators (distributed stocks). However, these figures do not include any details of stocks held by distributors of fuels or stocks held at retail sites, such as petrol stations. The figures for stocks in the balances also solely relate to those stocks currently present in the UK. As such they do not include details of any stocks that might be held by UK oil companies in other countries under bilateral agreements.

3.21 In order for the UK to be prepared for any oil emergency, the UK Government places an obligation on companies supplying oil products into final consumption in the UK to maintain a certain level of stocks of oil products used as fuels. As part of this, oil companies are allowed to hold stocks abroad under official governmental bilateral agreements that can count towards their stocking obligations. As such, to give a true picture of the amount of stocks available to the UK, i.e. that are owned by UK companies, the stocks figures in Table 3.10 take account of these bilateral stocks (see paragraphs 3.73 to 3.76).

Consumption of petroleum products

3.22 To help users gain the maximum information from the commodity balances, this section of the text will go through the data given on the consumption of oil products in the period 1997 to 1999. The main sectors of consumers will be looked at first (going down the tables) and then the data for individual products will be looked at (going across the tables).

3.23 Table 3.4 shows how overall deliveries of petroleum products into consumption in the UK in 1999, including those used by the UK refining industry as fuels within the refining process and all other uses, totalled 77 million tonnes. This was 2 per cent lower than in 1998 and 3 per cent lower than in 1997. As such, deliveries are following the trend of a decline seen since 1990, barring the slight increase seen in 1996.

3.24 As can be seen from the tables, one of the most significant changes in deliveries of products in recent years has been the decline in use for electricity generation. In 1999 only 1.3 million tonnes of oil products were used for electricity generation by major power producers and autogenerators of electricity, compared with 2.1 million tonnes in 1997 and much higher levels in earlier years (see Table 3.12). This change is primarily a result of the move by major

electricity producers away from oil based fuels towards using natural gas as their fuel for electricity generation. The fact that the level of usage by auto-producers of electricity has risen between 1997 and 1999 (up by 22 per cent) is due to the growth in auto-generation of electricity by industry as a whole rather than any specific move towards the use of oil products as fuels. In fact there has been a general move by industry away from oil products as fuels for all uses, not just for generation of electricity, which will be discussed later in this chapter. The data for fuels used in autogeneration of electricity in 1997 and 1998 have been revised in light of new information becoming available.

3.25 The data included against the blast furnaces heading in the Transformation sector represents fuel oil used in the manufacture of iron and steel which is directly injected into blast furnaces, as opposed to being used as a fuel to heat the blast furnaces. The fuel used for the latter (mostly gas oil/diesel oil) is included against the blast furnaces heading in the Energy Industry Use sector.

3.26 The other figures in the Energy Industry Use sector relate to uses within the UK refining industry in the manufacture of oil products. These are products either used as fuels during refining processes or products used by the refineries themselves as opposed to being sold to other consumers, but excluding any fuels used for the generation of electricity. These amounts are included in the Transformation sector totals. Given that there is a degree of interest in the total amounts of fuels used within refineries, Table 3.7 includes data on total refinery fuel usage (i.e. including that used in the generation of electricity) over the period 1995 to 1999. The data under the Other headings of the Energy Industry Use sector represent fuels used by the gas supply industry.

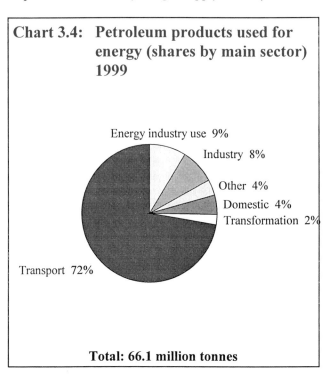

Chart 3.4: Petroleum products used for energy (shares by main sector) 1999

Energy industry use 9%

Industry 8%

Other 4%

Domestic 4%

Transformation 2%

Transport 72%

Total: 66.1 million tonnes

3.27 Final consumption of oil products in 1999, i.e. excluding any uses by the energy industries themselves or for transformation purposes, amounted to 69.7 million tonnes, 1 per cent lower than both 1998 and 1997. Chart 3.4 shows the breakdown of consumption for energy uses by each sector in 1999.

3.28 The total amount of oil products used by industry has declined in recent years, from 6 million tonnes in 1997 to 5½ million tonnes in 1999, a decrease of 5 per cent. The major decreases are seen for the chemical and food and drink industries, as well as declines being seen in most other sectors.

3.29 Transport sector consumption in 1999 was only slightly higher than in 1998, with increased usage for air transport (up 4½ per cent) being offset by a slight decrease in use for road transport (down 1 per cent). In 1999, transport usage totalled 48 million tonnes, and thus accounted for just over two-thirds of total final consumption of oil products in 1999. Consumption by other sectors decreased by 3 per cent in 1999 compared with 1998. Last year's edition of the Digest included estimates for the use of gas for road vehicles for the first time. These estimates were based on information on the amounts of duty received by HM Government from the tax on gas used as a road fuel. It has been possible to repeat this exercise, and it is estimated that some 8 thousand tonnes of gas (mostly butane or propane) was used in road vehicles in the UK in 1999. While a very small use when compared to overall consumption of these fuels, the consumption of these gases in road transport in 1999 has more than doubled since 1998. The DTI is considering, with DETR, the introduction of a voluntary survey of sellers of propane and butane for road vehicles to allow a more accurate estimate of the amounts of these fuels being used for transport purposes in the UK.

3.30 Consumption of non-energy products slightly increased in 1999, by 1½ per cent compared to 1998. In 1999, non-energy products made up 16 per cent of final consumption of oil products, compared with 15 per cent in 1998, and 15½ per cent in 1996. More detail on the non-energy uses of oil products, by product and by type of use where such information is available, is given in Table 3.D and paragraphs 3.64 to 3.70 later in this text.

3.31 Looking at the final consumption of individual products, three quarters of total final consumption in 1999 was made up of consumption of just three products; aviation turbine fuel, motor spirit and gas oil/diesel oil, i.e. transport products. Consumption of aviation turbine fuel increased by 4½ per cent in 1999 to 9.7 million tonnes, continuing the pattern seen in the 1990s of annual increases in the level of consumption. For motor spirit, consumption decreased by 1½ per cent in 1999. Total consumption of gas oil/diesel oil also fell by 1½ per cent in 1999, but that part of the total consumed as road fuel (i.e. as DERV)

slightly increased. More detailed information on consumption of motor spirit and gas oil/diesel oil over the period 1995 to 1999 is given in Table 3.8 and discussed in paragraphs 3.45 to 3.63.

3.32 As mentioned above, consumption of fuel oil in the UK has declined in recent years. In 1999 final consumption was down to just under 1½ million tonnes, two-thirds of its level in 1997. This is due to decreased use by industry (down 39 per cent) and decreased use by domestic and other premises for heating purposes and by other sectors (down by 37 per cent). Detail on the consumption of fuel oil broken down by grade is given in Table 3.8.

Supply and disposal of products (Table 3.7)

3.33 This table brings together the commodity balances for primary oils and for petroleum products into a single overall balance table. As such, whilst it follows the same general format as previously used for such balances (e.g. as used in Table 4.4 in the 1998 edition of the Digest), the headings have been revised so that the format of the commodity balances is followed.

3.34 The statistical difference for primary oils in the table is accounted for by own use in onshore terminals and gas separation plants, losses, platform and other field stock changes. Also a factor is the time lag that can exist between production and loading onto tankers being reported at an offshore field and the arrival of these tankers at onshore refineries and oil terminals. Whilst this gap is usually minimal and works such that any effect of this at the start of a month is balanced by a similar counterpart effect at the end of a month, there are instances where the length of this interval can be significant. It is thought that weather conditions in the period at the December 1995 and into the start of January 1996 were such that this interval was greater than usual. Thus some production was reported as occurring in December 1995 whilst the oil produced was not received until January 1996. This is thought to be the reason for the very low difference seen in 1995 (-95 thousand tonnes) compared to the large difference in 1996 (-1,409 thousand tonnes).

Table 3.A: Supply and disposal of petroleum adjusted for oil in transit 1995 and 1996

	1995	1996
Crude oil, NGLs and feedstocks		
Indigenous production	129,894	129,742
Offshore stock change (original)	- 160	- 2
Offshore stock change (adjusted)	- 960	+ 798
Statistical difference (original)	- 95	- 1,409
Statistical difference (adjusted)	- 895	- 609

3.35 Data is not available to properly correct the data for 1995 and 1996 to take account of this difference, but through analysis of the monthly data that is available, approximately 800 thousand tonnes of production in December 1995 was possibly in transit at the end of 1995, and thus not actually received onshore until January 1996. Using this as the basis for an adjustment, the statistical difference for primary oils for 1995 and 1996 would be specified in Table 3.A. The statistical differences seen in 1997 and 1998 are also large, and additional research has been carried out to try to identify the reasons for these large differences, although it has not proved possible to identify any specific reason for the differences.

3.36 As well as the known possible causes of error, such as clerical errors being made in the reporting of data at terminals and errors due to non-standard conditions being used to meter the flow of oil, there are several other technical factors that could cause errors in recording.

3.37 There are issues related to the accuracy of the metering process, in that meters are only required to be accurate to within a certain percentage range. There is also a technical issue related to the recording of quantities at the producing field (which is the input for the production data) and at oil terminals and refineries, since they are in effect measuring different types of oil. Terminals and refineries are able to measure a standardised, stabilised crude oil, i.e. with its water content and content of NGLs at a standard level and with the amounts being measured at standard conditions. However, at the producing field they are dealing with a "live" crude oil that can have a varying level of water and NGLs within it. Producing companies are asked to make adjustments so that production is recorded in terms of stabilised crude oil, but it is known that this estimation is very difficult to carry out. As such the quantities reported by a terminal for its receipts from any individual field will differ from the data reported by the field itself.

3.38 Part of the overall statistical difference may also be due to problems with the reporting of individual NGLs correctly at the production site and at terminals and refineries. It is known that there is some mixing of condensate and other NGLs in with what might otherwise be stabilised crude oil before it enters the pipeline. This mixing happens for several reasons. It saves having to have separate pipeline systems for transporting the NGLs. It also allows the viscosity of the oil passing down the pipeline to be varied as necessary. As said above, the quantity figures recorded by terminals are in terms of stabilised crude oil, with the NGL component removed, but there may be situations where what is being reported does not comply with this requirement.

3.39 It is known that there are some quantities of condensate extracted at gas terminals from the stream of gas extracted from some gas fields on the UKCS. Whilst of small quantity, these amounts will be recorded by terminals and refineries as UK indigenous receipts but will not be recorded as UK production.

3.40 Finally, refinery data is collated from details of individual shipments received and made by each refinery and terminal operating company. There are thousands of such shipments each year, and it is an immense task to cross-reference each shipment, which may be reported separately by two or three different companies involved in the movement. Whilst intensive work is carried out to check these returns, it is possible that some double counting of receipts might be occurring.

3.41 A recent development in the provision of data is that the DTI has taken over the role of collecting and collating data on the operation of oil companies from UKPIA. This has allowed an extra level of comparisons to be made, as it has been possible to look at all parts of the chain of movements of oil, from production right through to receipt at refineries. As such, this will allow thorough investigations to be made into the impact of the above mentioned factors on the data for 1997 and 1998, and will also allow measures to be put in place to identify and minimise their impact in the future.

3.42 With the downstream sector, the statistical differences can similarly be used to assess the validity and consistency of the data. As can be seen in the tables, these differences are generally a very small proportion of the totals involved. However in 1998 and 1999 their size has increased, with calculated deliveries to final consumption being some 1-1½ per cent higher than observed deliveries. More information on the reasons why these differences exist are given in paragraphs 3.97 to 3.102. A full review of the system for reporting the data on the downstream sector is under way, with revisions to the system planned to be introduced from the beginning of 2001. As part of this exercise, individual company reporting practices will be investigated to improve the quality of the data being reported.

Refinery capacity

3.43 Previous editions of the Digest have included a table showing distillation capacity (total, and by refinery); reforming and cracking/conversion capacity (totals). The data for refinery capacity as at the end of 1999 is now presented in Table 3.B below. These figures are collected annually by the Department of Trade and Industry from individual oil companies. Capacity per annum for each refinery is derived by applying the rated capacity of the plant per day when on-stream by the number of days the plant was on stream during the year. Fluctuations in the number of days the refinery is active is usually the main reason for annual changes in the level of capacity. Reforming capacity covers catalytic reforming, and Cracking/Conversion capacity covers processes for upgrading residual oils to lighter products, e.g. catalytic, thermal or hydro-cracking, visbreaking and coking.

3.44 At the end of 1999 the UK had 9 major refineries operating, with three minor refineries in existence. Distillation capacity in the UK at the end of 1999 was 3.5 million tonnes lower than at the end of 1998. This is due to the loss of 4.3 million tonnes of processing capacity through the closure of Shell's Shellhaven refinery at the end of November 1999. Total UK reforming capacity at the end of 1999 was 13.1 million tonnes, 16 per cent lower than at the end of 1998. Cracking and conversion capacity was only 1 per cent lower than at the end of 1998.

Table 3.B: UK refinery processing capacity as at end 1999 [1]

| | Million tonnes per annum | | |
	Distillation	Reforming	Cracking & Conversion
Shell UK Ltd			
Stanlow	11.5	1.4	3.8
Shellhaven (2)	0.0	0.0	0.0
Total (Shell)	11.5	1.4	3.8
Esso Petroleum Co. Ltd			
Fawley	15.6	2.8	4.6
BP Amoco Ltd			
Coryton (3)	9.6	1.6	3.3
Grangemouth	10.2	1.6	3.2
Total (BP Amoco)	19.8	3.2	6.5
TotalFina Ltd.			
Lindsey Oil Refinery Ltd South Killingholme	9.5	1.4	4.1
Texaco Refining Co. Ltd			
Pembroke (4)	10.1	1.5	6.1
Conoco Ltd			
Killingholme	9.4	2.1	9.0
Gulf Oil Refining Ltd			
Milford Haven (4)	0.0	0.0	0.0
Elf Oil Ltd / Murco Pet. Ltd			
Milford Haven	5.3	0.8	1.7
Phillips-Imperial Pet. Ltd			
North Tees	5.0	0.0	0.0
Carless Solvents Ltd			
Harwich	0.6	0.0	0.0
Eastham Refinery Ltd			
Eastham	1.0	0.0	0.0
Nynas UK AB			
Dundee (Camperdown)	0.7	0.0	0.0
Total all refineries	88.5	13.1	35.8

(1) Rated design capacity per day on stream multiplied by the average number of days on stream.
(2) Shellhaven closed in December 1999.
(3) Prior to 1996 owned by Mobil.
(4) Gulf Oil's refinery at Milford Haven closed in November 1997. The cracking facilities jointly owned by Gulf and Texaco as the Pembroke Cracking Company are now wholly owned by Texaco.

Additional information on inland deliveries of selected products (Table 3.8)

3.45 This table gives details for consumption of motor spirit, gas oil/diesel oil and fuel oils given in the main commodity balance tables for the period 1995 to 1999. This includes information on retail and commercial deliveries of motor spirit and DERV fuel that cannot be accommodated within the structure of the commodity balances, but which are of interest.

Also included in the table are details of the quantities of motor spirit and DERV fuel sold collectively by hypermarket and supermarket companies in the UK.

3.46 Motor spirit deliveries in 1999 were 1½ per cent down compared to 1998, and 2 per cent lower than in 1995. Deliveries of DERV fuel were only slightly higher (by ¼ per cent) in 1999, but they were 13 per cent higher than in 1995.

3.47 Several factors are behind the differing trends seen for motor spirit and DERV fuel. There has been an increase in the number of diesel-engined vehicles in use in the UK. Improved technology has resulted in the development of vehicles with performance and characteristics that are more acceptable to the motorist. While diesel vehicles have also been priced at levels comparable with their petrol equivalents, they deliver more miles per gallon. In the National Travel Survey for 1996 to 1998 carried out by the Department for the Environment, Transport and the Regions, diesel-engined cars averaged 51 miles per gallon of fuel, compared with 33 miles per gallon for petrol-engined cars.

3.48 In addition, during the early 1990s there was a significant differential in prices between DERV fuel and motor spirit prices that worked in favour of using DERV fuel. For example, in 1990, a typical retail price for a litre of 4-star petrol would have been 44.87 pence, compared with 40.48 pence for a litre of DERV fuel, representing a 10 per cent saving. This difference was seen at a time when fuel prices were rising as the Gulf crisis had an adverse effect on world markets, increasing the attraction that the efficiency gains of diesel vehicles had for the customer. In December 1999, average retail prices for a litre of the most common grade of motor spirit purchased (premium unleaded), and DERV fuel were 75.42 and 77.65 pence per litre respectively. It is thought that the removal of the favourable differential, and government policy to increase the level of taxation on DERV fuel for environmental reasons, has had a significant effect on slowing down in recent years the numbers of persons switching from petrol to diesel engined vehicles.

3.49 Chart 3.5 shows how the share of total motor spirit deliveries accounted for by unleaded fuel increased from 63 per cent in 1995 to 87 per cent in 1999. In the first quarter of 2000 the share was 91 per cent. Super premium unleaded reached a 4 per cent share of total motor spirit deliveries in 1995, but this has since fallen to 2 per cent in 1999. The current high level of duty applied to this fuel compared to the other grades of motor spirit has done much to reduce demand.

3.50 Since 1990 there has been an overall trend of a reduction in the consumption of motor spirit in the UK. Consumption in 1999 was 11½ per cent lower than the peak of 24 million tonnes of motor spirit consumed in 1990. There are several reasons for this. Firstly, as Chart 3.5 shows, there has been a drop in the level of consumption of leaded fuel in particular. However, this is not the cause of the overall drop in total consumption of motor spirit as the main reason for this decrease is motorists switching to using unleaded petrol in their vehicles.

3.51 As mentioned above (paragraph 3.48) and as illustrated in Chart 3.6, there is a large differential between the price of a litre of leaded and unleaded motor spirit. This factor has helped encourage motorists to switch from leaded to unleaded petrol. In addition, as part of a European strategy to reduce pollution from road traffic (known as the Auto-Oil Directive), leaded petrol (4-star) was banned from general sale from 1st January 2000, which gave an additional push, particularly during the second half of 1999, towards motorists switching to unleaded fuels.

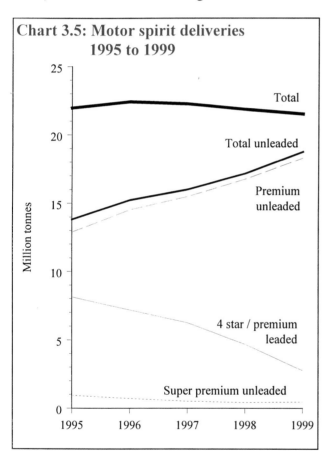

Chart 3.5: Motor spirit deliveries 1995 to 1999

3.52 Although some vehicles originally designed to use leaded fuel can use unleaded fuel, quite frequently the desire on the part of motorists to change to using unleaded fuel has provided an impetus for people to change their vehicles to either new ones or one of more recent manufacture. As a result, studies have shown that the percentage of the UK car fleet that can run on unleaded fuel has increased from 45 per cent in 1990 to 75 per cent in 1997, with its share increasing by around 4 per cent each year in recent years.

3.53 With regards to the phase-out of sales of leaded petrol, some vehicles that are currently using leaded petrol can safely use unleaded petrol in their cars with no adjustments being necessary. Other vehicles can

use unleaded petrol but require adjustments to the ignition timing in order to use the Premium grade unleaded fuel. Some older cars do rely on the lead in petrol to protect their engines from premature wear, and for these a variety of solutions exist to allow their continued use.

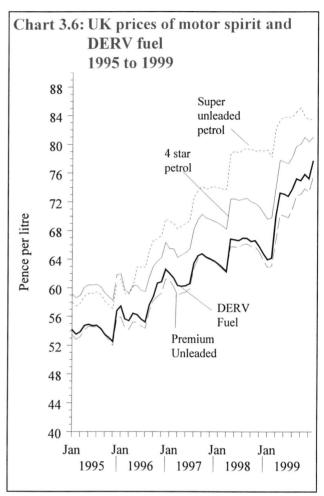

Chart 3.6: UK prices of motor spirit and DERV fuel 1995 to 1999

3.54 Lead in petrol does two things. It increases the fuel's octane rating making it less prone to 'knock' or 'pinking' caused by the fuel in the engine burning in an uncontrolled manner, potentially causing damage to the engine. Secondly, it protects the engine's exhaust valve seats from wear. Cars designed to run on unleaded fuel have very hard valve seats that resist wear. Vehicles that require leaded petrol for its high octane rating can use Super Unleaded petrol where it is available or LRP (Lead Replacement Petrol), which contains an AWA. An AWA (Anti-Wear Additive) is an alternative to lead to protect the engine's exhaust valve seats from excessive wear. Vehicles needing the protection from wear that leaded petrol provides can use LRP or buy an AWA for mixing with unleaded fuel of the correct octane rating or can have the engine modified to run on unleaded fuel.

3.55 LRP began to be widely available as a direct substitute for leaded petrol from the autumn of 1999. Where petrol stations offer LRP, it has replaced 4 star at leaded petrol pumps. Pumps dispensing the new fuel are clearly labelled and have a wide nozzle that does not fit the fuel filler of cars equipped with catalytic converters. It was thought possible that, as an alternative to LRP, some petrol stations might offer an

AWA for mixing with unleaded petrol. These additives would have been available in bottles or syringe-like injection applicators. However, very few such additives are on the market at the moment. It is expected that there will continue to be a decline in sales of LRP over the next few years. It is thus likely that its sale may cease from some petrol stations. If this happens it is possible that bottled additives may start to be made available. For further information on this subject, a free leaflet is available from the Department of the Environment, Transport and the Regions' free publications service (tel. 0870-1226-236), which can be viewed on the DETR Internet site at www.environment.detr.gov.uk/unleaded.

3.56 Chart 3.6 also illustrates the large differential that has existed for some time between the price of 4-star leaded petrol and DERV fuel, with the latter being priced at similar levels to premium grade unleaded petrol. As with the differential between leaded and unleaded petrol, this price differential also worked to encourage motorists to convert to diesel-engined vehicles. Chart 3.7 contains details of vehicle licence registrations for private cars during each year for the period 1990 to 1998, the latest year for which data is available, broken down by type of engine. Whilst the number of petrol engined vehicles licensed only grew by 1 per cent, the number of diesel-engined vehicles licensed has increased nearly four-fold in the same period.

3.57 In 1997, a differential rate of duty was introduced for Ultra Low Sulphur Diesel fuel (ULSD). Initially set at 1 pence per litre, this differential was increased in the March 1998 Budget to 2 pence per litre, and up to 3 pence per litre in the March 1999 Budget. This extra differential has mostly been used to allow producers to cover the additional costs of providing the ULSD, either through funding changes in refinery processes or through covering the extra cost of importing these low sulphur products. The introduction of this duty differential has had a significant effect on moving consumers over to what is regarded as a more environmentally friendly fuel. As such, HM Customs & Excise estimate that, as at the end of 1999, all sales of DERV fuel are now of ULSD.

3.58 ULSD has a maximum sulphur content of 50 parts per million by weight (0.005 per cent). This compares with the previous limit of 500 parts per million (0.05 per cent) in the UK, and the limit of 350 parts per million (0.035 per cent) which came into place from 1/1/2000 as part of the European strategy to reduce pollution from road traffic mentioned in paragraph 3.51. More information on the Auto-Oil Directive can be obtained from the European Commission at www.europa.eu.int.

3.59 The European strategy calls for a further reduction is sulphur levels in road fuels in 2005. This will require a reduction in the level of sulphur in motor spirit from the 150 parts per million in place from 1

January 2000 to 50 parts per million. This ultra low sulphur petrol (ULSP) offers environmental benefits over conventional unleaded petrol. The Department of the Environment, Transport and the Regions has carried out research into the potential benefits of ULSP. Its use can lead to reductions in emissions of nitrogen oxides of up to 6 per cent, carbon monoxide of between 6 and 18 per cent and reductions in hydrocarbon emissions of up to 15 per cent. There are also substantial reductions in emissions of non-regulated pollutants such as benzene, 1,3 butadiene, acetaldehyde and formaldehyde. The availability of ULSP would also allow the introduction of new engine technologies such as gasoline direct injection (GDI) that can improve fuel efficiency by up to 20 per cent. The reduction in fuel consumed would clearly lead to a reduction in carbon dioxide emissions.

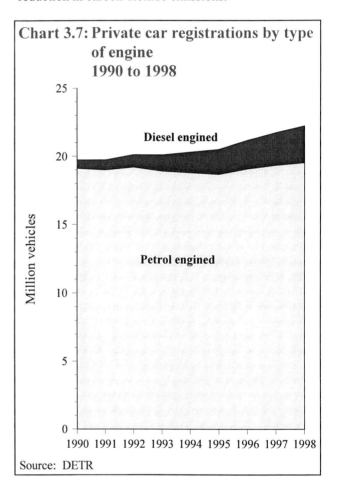

Chart 3.7: Private car registrations by type of engine 1990 to 1998

Source: DETR

3.60 To encourage an early move over to the production and use of ULSP in the UK, HM Government announced in the March 2000 budget its intention to introduce a duty incentive in favour of ULSP, in a similar way to the above mentioned incentive for the use of ULSD. This duty incentive of 1 pence per litre over conventional unleaded petrol is to be introduced during the second half of 2000. This incentive should encourage production and use of ultra low sulphur petrol in advance of the 2005 deadline. This will build upon the substantial success of the duty incentive for ultra low sulphur diesel that has meant that this cleaner fuel is now 100 per cent of the UK diesel market, more than 5 years ahead of the EC deadline.

3.61 Sales by super/hypermarkets have taken an increasing share of retail deliveries (i.e. deliveries to dealers) of motor spirit and DERV fuel in recent years as Table 3.C shows. These figures have been derived from a survey of super/hypermarket companies to collect details of their sales of motor spirit and DERV fuel. The share of total deliveries (i.e. including deliveries direct to commercial consumers) is shown in brackets.

3.62 The flattening of the upwards trend in super/hypermarkets share of retail deliveries in 1996 reflects increased action from other retailers to preserve their market shares (for example, the *Pricewatch* campaign operated by ESSO). The increases seen in subsequent years represent some slight increase in sales by super/hypermarket companies, but the percentage shares are also affected by the decline in the overall deliveries of motor spirit in the UK seen in these years as mentioned earlier.

Table 3.C: Super/hypermarkets share of retail deliveries

per cent

	Motor spirit		DERV fuel	
1995	21.9	(21.5)	14.9	(5.3)
1996	21.8	(21.3)	15.4	(6.0)
1997	22.7	(22.3)	16.7	(6.8)
1998	24.0	(23.6)	17.5	(7.6)
1999	25.6	(25.2)	18.7	(8.6)

3.63 The share of unleaded in total motor spirit deliveries by super/hypermarkets, 88 per cent in 1999, continues to be higher than for other types of outlet (87 per cent in 1999).

Additional information on inland deliveries for non-energy uses

3.64 Table 3.D below summarises additional data on the uses made of the total deliveries of oil products for non-energy uses included as the bottom line in the commodity balances in Tables 3.4 to 3.6. In it, extra information on the uses of lubricating oils and greases by use, and details of products used as petro-chemical feedstocks are given.

3.65 All inland deliveries of lubricating oils and petroleum coke have been classified as going for non-energy uses only. However, a certain part of each does go for energy uses, but it is difficult to estimate figures for energy use for these products with a great degree of accuracy, hence no estimates for energy use appear in the commodity balance tables.

3.66 For lubricating oils, work done by the International Energy Agency suggests that some 50 per cent of inland deliveries each year are re-used as a fuel, through either being burnt whilst being used as lubricants or by being recycled by being re-refined into fuel oils which are then burnt. The available data for

the UK suggest a slightly lower figure of 40 per cent (equal to around 290 to 330 thousand tonnes per year).

Table 3.D: Additional information on inland deliveries for non-energy uses 1997 to 1999

Thousand tonnes

	1997	1998	1999
Feedstock for petroleum chemical plants:			
Propane	830	831	679
Butane	402	383	415
Other gases	1,488	1,465	1,755
Total gases	2,721	2,679	2,849
Naphtha (L.D.F.)	2,640	2,882	2,957
Middle Distillate Feedstock (M.D.F.)	727	760	844
Other products	-	-	-
Total feedstock	6,088	6,321	6,650
Lubricating oils and grease:			
Aviation	3	2	2
Industrial	530	502	470
Marine	49	40	38
Motors	274	254	265
Agricultural	16	15	15
Fuel oil sold as lubricant	-	-	-
Total lubricating oils & grease	872	813	790
Other non-energy products:			
Industrial spirit	87	83	70
White spirit	108	96	95
Bitumen	2,015	1,967	1,963
Petroleum wax	44	18	19
Petroleum coke	1,095	887	939
Miscellaneous products	644	538	354
Total non-energy use	10,953	10,724	10,881

3.67 For petroleum coke, more information is available which has allowed more accurate estimates to be made. It has been possible to analyse the data available for the imports of petroleum coke to identify which type of company is importing the product. This work has shown that a significant proportion of petroleum coke imports each year are made by energy companies, such as power generators or fuels merchants, with a significant proportion also being imported by cement manufacturers. Whilst it cannot be certain that these imports are being used as a fuel, information on the use of petroleum coke in cement manufacture does suggest that it is being used as a fuel.

3.68 Using the data available for imports, estimates have been constructed for 1997, 1998 and 1999 which show that possibly up to 50 per cent of inland deliveries in some recent years (equal to around 550 thousand tonnes) of petroleum coke are imported by these companies for energy uses. In 1998, approximately 20 per cent of these supplies were for electricity generation, with 30 per cent for use as a fuel in the manufacture of cement, and the remainder being imported for sale as a solid fuel or to be used as an input into the manufacture of other solid fuels. In

1999, these proportions remained the same, but there was a reduction in the level of imports of petroleum coke, such that only around 280 thousand tonnes of petroleum coke is estimated to have gone for energy uses.

3.69 Data on the prices paid by industry for petroleum coke on a basis comparable with the other price data in Chapter 9 of the Digest are not available. However, analysis of the data on the quantity and value of imports of petroleum coke into the UK from HM Customs & Excise provides some estimates for the cost of imports and gives some indication of the prices being paid. These are only indicative of the prices being paid in the port of importation, and do not include the extra transport costs from the port to the final destination that would be part of more rigorous price estimate. Details of these estimates are included in Chapter 8 on trade in fuels, as part of Table 8.3. A breakdown has been made by grade of petroleum coke and type of use for imports into the UK, which is given in Table 3.E below. Calcined petroleum coke is virtually pure carbon, and as such is more valuable than non-calcined (otherwise known as "green") petroleum coke, as shown by the higher £ per tonne it commands and the fact that it is not used simply as a fuel.

3.70 Petroleum coke is a relatively low energy content fuel, having a calorific value of 39.5 GJ per tonne, compared with an average for petroleum products of 45.2 GJ per tonne, and 43.2 GJ per tonne for fuel oil. It is however higher than coal (27.3 GJ per tonne) and in certain areas is competing with coal as a fuel. It has the advantage of being a very cheap fuel, since it is often regarded as a waste product rather than a specific output from the refining process. Compared to imports of coal, prices per GJ are about 25 per cent lower.

Table 3.E: Estimated £ per tonne for imports of petroleum coke into the UK

| | Non-calcined ("green") petroleum coke | | | Calcined petroleum coke |
	Energy	Non-energy	Total	Non-energy
1997	35.7	39.1	37.0	142.5
1998	26.7	61.1	37.4	117.1
1999	20.0	31.7	25.3	133.2

Inland deliveries of gas oil/diesel fuel and fuel oils for energy use

3.71 As mentioned in the 1999 edition of the Digest, the table showing inland deliveries of these fuels, with a detailed breakdown by final user, included as Table 3.11 in last years Digest has been under review. Most of the industrial breakdown previously given in this table is already included in the main commodity balances (Tables 3.4 to 3.6). The additional information that was given in this table under each industry heading was of limited quality. Oil companies have reported as accurately as possible their deliveries

to the detailed industries. However, due to the fact that a certain percentage of their deliveries into the UK market of these fuels are sold to distributors for onward sale rather than being sold direct to the consuming company, the very detailed figures in this table are of limited accuracy. As such it has been decided not to include this table in the current edition of the Digest. However, users may still obtain a copy of the table direct from the contact details listed at the end of this chapter.

Inland deliveries by country (Table 3.9)

3.72 This table shows deliveries in England and Wales, Scotland, and Northern Ireland. The figures for deliveries for energy use between 1997 and 1999 show a decline in all three regions. The fall in the use of fuel oils over the period is 31 per cent in Scotland, 47 per cent in England and Wales and 37 per cent in Northern Ireland, reflecting the switch in fuel use at the Ballylumford power station in 1997 from fuel oil to gas. Added to this, the fall in use of fuel oil in England and Wales over the period was more than offset by increased deliveries of aviation turbine fuel and DERV fuel (by 15 and 2 per cent respectively). Feedstocks for use in petrochemical plants in Scotland increased in 1999 compared to 1998 (up by 9 per cent) despite an overall fall in non-energy use, related to an increase in capacity of the BP Grangemouth refinery and petro-chemical plant during the period.

Stocks of oil (Table 3.10)

3.73 This table shows stocks of crude oil, feedstocks (including partly processed oils) and products (in detail) at the end of each year. Stocks of crude oil and feedstocks increased in 1999, with increases in stocks held at terminal and offshore offsetting a decrease in stocks held at refineries. The increase in stocks held at terminals is related to increased throughput as a result of the increased level of output from UK production installations. Offshore stocks have increased due to the fact that more recent discoveries of oil are of smaller sized fields than previously discovered (see the Annex C, paragraph C.15). As such, the economics of producing the oil from these fields are such that the fields tend to hold crude oil in storage facilities at the field itself. It is then offloaded into tankers as opposed to being transported to shore via pipeline systems. The decrease in refinery stocks reflects to some extent refinery closures and also companies working to reduce cost levels as much as possible.

3.74 As stated in paragraphs 3.20 and 3.21, the details of stocks of petroleum products (and crude and process oils) included in Table 3.10 are all stocks that are owned by UK companies, and thus include details of any stocks owned but held abroad, e.g. in Rotterdam, under bilateral government agreements. As such, the level of stocks in this table represents the full availability of stocks to the UK in case of any oil emergency occurring from a disruption to international supplies.

3.75 Stocks of petroleum products at the end of 1999 were 15 per cent lower than a year earlier. From July 1999 the UK Government implemented a revised EU Directive (EU Directive 93/98) setting obligations on EU member states to hold stocks of oil for emergency situations. Part of the changes was a reduction in the level of stocking obligation placed on the UK due to it being a producer of oil. The benefits of this reduction in obligation were passed on to oil companies in the UK as soon as possible, and so a reduction in the level of oil stocks held by UK companies, both in terms of products and also crude oil held at refineries, was seen during the middle part of 1999. The total stocks of crude oil and products held by UK companies at the end of 1999 would be sufficient to meet the UK's needs for approximately 75 days.

3.76 More information on these changes was provided in an article published in the September 1999 edition of *Energy Trends*, a copy of which can be obtained from the contact persons listed at the end of this chapter.

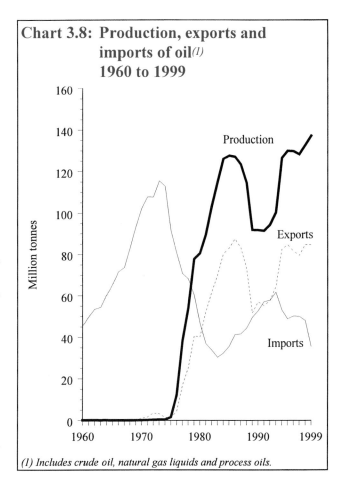

Chart 3.8: Production, exports and imports of oil[(1)] **1960 to 1999**

(1) Includes crude oil, natural gas liquids and process oils.

Long term trends

3.77 Tables 3.11 and 3.12 present extended time series of selected more aggregated data from the preceding tables, with a view to giving additional background on the historic development of the crude oil and petroleum sectors.

Crude oil and petroleum products: production, imports and exports (Table 3.11)

3.78 Table 3.11 shows data from 1960 to 1999 for production, imports and exports of crude oil (including natural gas liquids and feedstocks) and oil products. It also shows United Kingdom refinery throughput of crude oil, and the inland deliveries of oil products. Indigenous production of crude oil is shown in total with landward production shown separately.

3.79 The second (right-hand) part of the table consists of time series showing key aggregates, net exports figures, shares, etc. It should be noted that exports of crude oil include some imports that have been re-exported. In years of significant indigenous production these have little effect on exports as a proportion of indigenous production, but in the earlier years (approximately pre-1975) the re-exports exceeded indigenous production and thus the ratio of exports to indigenous production was over 1.

3.80 Chart 3.8 illustrates the trends in the production, exports and imports of crude oil. It shows that indigenous production of crude oil, etc. was negligible up to 1974 and then increased rapidly as North Sea production came on stream. Imports peaked in 1973, immediately prior to the first OPEC price 'hike'. The chart shows the rapid decline of net imports thereafter as imports fell and indigenous production rose, until 1981 when the surplus turned from net imports to net exports. Net exports peaked in 1984, one year before the peak for North Sea production in 1985.

3.81 The large fall in production in 1988 and, particularly 1989, reflects the effects of the Piper Alpha accident and subsequent incidents, and the continued 'low' production in 1990 and 1991 reflects the consequent safety work. Production increased in 1992 and again in 1993 topping 100 million tonnes for the first time since 1988. In 1995 production at 127 million tonnes returned to previous peak levels, and in 1999 set a new record at 137 million tonnes.

3.82 Table 3.11 shows that the imports share of refinery throughput of crude oil fell from around 100 per cent, before North Sea oil production started, to a low of 39 per cent in 1983 (the lowest year for imports), before rising to 64 per cent in 1993. Since then, with the significant increase in indigenous production, the imports share has fallen back to 51 per cent in 1999. These developments are mirrored by the changes in the ratio of indigenous production to refinery throughput. Ignoring pre-1976 figures, the proportion of indigenous production exported increased from 35 per cent to around two-thirds towards the end of the 1980s. Although the decreases in production in the late 1980s (paragraph 3.81) did lead to some reduction in the level of exports, the proportion of production exported has continued at roughly this level during the 1990s.

3.83   Net exports of oil products increased during the early 1990s to a record high in 1993, but fell in 1995 and remained broadly unchanged in 1996, partly as a result of the fire at Texaco's refinery in July 1995. The increases in net exports of products reflects the increased throughput from refineries mainly feeding through to increased exports of product rather than increases in deliveries to the domestic market. With the closure of the Gulf Oil refinery from December 1997, net exports of products decreased in both 1998 and 1999. Imports of crude oil in 1991 (and marginally again in 1992) exceeded exports for the first time since 1980. Net exports of crude oil resumed in 1993, and have continued to rise since that time. In 1999, at 47 million tonnes, net exports of crude oil were the highest since 1984 and overall net exports of crude oil and products at 56 million tonnes were at a new record level.

3.84   Refinery throughput peaked in 1973, and subsequently both this and refinery output (the difference is refinery use of fuel and gains/losses) fell to pre-1970 levels. Since the low point of 1983, both refinery throughput and output have increased by over 25 per cent. However, both fell in 1994 and again in 1995, partly as a result of the fire mentioned above. Refinery throughput and output in 1997 were both the largest seen since 1979, but with the closure of the Gulf Oil refinery in December 1997, refinery output fell by 3½ per cent in 1998 and then by another 6 per cent in 1999 to the lowest level seen since 1989.

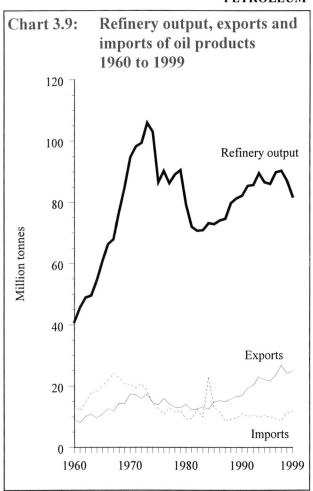

Chart 3.9:   Refinery output, exports and imports of oil products 1960 to 1999

3.85   Exports of oil products increased in 1991, 1992 and 1993, comfortably exceeding the earlier peak at the beginning of the 1970s, but fell in 1994 and 1995. In 1997 at 26.8 million tonnes they were the highest ever recorded. Imports of oil products were at their highest in 1967 and, apart from a 'blip' in 1984 as a result of the miners' strike, have been less than half this peak in recent years. As a result, 1984 apart, exports of oil products have exceeded imports in every year since 1974. In 1999, imports made up 17 per cent of inland deliveries compared with over 30 per cent in the early to mid-1960s. Chart 3.9 summarises the trend in refinery output, exports and imports of oil products over the period.

Inland deliveries of petroleum products (Table 3.12)

3.86 Table 3.12 shows data from 1960 to 1999 of deliveries of petroleum products split between non-energy uses in total and the major products delivered for energy use. The right-hand half of the table shows trends over the same period in petroleum products delivered to the main energy industries and to different categories of final user. It should be noted that whilst data for deliveries are considered to be a good proxy for consumption, differences can occur mainly because in the former no account is taken of the effect of stock changes at points further along the chain of consumption. The total of deliveries for energy use shown in the first (left-hand) half of the table excludes 'own use' by refineries shown in the right-hand part of the table.

3.87 Deliveries of petroleum products, in common with many other aggregate figures, (see Table 3.11) peaked in 1973. The 'blip' in 1984 reflects the increased deliveries of fuel oil, in particular, during the miners' strike. Fuel oil deliveries are now about 10 per cent of, and gas oil (other than DERV fuel) about half of the levels reached in the early 1970s. In contrast, deliveries of motor spirit and aviation turbine fuel have grown virtually continuously throughout the period. After limited growth during the 1970s and early 1980s, deliveries of DERV fuel resumed the high growth rates apparent in the 1960s, and have nearly doubled over the last 10 years. The upward surge of deliveries of transport fuels slowed in 1990 and ceased in 1991 with the twin impacts of the Gulf crisis and recession, with some recovery being seen in 1992.

3.88 Since 1990, deliveries of motor spirit have been decreasing each year, with the exception of 1996 which is 2 per cent higher than 1995, but is still 8 per cent lower than the level seen in 1990, and deliveries in 1995 were at their lowest level since 1986. 1999 has seen a 1½ per cent reduction over 1998. These changes reflect the switch to diesel-engined cars during the late 1980s and early 1990s, although this switch has lessened in recent years. As such, they are mirrored by the consistent pattern of increases in deliveries of DERV fuel each year since 1990, although this rate of increase has been slowing in recent years. Deliveries of aviation turbine fuel have also consistently increased each year since 1990.

In 1999, deliveries of DERV fuel were 43 per cent higher than in 1990, and deliveries of aviation turbine fuel had increased by 47 per cent over the same period. Chart 3.10 shows the trends in transport fuels from 1960 to 1999.

3.89 By the end of the 1980s and so far in the 1990s deliveries for non-energy uses were not far off their peak of the early to mid-1970s, in contrast to energy uses, which, despite the growth for transport, remains at around three-quarters of their peak levels.

3.90 The right-hand half of the table illustrates the growth in transport use - this includes the use of the fuels as given in the left hand side of the table plus other energy uses of petroleum products. Total uses by the transport industry are now well over three times the amount delivered in 1960. Deliveries to every other major sector are below 1973 levels - well below for electricity generators, gas works, iron and steel and 'other industries', and other final users (mainly agriculture, public administration and commerce).

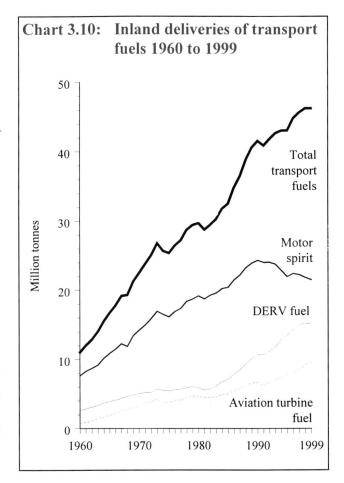

Chart 3.10: Inland deliveries of transport fuels 1960 to 1999

Technical notes and definitions

Indigenous production

3.91 The term indigenous is used throughout this section and includes oil from the UK Continental Shelf both offshore and onshore.

Deliveries

3.92 These are deliveries into consumption, as opposed to being estimates of actual consumption or use. They are split between inland deliveries and deliveries to marine bunkers. Inland deliveries will not necessarily be consumed in the United Kingdom (e.g. aviation fuels).

Sources of data

3.93 The majority of the data included in the text and tables of this chapter are derived from the UK Petroleum Industry Association (UKPIA) data collection system. In this, UKPIA collect data relating to the inland operations of the UK oil industry (i.e. information on the supply, refining and distribution of oil in the UK). The format and coverage of the data is such that it meets most of the needs of both government and the industry itself. As such, it operates by each member of UKPIA providing returns on its refining activities and deliveries of various products to the internal UK market. This information is supplemented whenever necessary to allow for complete coverage within the statistics, with separate exercises carried out on special topics (for example, the work on super/hypermarkets referred to in paragraph 3.61).

Statistical differences

3.94 In Tables 3.1 to 3.7, there are headings titled "statistical differences". These are differences that are seen between the separately observed figures for production and deliver of crude oil and products during the path of their movement from the point of production to the point of consumption.

3.95 These headings listed in the primary oil commodity balances (Tables 3.1 to 3.3) are differences that are seen between the separately observed and reported figures for production from onshore or offshore fields and supply to the UK market that cannot be accounted for by any specific factors. As such, they are primarily the result of the various inaccuracies in the meters at various points along offshore pipelines. These meters vary slightly in their accuracy within accepted tolerances, giving rises to both losses and gains when the volumes of oil flowing are measured. Temperature, pressure and natural leakage also contribute to the statistical differences as well. In addition, where data are shown on an energy basis, small discrepancies can occur between the estimated calorific values used at the field and the more accurately measured calorific value at the onshore terminal. There are also factors such as errors

due to rounding or unrecorded losses, such as leakage.

3.96 Other contributory factors in these balances are inaccuracies in the recording of the amounts of these substances reported as being disposed of to the various activities listed, including differences between the amounts reported as going to refineries and the actual amounts that pass through refineries.

3.97 Similar to this, the data under these headings in Tables 3.4 to 3.6 are the differences between the deliveries of petroleum products to the inland UK market reported by the supplying companies and estimates for such deliveries. These estimates are calculated by taking the output of products reported by refineries and then adjusting it by the relevant factors (such as imports and exports of the products, changes in the levels of stocks etc.).

3.98 As the data underlying both the observed deliveries into the UK market and the individual components of the estimates (i.e. production, imports, exports, stocks) comes from the same source (the oil companies), it may be thought that such differences should not exist. While it is true that each oil company provides data on its own activities in each area, there are separate areas of operation within the companies that report their own part of the overall data. The table below illustrates this.

Table 3.F: Sources of data within oil companies

Area covered	Source
Refinery production	Refinery
Imports & Exports	Refinery, logistics departments, oil traders
Stocks	Refinery, crude & product terminals, major storage and distribution sites
Final deliveries	Sales, marketing & accounts depts.

3.99 Each individual reporting source will have a direct interest in the data it is reporting. Refineries will be clear on what they produce and how much leaves the refinery gate as part of routine monitoring of the refinery operations. Sales to final consumers represent a company's main source of income, and so they will similarly be monitored closely, as will imports and exports. Companies will ensure that each component set of data reported is as accurate as possible, but in some cases the systems used are not integrated, which means that internal consistency checks to ensure consistency across all of the data reported cannot be made. Each part of a company may also work to different timings as well, which may further add to the degree of differences seen, but the main area where there is known to be a problem is with the "Transfers" heading in the commodity balances.

3.100 The data reported under this heading has two

components. Firstly there is an allowance for reclassification of products within the refining process. For example, butane is added to motor spirit to improve the octane rating, aviation turbine fuel could be reclassified as gas diesel oil if its quality deteriorates, and much of the fuel oil imported into the UK is further refined into other petroleum products. In addition to these inter-product transfers the data also includes an allowance to cover the receipt of backflows of products from petrochemical plants. Such plants are often very closely integrated with refineries (for example, BP's refinery at Grangemouth is right next to the petrochemical plant). A deduction for these backflows thus needs to be included under the "Transfers" heading so that calculated estimates reflect net output and are thus more comparable with the basis of the observed deliveries data.

3.101 However, there is scope for error in the recording of these two components. With inter-product transfers, the data is recorded within the refinery during the refining and blending processes. However, during these processes the usual units used to record the changes are volume rather than masses. As shown by the conversion factors given in Annex A, different factors apply for each product when converting from a volume to mass basis. What might be a balanced transfer in volume terms is thus not when converted to a mass basis. This is thought to be the main source of error within the individual product balances.

3.102 With the backflows data, as the observed data for deliveries derived from sales data are on a "net" basis they will exclude the element of backflows. However, it is thought that there is significant scope for error in the recording of the backflows when they are received at a refinery. For example, these could be seen simply as an input of fuel oils to be used as a feedstock, and thus recorded as an input without their precise nature being recorded – in effect a form of double-counting. It is this relationship between the petrochemical sector and refineries that is thought to be the main source of error in the overall oil commodity balances, and one which will be looked at during the review of the reporting system mentioned in paragraph 3.42.

Imports and exports
3.103 The information given under the headings "imports" and "exports" in this chapter are the figures recorded by importers and exporters of oil. They thus differ in some cases from the import and export figures provided by HM Customs and Excise that are given in Chapter 8 of this Digest. These differences may arise since whilst the trader's figures are a record of actual movements in the period, for non-EU trade, HM Customs and Excise figures show the trade as declared by exporters on documents received during the period stated. The Customs figures also include re-exports. These are products that may have originally entered the UK as imports from another country and been

stored in the UK prior to being exported back out of the UK, as opposed to having been actually produced in the UK.

3.104 In previous versions of the Digest, these imports and exports were called "arrivals" and "shipments" in an attempt to highlight their difference from the other sources of trade data. However, their name has now been changed to more clearly represent what the movements actually are; movements in and out of the United Kingdom.

Marine bunkers
3.105 This covers deliveries to ocean going and coastal vessels under international bunker contracts. Other deliveries to fishing, coastal and inland vessels are excluded.

Crude and process oils
3.106 These are all feedstocks, other than distillation benzene, for refining at refinery plants. Gasoline feedstock is any process oil whether clean or dirty which is used as a refinery feedstock for the manufacture of gasoline or naphtha. Other refinery feedstock is any process oil used for the manufacture of any other petroleum products.

Refineries
3.107 Refineries distilling crude and process oils to obtain petroleum products. This excludes petrochemical plants, plants only engaged in re-distilling products to obtain better grades, crude oil stabilisation plants and gas separation plants.

Products used as fuel (energy use)
3.108 The following paragraphs define the product headings used in the text and tables of this chapter, which are used for energy in some way, either directly as a fuel or as an input into electricity generation.

Refinery fuel - Petroleum products used as fuel at refineries.

Ethane - An ethane (C_2H_6) rich gas in natural gas and refinery gas streams. Primarily used, or intended to be used, as a chemical feedstock.

Propane - Hydrocarbon containing three carbon atoms, gaseous at normal temperature but generally stored and transported under pressure as a liquid. Used mainly for industrial purposes and some domestic heating and cooking.

Butane - Hydrocarbon containing four carbon atoms, otherwise as for propane. Additional uses - as a constituent of motor spirit to increase vapour pressure and as a chemical feedstock.

Other gases for gasworks - Ethane and other refinery gases resulting from the processing of crude petroleum.

Naphtha (Light distillate feedstock) - Petroleum

distillate boiling predominantly below 200°C.

Aviation spirit - All light hydrocarbon oils intended for use in aviation piston-engine power units, whether in the air, on land, or on water, including bench testing of aircraft engines.

Motor spirit - Blended light petroleum components used as fuel for spark-ignition internal-combustion engines other than aircraft engines:

(i) 4 star grade - all finished motor spirit with an octane number (research method) not less than 97, conforming to British Standard 4040. This can include leaded petrol or unleaded petrol containing an alternative to lead as an anti-wear additive (lead replacement petrol – LRP).

(ii) Premium unleaded grade - all finished motor spirit, with an octane number (research method) not less than 95, conforming to British Standard 7070.

(iii) Super premium unleaded grade - all finished motor spirit, with an octane number (research method) not less than 97, conforming to British Standard 7800.

Aviation turbine fuel - All other turbine fuel intended for use in aviation gas-turbine power units, whether in the air, on land or on water, including bench testing of aircraft engines.

Burning oil (kerosene) - Refined petroleum fuel, intermediate in volatility between motor spirit and gas oil, used for lighting and heating. White spirit and kerosene used for lubricant blends are excluded.

Gas oil/automotive diesel - Petroleum fuel having a distillation range immediately between kerosene and light-lubricating oil.

(i) **DERV (Diesel Engined Road Vehicle) fuel** - automotive diesel fuel for use in high speed, compression ignition engines in vehicles subject to Vehicle Excise Duty.

(ii) **Gas oil** - used as a burner fuel in heating installations, for industrial gas turbines and as for DERV (but in vehicles not subject to Vehicle Excise Duty e.g. Agriculture vehicles, fishing vessels, construction equipment).

(iii) **Marine diesel oil** - heavier type of gas oil suitable for heavy industrial and marine compression-ignition engines.

Fuel oil - Heavy petroleum residue blends used in atomising burners and for heavy duty marine diesel engines (marine bunkers, etc.) normally requiring pre-heating before combustion, but excluding fuel oil for grease making or lubricating oil and fuel oil sold as such for road making.

Orimulsion - An emulsion of bitumen in water used as a fuel primarily in power stations. In the tables Orimulsion is normally excluded from fuel oil, but where it is not shown separately it will be designated as part of fuel oils. From May 1995 Orimulsion has

been classified as a type of bitumen for overseas trade purposes (see below). It was last imported in February 1997. Since that time the sole power station in the UK that was using it as a fuel has changed to alternative sources of energy.

Products not used as fuel (non-energy use)

3.109 The following paragraphs define the product headings used in the text and tables of this chapter, which are used for non-energy purposes.

Feedstock for petroleum chemical plants - All petroleum products intended for use in the manufacture of petroleum chemicals. This includes middle distillate feedstock of which there are several grades depending on viscosity. The boiling point ranges between 200°C and 400°C. (A deduction has been made from these figures equal to the quantity of feedstock used in making the conventional petroleum products that are produced during the processing of the feedstock. The output and deliveries of these conventional petroleum products are included elsewhere as appropriate.)

White spirit - A highly refined distillate with a boiling range of about 150°C to 200°C used as a paint solvent and for dry cleaning purposes etc.

Industrial spirit - Refined petroleum fractions with boiling ranges up to 200°C dependent on the use to which they are put - e.g. seed extraction, rubber solvents, perfume etc.

Lubricating oils (and grease) - Refined heavy distillates obtained from the vacuum distillation of petroleum residues. Includes liquid and solid hydrocarbons sold by the lubricating oil trade, either alone or blended with fixed oils, metallic soaps and other organic and/or inorganic bodies. A certain percentage of inland deliveries are re-used as a fuel (see paragraphs 3.64 to 3.70).

Bitumen - The residue left after the production of lubricating oil distillates and vacuum gas oil for upgrading plant feedstock. Used mainly for road making and building construction purposes. Includes other petroleum products, creosote and tar mixed with bitumen for these purposes and fuel oil sold as such for road making. In May 1995 harmonisation of EU trade categories, resulted in Orimulsion being reclassified as a bitumen, but for this chapter Orimulsion is still included under products used as fuel (energy use).

Petroleum wax - Includes paraffin wax, which is a white crystalline hydrocarbon material of low oil content normally obtained during the refining of lubricating oil distillate, paraffin scale, slack wax, microcrystalline wax and wax emulsions. Used for candle manufacture, polishes, food containers, wrappings etc.

Petroleum cokes - Carbonaceous material derived

from hydrocarbon oils, uses for which include metallurgical electrode manufacture. Quantities of imports of this product are used as a fuel, primarily in the manufacture of cement (see paragraphs 3.64 to 3.70).

Miscellaneous products - Includes aromatic extracts, defoament solvents and other minor miscellaneous products.

Main classes of consumer

3.110 The following are definitions of the main groupings of users of petroleum products used in the text and tables of this chapter.

Gas works - Deliveries of petroleum products to establishments producing gas for public supply.

Electricity generators - Petroleum products delivered for use by major power producers and other companies for electricity generation including those deliveries to the other industries listed below which are used for autogeneration of electricity (Tables 3.4 to 3.6). This includes petroleum products used to generate electricity at oil refineries and is recorded in the Transformation sector, as opposed to other uses of refinery fuels which are recorded in the Energy Industry Use sector.

Agriculture - Deliveries of fuel oil and gas oil/diesel oil for use in agricultural power units, dryers and heaters. Burning oil for farm use.

Iron and steel - Deliveries of petroleum products to steel works and iron foundries.

Other industries - The industries covered correspond to the industrial groups shown in Table 1.E excluding Iron and Steel of Chapter 1.

Marine - Fuel oil and gas oil/diesel oil delivered, other than under international bunker contracts, for fishing vessels, UK oil and gas exploration and production, coastal and inland shipping and for use in ports and harbours.

Railways - Deliveries of fuel oil, gas oil/diesel oil and burning oil to railways, excluding deliveries to railway power stations.

Air transport - Total inland deliveries of aviation turbine fuel and aviation spirit. The figures cover deliveries of aviation fuels in the United Kingdom to international and other airlines, British and foreign governments (including armed services) and for private flying.

Road transport - Deliveries of motor spirit and DERV fuel for use in road vehicles of all kinds. Petroleum industry estimates for the consumption of road transport fuels by different vehicle classes, formerly provided in this Digest, are no longer available to the Department. The Department of the Environment, Transport and the Regions has provided alternative estimates. These are based on details of average vehicle mileages and assumed miles per gallon. The methodology behind these analyses and the estimates themselves are currently in the process of being updated. As such listed below is an indicative breakdown of road fuel consumption. More up-to-date information should be available from the DETR later in the year.

Table 3.G: Estimated consumption of road transport fuels by vehicle class

Motor spirit:	
Cars and taxis	95%
Goods vehicles, mainly light vans	4%
Remainder, mainly motor cycles,	
Mopeds, etc.	1%
DERV:	
Goods vehicles	68%
Buses and coaches	7%
Remainder, mainly diesel-engined cars	
And taxis.	25%

Source: DETR

Domestic - Fuel oil and gas oil/diesel oil delivered for central heating of private houses and other dwellings and deliveries of kerosene (burning oil) and liquefied petroleum gases for domestic purposes (see Tables 3.4 to 3.6).

Public services - Deliveries to national and local government premises (including educational, medical and welfare establishments and British and foreign armed forces) of fuel oil and gas oil/diesel oil for central heating and of kerosene (burning oil).

Miscellaneous - Deliveries of fuel oil and gas oil/diesel oil for central heating in premises other than those classified as domestic or public services and fuel oil and gas oil/diesel oil used by the petroleum industry other than as refinery fuel.

Monthly and quarterly data

3.111 Monthly or quarterly aggregate data for certain series presented in this chapter are available in Tables 13 to 17 of the monthly DTI publication *Energy Trends*. See Annex F for more information about *Energy Trends*.

Contact: *Kevin Williamson (Statistician)*
              020 7215 5184
              Clive Evans
              020 7215 5189
              Ian Corrie
              020 7215 2714

3.1 Commodity balances 1999[(1)]

Primary oil

Thousand tonnes

	Crude oil	Ethane	Propane	Butane	Condensate	Total NGL	Feedstock	Total primary oil
Supply								
Production	128,262	2,022	2,853	2,005	1,957	8,837	-	137,099
Other sources	-	-	-	-	-	-	-	-
Imports	39,321	-	-	-	-	-	5,570	44,891
Exports	-85,052	-36	-1,980	-1,154	-700	-3,870	-2,875	-91,797
Marine bunkers	-	-	-	-	-	-	-	-
Stock change (2)	-347	..	..	..	..	+17	+132	-198
Transfers	-	-1,527	-865	-931	-	-3,323	+2,105	-1,218
Total supply	**82,184**	..	..	..	..	**1,661**	**4,932**	**88,777**
Statistical difference (3)(4)	+636	..	..	..	..	+42	-510	+168
Total demand (4)	**81,548**	..	..	..	..	**1,619**	**5,442**	**88,609**
Transformation(4)	**81,548**	..	..	..	..	**1,296**	**5,442**	**88,286**
Electricity generation	-	-	-	-	-	-	-	-
Major power producers	-	-	-	-	-	-	-	-
Autogenerators	-	-	-	-	-	-	-	-
Petroleum refineries	81,548	..	..	..	..	1,296	5,442	88,286
Coke manufacture	-	-	-	-	-	-	-	-
Blast furnaces	-	-	-	-	-	-	-	-
Patent fuel manufacture	-	-	-	-	-	-	-	-
Other	-	-	-	-	-	-	-	-
Energy industry use	-	**316**	**7**	-	-	**323**	-	**323**
Electricity generation	-	-	-	-	-	-	-	-
Oil & gas extraction(4)	-	316	7	-	-	323	-	323
Petroleum refineries	-	-	-	-	-	-	-	-
Coal extraction	-	-	-	-	-	-	-	-
Coke manufacture	-	-	-	-	-	-	-	-
Blast furnaces	-	-	-	-	-	-	-	-
Patent fuel manufacture	-	-	-	-	-	-	-	-
Pumped storage	-	-	-	-	-	-	-	-
Other	-	-	-	-	-	-	-	-
Losses	-	-	-	-	-	-	-	-

(1) As there is no use made of primary oils and feedstocks by industries other than the oil and gas extraction and petroleum refining industries, other industry headings have not been included in this table. As such, this table is a summary of the activity of what is known as the Upstream oil industry.

(2) Stock fall (+), stock rise (-).

(3) Total supply minus total demand.

(4) Figures for total demand for the individual NGLs (and thus for the statistical differences as well) are not available. While separate data are available on the use of individual NGL's in the extraction of oil and gas, details of inputs into refineries of NGLs are only available at aggregate level for total NGLs, resulting in accurate estimates for the total demand for each NGL not being possible. As they are available, details of the use of individual NGLs in the extraction of oil and gas are included in this table.

3.2 Commodity balances 1998[1]

Primary oil

Thousand tonnes

	Crude oil	Ethane	Propane	Butane	Condensate	Total NGL	Feedstock	Total primary oil
Supply								
Production	124,222	1,734r	3,031	1,646r	1,734	8,145r	-	132,367r
Other sources	-	-	-	-	-	-	-	-
Imports	39,447	-	-	-	-	-	8,510	47,957
Exports	-79,651	-41	-1,842	-856	-641	-3,380	-1,581	-84,612
Marine bunkers	-	-	-	-	-	-	-	-
Stock change (2)	-622	..	..	..	..	-29	+58	-593
Transfers	-	-1,215	-1,071	-1,171	-	-3,457	1,255	-2,202
Total supply	**83,396**	..	..	..	..	**1,279r**	**8,242**	**92,917r**
Statistical difference (3)(4)	-1,114	..	..	..	..	-431r	+312	-1,233r
Total demand (4)	**84,510**	..	..	..	..	**1,710**	**7,930**	**94,150**
Transformation(4)	**84,510**	..	..	..	..	**1,357**	**7,930**	**93,797**
Electricity generation	-	-	-	-	-	-	-	-
Major power producers	-	-	-	-	-	-	-	-
Autogenerators	-	-	-	-	-	-	-	-
Petroleum refineries	84,510	..	..	..	..	1,357	7,930	93,797
Coke manufacture	-	-	-	-	-	-	-	-
Blast furnaces	-	-	-	-	-	-	-	-
Patent fuel manufacture	-	-	-	-	-	-	-	-
Other	-	-	-	-	-	-	-	-
Energy industry use	-	**302r**	**38r**	**13**	-	**353**	-	**353**
Electricity generation	-	-	-	-	-	-	-	-
Oil & gas extraction(4)	-	302r	38r	13	-	353	-	353
Petroleum refineries	-	-	-	-	-	-	-	-
Coal extraction	-	-	-	-	-	-	-	-
Coke manufacture	-	-	-	-	-	-	-	-
Blast furnaces	-	-	-	-	-	-	-	-
Patent fuel manufacture	-	-	-	-	-	-	-	-
Pumped storage	-	-	-	-	-	-	-	-
Other	-	-	-	-	-	-	-	-
Losses	-	-	-	-	-	-	-	-

(1) As there is no use made of primary oils and feedstocks by industries other than the oil and gas extraction and petroleum refining industries, other industry headings have not been included in this table. As such, this table is a summary of the activity of what is known as the Upstream oil industry.

(2) Stock fall (+), stock rise (-).

(3) Total supply minus total demand.

(4) Figures for total demand for the individual NGLs (and thus for the statistical differences as well) are not available. While separate data are available on the use of individual NGL's in the extraction of oil and gas, details of inputs into refineries of NGLs are only available at aggregate level for total NGLs, resulting in accurate estimates for the total demand for each NGL not being possible. As they are available, details of the use of individual NGLs in the extraction of oil and gas are included in this table.

3.3 Commodity balances 1997[1]

Primary oil

Thousand tonnes

	Crude oil	Ethane	Propane	Butane	Condensate	Total NGL	Feedstock	Total primary oil
Supply								
Production	120,321	1,752	2,840	1,775	1,546	7,913	-	128,234
Other sources	-	-	-	-	-	-	-	-
Imports	41,333	-	-	-	-	-	8,661	49,994
Exports	-75,113	-76	-1,639r	-659	-568	-2,942r	-1,345	-79,400r
Marine bunkers	-	-	-	-	-	-	-	-
Stock change (2)	-311	..	..	..	..	+18	+85	-208
Transfers	-	-1,253r	-1,080	-1,083r	-	-3,416r	794	-2,622r
Total supply	**86,230**	..	..	..	..	**1,573r**	**8,195**	**95,998r**
Statistical difference (3)(4)	-1,598	..	..	..	..	-142r	+365	-1,375r
Total demand (4)	**87,828**	..	..	..	..	**1,715r**	**7,830**	**97,373r**
Transformation(4)	**87,828**	..	..	..	..	**1,365**	**7,830**	**97,023**
Electricity generation	-	-	-	-	-	-	-	-
Major power producers	-	-	-	-	-	-	-	-
Autogenerators	-	-	-	-	-	-	-	-
Petroleum refineries	87,828	..	..	..	..	1,365	7,830	97,023
Coke manufacture	-	-	-	-	-	-	-	-
Blast furnaces	-	-	-	-	-	-	-	-
Patent fuel manufacture	-	-	-	-	-	-	-	-
Other	-	-	-	-	-	-	-	-
Energy industry use	-	**307r**	**38r**	**5**	-	**350r**	-	**350r**
Electricity generation	-	-	-	-	-	-	-	-
Oil & gas extraction(4)	-	307r	38r	5	-	350r	-	350r
Petroleum refineries	-	-	-	-	-	-	-	-
Coal extraction	-	-	-	-	-	-	-	-
Coke manufacture	-	-	-	-	-	-	-	-
Blast furnaces	-	-	-	-	-	-	-	-
Patent fuel manufacture	-	-	-	-	-	-	-	-
Pumped storage	-	-	-	-	-	-	-	-
Other	-	-	-	-	-	-	-	-
Losses	-	-	-	-	-	-	-	-

(1) As there is no use made of primary oils and feedstocks by industries other than the oil and gas extraction and petroleum refining industries, other industry headings have not been included in this table. As such, this table is a summary of the activity of what is known as the Upstream oil industry.

(2) Stock fall (+), stock rise (-).

(3) Total supply minus total demand.

(4) Figures for total demand for the individual NGLs (and thus for the statistical differences as well) are not available. While separate data are available on the use of individual NGL's in the extraction of oil and gas, details of inputs into refineries of NGLs are only available at aggregate level for total NGLs, resulting in accurate estimates for the total demand for each NGL not being possible. As they are available, details of the use of individual NGLs in the extraction of oil and gas are included in this table.

3.4 Commodity balances 1999

Petroleum products

	Ethane	Propane	Butane	Other gases	Naptha	Aviation spirit	Motor spirit	Industrial spirit	White spirit	Aviation turbine fuel	Burning oil
Supply											
Production	33	1,526	477	2,855	2,485	16	25,587	20	110	7,352	3,603
Other sources	1,527	865	931	-	-	-	-	-	-	-	-
Imports	-	47	219	-	465	13	2,246	51	2	2,618	98
Exports	-10	-316	-126	-	-605	-1	-6,332	-13	-2	-739	-253
Marine bunkers	-	-	-	-	-	-	-	-	-	-	-
Stock change (1)	+1	-24	+3	+1	+113	+7	+125	+10	-1	+173	-82
Transfers	-28	-110	-557	-33	+181	+28	+143	+12	-	+274	+44
Total supply	**1,523**	**1,988**	**947**	**2,823**	**2,639**	**63**	**21,769**	**80**	**109**	**9,678**	**3,410**
Statistical difference (2)	-131	+389	+383	-31	-339	+19	+257	+10	+14	+19	-146
Total demand	**1,654**	**1,599**	**564**	**2,854**	**2,978**	**44**	**21,512**	**70**	**95**	**9,659**	**3,556**
Transformation	-	-	-	**222**	-	-	-	-	-	-	-
Electricity generation	-	-	-	222	-	-	-	-	-	-	-
Major power producers	-	-	-	-	-	-	-	-	-	-	-
Autogenerators	-	-	-	222	-	-	-	-	-	-	-
Petroleum refineries	-	-	-	-	-	-	-	-	-	-	-
Coke manufacture	-	-	-	-	-	-	-	-	-	-	-
Blast furnaces	-	-	-	-	-	-	-	-	-	-	-
Patent fuel manufacture	-	-	-	-	-	-	-	-	-	-	-
Other	-	-	-	-	-	-	-	-	-	-	-
Energy industry use	**33**	**25**	**25**	**2,424**	**21**	-	-	-	-	-	-
Electricity generation	-	-	-	-	-	-	-	-	-	-	-
Oil & gas extraction	-	-	-	-	-	-	-	-	-	-	-
Petroleum refineries	33	1	-	2,424	21	-	-	-	-	-	-
Coal extraction	-	-	-	-	-	-	-	-	-	-	-
Coke manufacture	-	-	-	-	-	-	-	-	-	-	-
Blast furnaces	-	3	-	-	-	-	-	-	-	-	-
Patent fuel manufacture	-	-	-	-	-	-	-	-	-	-	-
Pumped storage	-	-	-	-	-	-	-	-	-	-	-
Other	-	21	25	-	-	-	-	-	-	-	-
Losses	-	-	-	-	-	-	-	-	-	-	-
Final consumption	**1,621**	**1,574**	**539**	**208**	**2,957**	**44**	**21,512**	**70**	**95**	**9,659**	**3,556**
Industry	**74**	**724**	-	-	-	-	-	-	-	-	**1,049**
Unclassified	74	703	-	-	-	-	-	-	-	-	1,049
Iron & steel	-	21	-	-	-	-	-	-	-	-	-
Non-ferrous metals	-	-	-	-	-	-	-	-	-	-	-
Mineral products	-	-	-	-	-	-	-	-	-	-	-
Chemicals	-	-	-	-	-	-	-	-	-	-	-
Mechanical engineering etc	-	-	-	-	-	-	-	-	-	-	-
Electrical engineering etc	-	-	-	-	-	-	-	-	-	-	-
Vehicles	-	-	-	-	-	-	-	-	-	-	-
Food, beverages etc	-	-	-	-	-	-	-	-	-	-	-
Textiles, leather, etc	-	-	-	-	-	-	-	-	-	-	-
Paper, printing etc	-	-	-	-	-	-	-	-	-	-	-
Other industries	-	-	-	-	-	-	-	-	-	-	-
Construction	-	-	-	-	-	-	-	-	-	-	-
Transport	-	-	**8**	-	-	**44**	**21,512**	-	-	**9,659**	**12**
Air	-	-	-	-	-	44	-	-	-	9,659	-
Rail	-	-	-	-	-	-	-	-	-	-	12
Road	-	-	8	-	-	-	21,512	-	-	-	-
National navigation	-	-	-	-	-	-	-	-	-	-	-
Pipelines	-	-	-	-	-	-	-	-	-	-	-
Other	-	**171**	**116**	-	-	-	-	-	-	-	**2,495**
Domestic	-	171	116	-	-	-	-	-	-	-	2,471
Public administration	-	-	-	-	-	-	-	-	-	-	12
Commercial	-	-	-	-	-	-	-	-	-	-	-
Agriculture	-	-	-	-	-	-	-	-	-	-	12
Miscellaneous	-	-	-	-	-	-	-	-	-	-	-
Non energy use	**1,547**	**679**	**415**	**208**	**2,957**	-	-	**70**	**95**	-	-

(1) Stock fall (+), stock rise (-).
(2) Total supply minus total demand.

3.4 Commodity balances 1999 (continued)

Petroleum products

Thousand tonnes

Gas oil	Marine diesel oil	Fuel oils	Lubri -cants	Bitu -men	Petroleum wax	Petroleum coke	Orimul -sion	Misc. products	Total products	
										Supply
26,232	5	12,368	920	1,667	264	1,837	-	601	87,958	Production
-	-	-	-	-	-	-	-	-	3,323	Other sources
5,093	-	378	164	244	-	644	-	-	12,282	Imports
-6,667	-	-4,929	-673	-271	-17	-641	-	-61	-21,656	Exports
-1,061	-90	-1,179	-	-	-	-	-	-	-2,330	Marine bunkers
+232	-1	+65	+123	+4	-32	-40	-	-102	+575	Stock change (1)
-702	+12	-1,683	+110	+380	-228	-6	-	+58	-2,105	Transfers
23,127	-74	5,020	644	2,024	-13	1,794	-	496	78,047	**Total supply**
+285	-89	+756	-146	+61	-32	-335	-	+141	+1,085	**Statistical difference** (2)
22,842	15	4,264	790	1,963	19	2,129	-	355	76,962	**Total demand**
153	-	**1,176**	-	-	-	-	-	-	**1,551**	**Transformation**
153	-	906	-	-	-	-	-	-	1,281	Electricity generation
59	-	345	-	-	-	-	-	-	404	Major power producers
94	-	561	-	-	-	-	-	-	877	Autogenerators
-	-	-	-	-	-	-	-	-	-	Petroleum refineries
-	-	-	-	-	-	-	-	-	-	Coke manufacture
-	-	270	-	-	-	-	-	-	270	Blast furnaces
-	-	-	-	-	-	-	-	-	-	Patent fuel manufacture
-	-	-	-	-	-	-	-	-	-	Other
221	-	**1,775**	-	-	-	**1,190**	-	-	**5,714**	**Energy industry use**
-	-	-	-	-	-	-	-	-	-	Electricity generation
-	-	-	-	-	-	-	-	-	-	Oil & gas extraction
76	-	1,773	-	-	-	1,190	-	-	5,518	Petroleum refineries
-	-	-	-	-	-	-	-	-	-	Coal extraction
-	-	-	-	-	-	-	-	-	-	Coke manufacture
144	-	-	-	-	-	-	-	-	147	Blast furnaces
-	-	-	-	-	-	-	-	-	-	Patent fuel manufacture
-	-	-	-	-	-	-	-	-	-	Pumped storage
1	-	2	-	-	-	-	-	-	49	Other
-	-	-	-	-	-	-	-	-	-	**Losses**
22,468	15	1,313	790	1,963	19	939	-	355	69,697	**Final Consumption**
2,880	-	**841**	-	-	-	-	-	-	**5,568**	**Industry**
-	-	-	-	-	-	-	-	-	1,826	Unclassified
9	-	53	-	-	-	-	-	-	83	Iron & steel
26	-	9	-	-	-	-	-	-	35	Non-ferrous metals
168	-	46	-	-	-	-	-	-	214	Mineral products
136	-	231	-	-	-	-	-	-	367	Chemicals
167	-	18	-	-	-	-	-	-	185	Mechanical engineering etc
26	-	35	-	-	-	-	-	-	61	Electrical engineering etc
105	-	27	-	-	-	-	-	-	132	Vehicles
146	-	190	-	-	-	-	-	-	336	Food, beverages etc
38	-	41	-	-	-	-	-	-	79	Textiles, leather, etc
28	-	58	-	-	-	-	-	-	86	Paper, printing etc
1,537	-	130	-	-	-	-	-	-	1,667	Other industries
494	-	3	-	-	-	-	-	-	497	Construction
16,544	**15**	**70**	-	-	-	-	-	-	**47,864**	**Transport**
-	-	-	-	-	-	-	-	-	9,703	Air
452	-	-	-	-	-	-	-	-	464	Rail
15,188	-	-	-	-	-	-	-	-	36,708	Road
904	15	70	-	-	-	-	-	-	989	National navigation
-	-	-	-	-	-	-	-	-	-	Pipelines
2,200	-	**402**	-	-	-	-	-	-	**5,384**	**Other**
159	-	2	-	-	-	-	-	-	2,919	Domestic
869	-	256	-	-	-	-	-	-	1,137	Public administration
434	-	48	-	-	-	-	-	-	482	Commercial
597	-	90	-	-	-	-	-	-	699	Agriculture
141	-	6	-	-	-	-	-	-	147	Miscellaneous
844	-	-	790	1,963	19	939	-	355	10,881	**Non energy use**

3.5 Commodity balances 1998

Petroleum products

	Ethane	Propane	Butane	Other gases	Naptha	Aviation spirit	Motor spirit	Industrial spirit	White spirit	Aviation turbine fuel	Burning oil
Supply											
Production	36	1,551	428	2.948	2,352	-	27,392	2	134	7,942	3,471
Other sources	1,215	1,071	1,171	-	-	-	-	-	-	-	-
Imports	-	82	158	-	855	32	1,986	49	2	2,660	131
Exports	11r	-727	-142	-	-520	-1	-7,959r	-23	-9	-828	-267
Marine bunkers	-	-	-	-	-	-	-	-	-	-	-
Stock change *(1)*	+1	+38	+1	-1	-117	-6	+244	-3	-5	-60	+31
Transfers	-17	-19	-1,351	-44	-153	-3	+1,103	+89	-8	-131	+166
Total supply	**1,224r**	**1,996**	**265**	**2.903**	**2,417**	**22**	**22,766r**	**114**	**114**	**9,583**	**3,532**
Statistical difference *(2)*	-172	+191	-300	-19	-482	-14	+918r	+31	+18	+342	-42
Total demand	**1,396**	**1,805r**	**565r**	**2,922**	**2,899**	**36**	**21,848**	**83**	**96**	**9,241**	**3,574r**
Transformation	-	-	-	**228r**	-	-	-	-	-	-	-
Electricity generation	-	-	-	228r	-	-	-	-	-	-	-
Major power producers	-	-	-	-	-	-	-	-	-	-	-
Autogenerators	-	-	-	228r	-	-	-	-	-	-	-
Petroleum refineries	-	-	-	-	-	-	-	-	-	-	-
Coke manufacture	-	-	-	-	-	-	-	-	-	-	-
Blast furnaces	-	-	-	-	-	-	-	-	-	-	-
Patent fuel manufacture	-	-	-	-	-	-	-	-	-	-	-
Other	-	-	-	-	-	-	-	-	-	-	-
Energy industry use	**36**	**28**	**22**	**2,520r**	**17**	-	-	-	-	-	-
Electricity generation	-	-	-	-	-	-	-	-	-	-	-
Oil & gas extraction	-	-	-	-	-	-	-	-	-	-	-
Petroleum refineries	36	1	-	2,520r	17	-	-	-	-	-	-
Coal extraction	-	-	-	-	-	-	-	-	-	-	-
Coke manufacture	-	-	-	-	-	-	-	-	-	-	-
Blast furnaces	-	3	-	-	-	-	-	-	-	-	-
Patent fuel manufacture	-	-	-	-	-	-	-	-	-	-	-
Pumped storage	-	-	-	-	-	-	-	-	-	-	-
Other	-	24	22	-	-	-	-	-	-	-	-
Losses	-	-	-	-	-	-	-	-	-	-	-
Final consumption	**1,360**	**1,777r**	**543**	**174**	**2,882**	**36**	**21,848**	**83**	**96**	**9,241**	**3,574r**
Industry	**69**	**797r**	-	-	-	-	-	-	-	-	**840r**
Unclassified	69	776r	-	-	-	-	-	-	-	-	840r
Iron & steel	-	21r	-	-	-	-	-	-	-	-	-
Non-ferrous metals	-	-	-	-	-	-	-	-	-	-	-
Mineral products	-	-	-	-	-	-	-	-	-	-	-
Chemicals	-	-	-	-	-	-	-	-	-	-	-
Mechanical engineering etc	-	-	-	-	-	-	-	-	-	-	-
Electrical engineering etc	-	-	-	-	-	-	-	-	-	-	-
Vehicles	-	-	-	-	-	-	-	-	-	-	-
Food, beverages etc	-	-	-	-	-	-	-	-	-	-	-
Textiles, leather, etc	-	-	-	-	-	-	-	-	-	-	-
Paper, printing etc	-	-	-	-	-	-	-	-	-	-	-
Other industries	-	-	-	-	-	-	-	-	-	-	-
Construction	-	-	-	-	-	-	-	-	-	-	-
Transport	-	-	**4r**	-	-	**36**	**21,848**	-	-	**9,241**	**12**
Air	-	-	-	-	-	36	-	-	-	9,241	-
Rail	-	-	-	-	-	-	-	-	-	-	12
Road	-	-	4r	-	-	-	21,848	-	-	-	-
National navigation	-	-	-	-	-	-	-	-	-	-	-
Pipelines	-	-	-	-	-	-	-	-	-	-	-
Other	-	**149r**	**156r**	-	-	-	-	-	-	-	**2,722**
Domestic	-	149r	156r	-	-	-	-	-	-	-	2,698
Public administration	-	-	-	-	-	-	-	-	-	-	12
Commercial	-	-	-	-	-	-	-	-	-	-	-
Agriculture	-	-	-	-	-	-	-	-	-	-	12
Miscellaneous	-	-	-	-	-	-	-	-	-	-	-
Non energy use	**1,291**	**831**	**383**	**174**	**2,882**	-	-	**83**	**96**	-	-

(1) Stock fall (+), stock rise (-).
(2) Total supply minus total demand.

3.5 Commodity balances 1998 (continued)

Petroleum products

Thousand tonnes

Gas oil	Marine diesel oil	Fuel oils	Lubri -cants	Bitu -men	Petroleum wax	Petroleum coke	Orimul -sion	Misc. products	Total products	
										Supply
27,924	10	13,476	1,134	2,190	59	1,885	-	630	93,564	Production
-	-	-	-	-	-	-	-	-	3,457	Other sources
3,468	-	791	198	76	-	883r	-	-r	11,371r	Imports
-6,201	-	-5,834	-632	-334	-14	-831r	-	-	-24,333r	Exports
-1,204	-192	-1,684	-	-	-	-	-	-	-3,080	Marine bunkers
-215	-	+84	-5	+20	+5	-42	-	-63r	-93r	Stock change *(1)*
-103	+40	-949	+71	+57	-28	+8	-	+17	-1,255	Transfers
23,669	-142	5,884	766	2,009	22	1,903r	-	584r	79,631r	**Total supply**
+456r	-143	+539r	-47	+42	+4	-175r	-	+46	+1,193r	**Statistical difference** *(2)*
23,213r	1	5,345r	813	1,967	18	2,078	-	538	78,438r	**Total demand**
132r	-	**1,571**	-	-	-	-	-	-	**1,931r**	**Transformation**
132r	-	1,303r	-	-	-	-	-	-	1,663r	Electricity generation
56	-	700	-	-	-	-	-	-	756	Major power producers
76r	-	603r	-	-	-	-	-	-	907r	Autogenerators
-	-	-	-	-	-	-	-	-	-	Petroleum refineries
-	-	-	-	-	-	-	-	-	-	Coke manufacture
-	-	268r	-	-	-	-	-	-	268r	Blast furnaces
-	-	-	-	-	-	-	-	-	-	Patent fuel manufacture
-	-	-	-	-	-	-	-	-	-	Other
286r	-	**2,122r**	-	-	-	**1,191**	-	-	**6,222r**	**Energy industry use**
-	-	-	-	-	-	-	-	-	-	Electricity generation
-	-	-	-	-	-	-	-	-	-	Oil & gas extraction
159r	-	2,122r	-	-	-	1,191	-	-	6,046r	Petroleum refineries
-	-	-	-	-	-	-	-	-	-	Coal extraction
-	-	-	-	-	-	-	-	-	-	Coke manufacture
126r	-	-	-	-	-	-	-	-	129r	Blast furnaces
-	-	-	-	-	-	-	-	-	-	Patent fuel manufacture
-	-	-	-	-	-	-	-	-	-	Pumped storage
1	-	-	-	-	-	-	-	-	47	Other
-	-	-	-	-	-	-	-	-	-	**Losses**
22,795r	1	1,652r	813	1,967	18	887	-	538	70,285r	**Final Consumption**
2,866r	-	1,050r	-	-	-	-	-	-	5,622r	**Industry**
-	-	-	-	-	-	-	-	-	1,685r	Unclassified
31r	-	30r	-	-	-	-	-	-	82r	Iron & steel
22	-	17r	-	-	-	-	-	-	39r	Non-ferrous metals
176	-	48	-	-	-	-	-	-	224	Mineral products
152r	-	312r	-	-	-	-	-	-	464r	Chemicals
165r	-	33r	-	-	-	-	-	-	198r	Mechanical engineering etc
27	-	61	-	-	-	-	-	-	88	Electrical engineering etc
93r	-	32	-	-	-	-	-	-	125r	Vehicles
151r	-	242r	-	-	-	-	-	-	393r	Food, beverages etc
38	-	57	-	-	-	-	-	-	95	Textiles, leather, etc
34r	-	81r	-	-	-	-	-	-	115r	Paper, printing etc
1,482r	-	125r	-	-	-	-	-	-	1,607r	Other industries
495	-	12	-	-	-	-	-	-	507	Construction
16,593r	1	105r	-	-	-	-	-	-	47,840r	**Transport**
-	-	-	-	-	-	-	-	-	9,277	Air
469r	-	-	-	-	-	-	-	-	481r	Rail
15,143	-	-	-	-	-	-	-	-	36,995r	Road
981	1	105r	-	-	-	-	-	-	1,087r	National navigation
-	-	-	-	-	-	-	-	-	-	Pipelines
2,576r	-	497r	-	-	-	-	-	-	6,100r	**Other**
191	-	1	-	-	-	-	-	-	3,195r	Domestic
1,009r	-	362r	-	-	-	-	-	-	1,383r	Public administration
509r	-	46r	-	-	-	-	-	-	555r	Commercial
698	-	76	-	-	-	-	-	-	786	Agriculture
169	-	12r	-	-	-	-	-	-	181r	Miscellaneous
760	-	-	813	1,967	18	887	-	538	10,723	**Non energy use**

3.6 Commodity balances 1997

Petroleum products

	Ethane	Propane	Butane	Other gases	Naptha	Aviation spirit	Motor spirit	Industrial spirit	White spirit	Aviation turbine fuel	Burning oil
Supply											
Production	11	1,532	429	3,068	2,871	-	28,260	1	127	8,342	3,336
Other sources	1,253r	1,080	1,083r	-	-	-	-	-	-	-	-
Imports	-	104	248	-	684	36	1,391	76	7	1,417	126
Exports	-	-672	-145	-	-955	-1	-8,602	-24	-6	-913	-306
Marine bunkers	-	-	-	-	-	-	-	-	-	-	-
Stock change (1)	+1	-38	-31	+2	+35	+1	+392	-4	+5	-	-4
Transfers	-11	-12	-855	-28	-97	-2	+698	+56	-5	-83	+105
Total supply	**1,254r**	**1,994**	**729r**	**3,042**	**2,538**	**34**	**22,139**	**105**	**128**	**8,763**	**3,257**
Statistical difference(2)	-190r	+184	+105r	-15	-119	-3	-113	+18	+20	+352	-86
Total demand	**1,444**	**1,810**	**624r**	**3,057**	**2,657**	**37**	**22,252**	**87**	**108**	**8,411**	**3,343**
Transformation	-	-	-	**236r**	-	-	-	-	-	-	-
Electricity generation	-	-	-	236r	-	-	-	-	-	-	-
Major power producers	-	-	-	-	-	-	-	-	-	-	-
Autogenerators	-	-	-	236r	-	-	-	-	-	-	-
Petroleum refineries	-	-	-	-	-	-	-	-	-	-	-
Coke manufacture	-	-	-	-	-	-	-	-	-	-	-
Blast furnaces	-	-	-	-	-	-	-	-	-	-	-
Patent fuel manufacture	-	-	-	-	-	-	-	-	-	-	-
Other	-	-	-	-	-	-	-	-	-	-	-
Energy industry use	**11**	**34**	**25**	**2,693r**	**17**	-	-	-	-	..	-
Electricity generation	-	-	-	-	-	-	-	-	-	-	-
Oil & gas extraction	-	-	-	-	-	-	-	-	-	..	-
Petroleum refineries	11	11	-	2,693r	17	-	-	-	-	-	-
Coal extraction	-	-	-	-	-	-	-	-	-	-	-
Coke manufacture	-	-	-	-	-	-	-	-	-	-	-
Blast furnaces	-	3	-	-	-	-	-	-	-	-	-
Patent fuel manufacture	-	-	-	-	-	-	-	-	-	-	-
Pumped storage	-	-	-	-	-	-	-	-	-	-	-
Other	-	20	25	-	-	-	-	-	-	-	-
Losses	-	-	-	-	-	-	-	-	-	-	-
Final consumption	**1,433**	**1,776**	**599r**	**128**	**2,640**	**37**	**22,252**	**87**	**108**	**8,411**	**3,343**
Industry	**73**	**829**	-	-	-	-	-	-	-	-	**783**
Unclassified	73	812	-	-	-	-	-	-	-	-	783
Iron & steel	-	17	-	-	-	-	-	-	-	-	-
Non-ferrous metals	-	-	-	-	-	-	-	-	-	-	-
Mineral products	-	-	-	-	-	-	-	-	-	-	-
Chemicals	-	-	-	-	-	-	-	-	-	-	-
Mechanical engineering etc	-	-	-	-	-	-	-	-	-	-	-
Electrical engineering etc	-	-	-	-	-	-	-	-	-	-	-
Vehicles	-	-	-	-	-	-	-	-	-	-	-
Food, beverages etc	-	-	-	-	-	-	-	-	-	-	-
Textiles, leather, etc	-	-	-	-	-	-	-	-	-	-	-
Paper, printing etc	-	-	-	-	-	-	-	-	-	-	-
Other industries	-	-	-	-	-	-	-	-	-	-	-
Construction	-	-	-	-	-	-	-	-	-	-	-
Transport	-	-	**2**	-	-	**37**	**22,252**	-	-	**8,411**	**12**
Air	-	-	-	-	-	37	-	-	-	8,411	-
Rail	-	-	-	-	-	-	-	-	-	-	12
Road	-	-	2	-	-	-	22,252	-	-	-	-
National navigation	-	-	-	-	-	-	-	-	-	-	-
Pipelines	-	-	-	-	-	-	-	-	-	-	-
Other	-	**117**	**195r**	-	-	-	-	-	-	-	**2,548**
Domestic	-	117	195r	-	-	-	-	-	-	-	2,524
Public administration	-	-	-	-	-	-	-	-	-	-	12
Commercial	-	-	-	-	-	-	-	-	-	-	-
Agriculture	-	-	-	-	-	-	-	-	-	-	12
Miscellaneous	-	-	-	-	-	-	-	-	-	-	-
Non energy use	**1,360**	**830**	**402**	**128**	**2,640**	-	-	**87**	**108**	-	-

(1) Stock fall (+), stock rise (-).
(2) Total supply minus total demand.

3.6 Commodity balances 1997 (continued)

Petroleum products

Thousand tonnes

Gas oil	Marine diesel oil	Fuel oils	Lubri -cants	Bitu -men	Petroleum wax	Petroleum coke	Orimul -sion	Misc. products	Total products	
										Supply
28,829	-	14,062	1,231	2,258	65	1,781	-	734	96,937	Production
-	-	-	-	-	-	-	-	-	3,416r	Other sources
1,866	-	1,212	210	51	-	985r	182	111r	8,706	Imports
-6,586	-	-6,863	-722	-312	-17	-544	-	-87	-26,755r	Exports
-1,087	-70	-1,805	-	-	-	-	-	-	-2,962	Marine bunkers
+42	-	+141	+3	-25	+2	+44	+45	-88	+523	Stock change (1)
-65	+25	-599	+45	+36	-18	+5	-	+11	-794	Transfers
22,999	-45	6,148	767	2,008	32	2,271r	227	681r	79,071r	**Total supply**
-79	-46	-103r	-105	-7	-12	-7r	+45	-17r	-178r	**Statistical difference** (2)
23,078r	1	6,251r	872	2,015	44	2,278	182	698	79,249r	**Total demand**
150r	-	**2,126r**	-	-	-	-	182	-	**2,695r**	**Transformation**
150r	-	1,672r	-	-	-	-	182	-	2,241r	Electricity generation
66	-	1,137r	-	-	-	-	182	-	1,385r	Major power producers
84r	-	535r	-	-	-	-	-	-	856r	Autogenerators
-	-	-	-	-	-	-	-	-	-	Petroleum refineries
-	-	-	-	-	-	-	-	-	-	Coke manufacture
-	-	454r	-	-	-	-	-	-	454r	Blast furnaces
-	-	-	-	-	-	-	-	-	-	Patent fuel manufacture
-	-	-	-	-	-	-	-	-	-	Other
267r	-	**2,036r**	-	-	-	1,183	-	54	**6,320r**	**Energy industry use**
-	-	-	-	-	-	-	-	-	-	Electricity generation
-	-	-	-	-	-	-	-	-	-	Oil & gas extraction
151r	-	2,036r	-	-	-	1,183	-	54	6,156r	Petroleum refineries
-	-	-	-	-	-	-	-	-	-	Coal extraction
-	-	-	-	-	-	-	-	-	-	Coke manufacture
115r	-	-	-	-	-	-	-	-	118r	Blast furnaces
-	-	-	-	-	-	-	-	-	-	Patent fuel manufacture
-	-	-	-	-	-	-	-	-	-	Pumped storage
1	-	-	-	-	-	-	-	-	46	Other
-	-	-	-	-	-	-	-	-	-	**Losses**
22,661r	1	2,089r	872	2,015	44	1,095	-	644	70,235r	**Final Consumption**
2,768r	-	1,332r	-	-	-	-	-	-	5,785r	**Industry**
-	-	-	-	-	-	-	-	-	1,668	Unclassified
28r	-	60r	-	-	-	-	-	-	105r	Iron & steel
23	-	27r	-	-	-	-	-	-	50r	Non-ferrous metals
167	-	75	-	-	-	-	-	-	242	Mineral products
154r	-	378r	-	-	-	-	-	-	532r	Chemicals
181	-	50	-	-	-	-	-	-	231	Mechanical engineering etc
34	-	72	-	-	-	-	-	-	106	Electrical engineering etc
93	-	53	-	-	-	-	-	-	146	Vehicles
155r	-	302r	-	-	-	-	-	-	457r	Food, beverages etc
37	-	59	-	-	-	-	-	-	96	Textiles, leather, etc
33r	-	87	-	-	-	-	-	-	120r	Paper, printing etc
1,354r	-	153r	-	-	-	-	-	-	1,507r	Other industries
509	-	16	-	-	-	-	-	-	525	Construction
16,474r	1	127r	-	-	-	-	-	-	47,316r	**Transport**
-	-	-	-	-	-	-	-	-	8,448	Air
463r	-	-	-	-	-	-	-	-	475r	Rail
14,976	-	-	-	-	-	-	-	-	37,230	Road
1,035	1	127r	-	-	-	-	-	-	1,163	National navigation
-	-	-	-	-	-	-	-	-	-	Pipelines
2,692r	-	630r	-	-	-	-	-	-	6,182r	**Other**
216	-	3	-	-	-	-	-	-	3,055r	Domestic
1,115r	-	469r	-	-	-	-	-	-	1,596r	Public administration
523r	-	64r	-	-	-	-	-	-	587r	Commercial
663	-	81	-	-	-	-	-	-	756	Agriculture
175	-	13r	-	-	-	-	-	-	188r	Miscellaneous
727	-	-	872	2,015	44	1,095	-	644	10,952	Non energy use

3.7 Supply and disposal of petroleum[1]

Thousand tonnes

	1995	1996	1997	1998	1999
Primary oils (Crude oil, NGLs and feedstocks)					
Indigenous production *(2)*	129,894	129,742	128,234	132,367r	137,099
Imports	48,749	50,099	49,994	47,957	44,891
Exports *(3)*	-84,577	-81,563	-79,400r	-84,612	-91,797
Transfers - Transfers to products *(4)*	-3,023	-3,267	-3,416r	-3,457	-3,323
Product rebrands *(5)*	+1,110	+997	+794	+1,255	+2,105
Stock change *(6)* - Offshore	-160	-2	-200	-127	-83
Oil terminals	+1,044	-354	-8	-466	-115
Use during production *(7)*	-389	-401	-350r	-353	-323
Calculated refinery throughput *(8)*	92,648	95,251	95,648r	92,564r	88,454
Overall statistical difference *(9) (10)*	-95	-1,409	-1,375r	--1,233r	+168
Actual refinery throughput	**92,743**	**96,660**	**97,023**	**93,797**	**88,286**
Petroleum products					
Losses in refining process	129	152	86	233	324
Refinery gross production *(11)*	92,614	96,508	96,937	93,564	87,958
Transfers - Transfers to products *(4)*	+3,023	+3,267	+3,416r	+3,457	+3,323
Product rebrands *(5)*	-1,110	-997	-794	-1,255	-2,105
From other sources *(12)*	-	-	-	-	-
Imports	9,878	9,315	8,706	11,371r	12,282
Exports *(13)*	-21,614	-23,681	-26,755r	-24,333r	-21,656
Marine bunkers	-2,465	-2,665	-2,962	-3,080	-2,330
Stock changes *(6)* - Refineries	+46	-76	+417	-66	+399
Power generators	-26	+168	+106	-28	+176
Calculated total supply	80,346	81,839	79,071r	79,631r	78,047
Statistical difference *(9)*	+171	-176	-178	+1,193r	+1,085
Total demand *(4)*	**80,175**	**82,015**	**79,249r**	**78,438r**	**76,962**
Of which:					
Energy use	68,855	70,702	68,297r	67,715r	66,081
Of which, for electricity generation *(14)*	4,370r	4,028r	2,241r	1,663r	1,281
total refinery fuels *(14)*	6,481	6,623	6,572	6,468	5,976
Non-energy use	11,320	11,293	10,952	10,723	10,881

(1) *Aggregate monthly data on oil production, trade, refinery throughput and inland deliveries are published in Tables 13 to 17 of Energy Trends- see Annex F.*
(2) *Crude oil plus condensates and petroleum gases derived at onshore treatment plants.*
(3) *Includes NGLs, process oils and re-exports.*
(4) *Disposals of NGLs by direct sale (excluding exports) or for blending.*
(5) *Product rebrands (inter-product blends or transfers) represent petroleum products received at refineries/ plants as process oils for refinery or cracking unit operations.*
(6) *Impact of stock changes on supplies. A stock fall is shown as (+) as it increases supplies, and vice-versa for a stock rise (-).*
(7) *Own use in onshore terminals and gas separation plants.*
(8) *Equivalent to the total supplies reported against the upstream transformation sector in Tables 3.1 to 3.3.*
(9) *Supply greater than (+) or less than (-) recorded throughput or disposals.*
(10) *This total includes differences between the figures for indigenous production as recorded by individual fields and indigenous receipts, which is accounted for by own use in onshore terminals and gas separation plants, losses, platform and other field stock changes and the time lag between production on offshore loaders and tankers arrival at refineries. The size of this component of the overall statistical difference was previously given separately, and is given in the table below for information. See Chapter 3, paragraphs 3.33 to 3.41 for information on the large 1995 and 1996 differences.*

	1995	1996	1997	1998	1999
Indigenous receipts	129,534	130,792	129,037r	133,125r	133,693
Statistical difference - upstream production sector	*-189*	*-1,453*	*-1,353r*	*-1,238r*	*-527*

(11) *Includes refinery fuels.*
(12) *Petroleum products derived from other sources, mainly bitumen and lubricating oils.*
(13) *Excludes NGLs.*
(14) *Figures cover petroleum used to generate electricity by all major power producers and by all other generators, including petroleum used to generate electricity at refineries. These quantities are also included in the totals reported as used as refinery fuel, so there is thus some overlap in these figures.*

3.8 Additional information on inland deliveries of selected products[1][2][3]

Thousand tonnes

	1995	1996	1997	1998	1999
Motor spirit					
Retail deliveries (4)					
Hypermarkets(5)					
4 star / Leaded premium	1,608	1,451	1,253	1,001	641
Super premium unleaded	108	55	24	18	13
Premium unleaded	3,003	3,278	3,680	4,130	4,775
Total hypermarkets	4,719	4,784	4,957	5,149	5,429
Refiners/other traders					
4 star / Leaded premium	6,365	5,592	4,885	3,594	2,061
Super premium unleaded	817	643	482	391	442
Premium unleaded	9,601	10,950	11,508	12,302	13,205
Total Refiners/other traders	16,783	17,185	16,876	16,287	15,708
Total retail deliveries					
4 star / Leaded premium	7,973	7,043	6,138	4,595	2,702
Super premium unleaded	925	698	506	409	455
Premium unleaded	12,604	14,228	15,188	16,432	17,980
Total retail deliveries	21,502	21,969	21,833	21,436	21,137
Commercial consumers (6)					
4 star / Leaded premium	149	135	112	91	59
Super premium unleaded	17	11	9	4	6
Premium unleaded	285	294	298	317	310
Total commercial consumers	451	440	419	412	375
Total motor spirit	**21,953**	**22,409**	**22,252**	**21,848**	**21,512**
Unleaded as % of Total motor spirit	63.0	68.0	71.9	78.6	87.2
Gas oil/diesel oil					
DERV fuel:					
Retail deliveries (4):					
Hypermarkets (5)	717	855	1,023	1,153	1,306
Refiners/other traders	4,097	4,682	5,104	5,449	5,684
Total retail deliveries	4,814	5,537	6,127	6,602	6,990
Commercial consumers (6)	8,643	8,828	8,849	8,541	8,198
Total DERV fuel	13,457	14,365	14,976	15,143	15,188
Gas oil	7,227	7,631	7,325	7,244	6,805
Marine diesel oil	-	-	1	1	15
Total Gas oil/diesel oil	**20,684**	**21,996**	**22,302**	**22,388**	**22,008**
Fuel oils (7)					
Light	110	108	135	76	74
Medium	559	484	381	259	251
Heavy	6,031	5,390	3,238	2,600	1,861
Orimulsion (8)	1,276	872	182	-	-
Total fuel oils	**7,975**	**6,854**	**3,936**	**2,935**	**2,186**

(1) Aggregate monthly data for inland deliveries of oil products are published in Table 16 of Energy Trends - see Annex F.

(2) The end use section analyses are based partly on recorded figures and on estimates made by the Institute of Petroleum and the Department of Trade and Industry and are intended to be for general guidance only. See also the notes in the main text of this chapter.

(3) For a full breakdown of the end-uses of all oil products, see Commodity Balances in Tables 3.4 to 3.6.

(4) Retail deliveries - deliveries to garages, etc. mainly for resale to final consumers.

(5) Data for sales by super and hypermarket companies are collected via a separate reporting system, but are consistent with the main data collected by UKPIA - see the notes in the main text of this chapter.

(6) Commercial consumers - direct deliveries for use in consumer's business.

(7) Inland deliveries excluding that used as a fuel in refineries, but including that used for electricity generation by major electricity producers and other industries.

(8) Deliveries of Orimulsion ceased in February 1997.

3.9   Inland deliveries by country[1]

Thousand tonnes

	England and Wales [2]			Scotland			Northern Ireland		
	1997	1998	1999	1997	1998	1999	1997	1998	1999
Energy use									
Gases for gasworks and other uses									
Butane and propane	1,046	974	914	112	146	123	36	34	30
Other gases	-	-	-	73	69	74	-	-	-
Aviation spirit	33	32	36	3	3	5	1	1	3
Motor spirit:									
Dealers	19,981	19662	19,444	1,407	1,373	1,364	444	401	329
Commercial consumers	367	368	334	30	29	30	22	16	11
Total motor spirit	20,349	20,030	19,778	1,437	1,402	1,394	466	417	340
Kerosenes									
Aviation turbine fuel	7,905	8,742	9,108	440	429	479	66	70	72
Burning oil	2,672	2,665	2,604	181	221	247	491	688	706
Gas oil/diesel oil									
DERV fuel	13,527	13,680	13,802	1,111	1,189	1,174	338	274	212
Other [3]	5,672	5,554	5,262	1,313	1,301	1,138	340	390	420
Fuel oils [4][5]	3,069	2,292	1,617	443	332	304	423	310	265
Total products used as energy	**54,274**	**53,968**	**53,120**	**5,113**	**5,093**	**4,938**	**2,161**	**2,183**	**2,048**
Non-energy use									
Feedstock for petroleum chemical plants	3,422	3,871	3,979	2,666	2,452	2,671	-	-	-
Industrial spirit	86	82	68	2	2	2	-	-	-
White spirit	108	96	95	-	-	-	-	-	-
Lubricating oils	832	770	749	31	36	33	9	7	8
Bitumen	1,692	1,634	1,646	229	242	205	93	90	112
Petroleum wax	42	12	12	3	6	7	-	-	-
Total products used as non-energy [6]	**7,920**	**7,884**	**7,841**	**2,931**	**2,742**	**2,920**	**102**	**97**	**120**
Total all products	**62,194**	**61,852**	**60,961**	**8,044**	**7,835**	**7,858**	**2,263**	**2,281**	**2,168**

(1) Excludes products used as a fuel within refineries that are included in Tables 3.4 to 3.6.
(2) Includes the Channel Islands and the Isle of Man.
(2) Includes deliveries of marine diesel oil.
(3) Includes Orimulsion.
(4) Deliveries of orimulsion ceased in February 1997.
(5) Includes deliveries of miscellaneous products and petroleum coke.

3.10 Stocks of crude oil and petroleum products at end of year[1]

Thousand tonnes

	1995	1996	1997	1998	1999
Crude and process oils					
Refineries (2)	5,075	4,971	4,977	5,074	4,560
Terminals (3)	1,003	1,461	1,463	1,832	2,461
Offshore (4)	588	590r	790	917	1,000
Total crude and process oils (5)	**6,741**	**7,065r**	**7,390**	**7,883**	**8,080**
Petroleum products					
Ethane	6	8	7	6	6
Propane	156	120	157	120	144
Butane	89	61	92	92	89
Other petroleum gases	4	4	2	3	3
Naphtha	406	379	344	461	349
Aviation spirit	9	7	6	12	5
Motor spirit	2,473	2,502	2,218	1,984	1,425
Industrial spirit	14	11	15	17	7
White spirit	14	15	10	14	15
Aviation turbine fuel	489	573	573	637	461
Burning oil	250	284	288	257	339
Gas oil (6) (7)	1,705	1,678	1,639	3,703	2,984
Marine diesel oil	-	-	-	-	1
Fuel oils (7)	2,974	2,963	2,880	1,466	1,401
Lubricating oils	397	289	287	292	169
Bitumen	198	189	214	194	189
Petroleum wax	8	9	8	3	36
Petroleum coke	245	289	245	287	327
Miscellaneous products	74	66	154	217	320
Total all products	**9,511**	**9,447**	**9,139**	**9,765**	**8,269**
Of which : net bilateral stocks (8)	1,459	1,484	1,698	2,228	1,307

(1) Aggregate monthly data on the level of stocks of crude oil and oil products are published in Table 14 of Energy Trends - see Annex F.
(2) Stocks of crude oil, NGLs and process oils at UK refineries.
(3) Stocks of crude oil and NGLs at UKCS pipeline terminals.
(4) Stocks of crude oil in tanks and partially loaded tankers at offshore fields.
(5) Includes process oils held abroad for UK use approved by bilateral agreements.
(6) Includes middle distillate feedstock.
(7) The increase in gas oil stocks and the decrease in fuel oil stocks can be attributed to the change in patterns of stocks held abroad, under bilateral agreements, by UK companies as part of their national stocking obligation.
(8) The difference between stocks held abroad for UK use under approved bilateral agreements and the equivalent stocks held in the UK for foreign use.

3.11 Crude oil and petroleum products: production, imports and exports[1][2], 1960 to 1999

Thousand tonnes

	Crude oil [3]					Oil products			
	Imports	Indigenous production		Exports	Refinery throughput	Refinery Output [4]	Exports	Imports	Inland Deliveries [4]
		Total	Landward						
1960	45,437	148	148	-	45,176	40,931	9,131	13,951	39,982
1961	49,702	151	151	-	49,891	45,627	7,947	11,809	43,180
1962	53,461	129	129	80	53,134	48,901	10,187	14,802	47,956
1963	54,406	125	125	34	54,170	49,591	10,871	17,929	52,645
1964	60,177	129	129	54	59,414	54,573	9,612	18,163	58,017
1965	65,409	84	84	87	66,111	60,900	10,892	20,141	63,762
1966	71,544	78	78	107	71,722	66,311	12,645	21,808	69,044
1967	73,704	89	89	726	73,531	67,966	11,865	24,125	74,926
1968	83,195	84	84	59	83,100	77,049	14,453	22,816	79,236
1969	93,153	85	85	334	91,699	85,087	14,344	20,758	85,425
1970	102,155	156	83	1,182	101,911	94,696	17,424	20,428	91,151
1971	107,736	212	85	1,569	105,342	98,245	17,166	19,369	91,991
1972	107,706	333	85	3,558	106,980	99,368	15,979	20,827	98,469
1973	115,472	372	88	3,235	114,338	105,954	17,404	18,300	99,786
1974	112,822	410	107	1,404	111,217	103,060	14,631	14,537	93,409
1975	91,366	1,564	99	1,524	93,597	86,647	13,924	12,786	82,824
1976	80,466	12,169	99	4,285	97,784	90,284	15,988	10,709	81,579
1977	70,697	38,265	99	16,793	93,615	86,338	14,160	13,050	82,759
1978	68,144	54,006	88	25,200	96,390	89,156	13,194	11,586	84,141
1979	60,380	77,748	121	40,569	97,806	90,583	12,988	12,035	84,554
1980	46,717	80,467	237	40,180	86,341	79,227	14,110	9,245	71,177
1981	36,855	89,454	232	52,206	78,287	72,006	12,256	9,402	66,256
1982	33,754	103,211	253	61,670	77,130	70,747	12,637	12,524	67,246
1983	30,324	114,960	316	69,923	76,876	70,927	13,331	9,907	64,464
1984	32,272	126,065	345	80,143	79,117	73,187	12,478	23,082	81,435
1985	35,576	127,611	380	82,980	78,431	72,904	14,828	13,101	69,781
1986	41,209	127,068	504	87,437	80,155	74,089	15,283	11,767	69,227
1987	41,541	123,351	578	83,220	80,449	74,656	14,980	8,570	67,701
1988	44,272	114,459	761	73,330	85,662	79,837	15,802	9,219	72,317
1989	49,500	91,710	722	51,664	87,669	81,392	16,683	9,479	73,028
1990	52,710	91,604	1,758	56,999	88,692	82,286	16,899	11,005	73,943
1991	57,084	91,261	3,703	55,131	92,001	85,476	19,351	10,140	74,506
1992	57,683	94,251	3,962	57,627	92,334	85,783	20,250	10,567	75,470
1993	61,701	100,189	3,737	64,415	96,273	89,584	23,031	10,064	75,790
1994	53,096	126,542	4,649	82,393	93,161	86,644	22,156	10,441	74,957
1995	48,749	129,894	5,051	84,577	92,743	86,133	21,614	9,878	73,694
1996	50,099	129,742	5,251	81,563	96,660	89,885	23,681	9,315	75,390
1997	49,994	128,234	4,981	79,400	97,023	90,366	26,755	8,706	72,501
1998	47,957	132,367r	5,161	84,612	93,797	87,096	24,333	11,371	71,969
1999	44,891	137,099	4,285	91,797	88,286	81,987	21,656	12,282	70,987

(1) *Aggregate monthly data on crude oil production and trade in oil and oil products are published in Table 13 of Energy Trends - see Annex F.*
(2) *See paragraphs 3.77 to 3.85.*
(3) *Includes natural gas liquids and feedstocks.*
(4) *Excludes products used as fuels within refinery processes.*

3.11 Crude oil and petroleum products: production, imports and exports[1][2], 1960 to 1999 (continued)

	Net exports			Crude oil			Oil products
	Crude oil *(5)*	Oil products *(5)*	Total *(5)*	Ratio of imports to ref. throughput	Ratio of indigenous production to ref. throughput	Ratio of exports to indigenous production	Imports: Share of inland deliveries
	Thousand tonnes			Ratio			Percentage
1960	(45,437)	(4,820)	(50,257)	1.006	0.003	-	34.9
1961	(49,702)	(3,862)	(53,564)	0.996	0.003	-	27.3
1962	(53,381)	(4,615)	(57,996)	1.006	0.002	0.620	30.9
1963	(54,372)	(7,058)	(61,430)	1.004	0.002	0.272	34.1
1964	(60,123)	(8,551)	(68,674)	1.013	0.002	0.419	31.3
1965	(65,322)	(9,249)	(74,571)	0.989	0.001	1.036	31.6
1966	(71,437)	(9,163)	(80,600)	0.998	0.001	1.372	31.6
1967	(72,978)	(12,260)	(85,238)	1.002	0.001	8.157	32.2
1968	(83,136)	(8,363)	(91,499)	1.001	0.001	0.702	28.8
1969	(92,819)	(6,414)	(99,233)	1.016	0.001	3.929	24.3
1970	(100,973)	(3,004)	(103,977)	1.002	0.001	7.577	22.4
1971	(106,167)	(2,203)	(108,370)	1.023	0.001	7.401	21.1
1972	(104,148)	(4,848)	(108,996)	1.007	0.002	10.685	21.2
1973	(112,237)	(896)	(113,133)	1.010	0.002	8.696	18.3
1974	(111,418)	94	(111,324)	1.014	0.002	3.424	15.6
1975	(89,842)	1,138	(88,704)	0.976	0.012	0.974	15.4
1976	(86,181)	5,279	(80,902)	0.925	0.118	0.352	13.1
1977	(53,904)	1,110	(52,794)	0.755	0.409	0.439	15.8
1978	(42,944)	1,608	(41,336)	0.707	0.560	0.467	13.8
1979	(19,811)	953	(18,858)	0.617	0.796	0.522	14.2
1980	(6,537)	4,865	(1,672)	0.541	0.932	0.499	13.0
1981	15,351	2,854	18,205	0.471	1.143	0.583	14.2
1982	27,916	113	28,029	0.438	1.338	0.597	18.6
1983	39,599	3,424	43,023	0.394	1.497	0.608	15.4
1984	48,141	(10,604)	37,537	0.408	1.593	0.638	28.3
1985	47,404	1,727	49,131	0.454	1.627	0.650	18.8
1986	46,228	3,516	49,744	0.514	1.585	0.688	17.0
1987	41,679	6,410	48,089	0.516	1.533	0.675	12.7
1988	29,057	6,583	35,640	0.517	1.336	0.641	12.7
1989	2,164	7,204	9,368	0.565	1.046	0.563	13.0
1990	4,289	5,894	10,183	0.594	1.033	0.622	14.9
1991	(1,953)	9,211	7,258	0.620	0.992	0.604	13.6
1992	(56)	9,683	9,627	0.625	1.021	0.611	14.0
1993	2,714	12,967	15,681	0.641	1.041	0.643	13.3
1994	29,297	11,715	41,012	0.570	1.359	0.651	13.9
1995	35,828	11,736	47,564	0.526	1.401	0.651	13.4
1996	31,464	14,366	45,830	0.518	1.342	0.629	12.1
1997	29,406	18,037	47,443	0.515	1.322	0.619	12.0
1998	36,655	12,962	49,617	0.511	1.414	0.638	15.7
1999	46,906	9,375	56,281	0.508	1.553	0.670	17.3

(5) Figures in brackets signify that in that particular year imports were greater than exports, producing a negative net exports figure.

3.12 Inland deliveries of petroleum, 1960 to 1999[1]

Million tonnes

	Total	Deliveries for energy uses								Deliveries for non-energy uses
		Motor spirit	DERV fuel	Aviation turbine fuel	Burning oil	Gas oil (2)	Fuel oils (3)	Petroleum gases	Total for energy uses (4)	
1960	**39.98**	7.65	2.62	0.79	1.32	3.41	17.72	0.40	**35.89**	**4.09**
1961	**43.18**	8.27	2.88	0.89	1.25	3.80	19.12	0.48	**38.98**	**4.20**
1962	**47.96**	8.70	3.09	1.10	1.45	4.38	21.67	0.65	**43.34**	**4.62**
1963	**52.65**	9.19	3.36	1.51	1.70	5.03	23.07	0.91	**47.14**	**5.50**
1964	**58.02**	10.17	3.69	1.69	1.49	5.56	25.20	1.27	**51.79**	**6.23**
1965	**63.76**	10.91	3.91	1.93	1.61	6.33	27.73	1.56	**57.32**	**6.44**
1966	**69.04**	11.50	4.11	2.19	1.68	7.04	29.56	1.90	**62.23**	**6.81**
1967	**74.63**	12.28	4.36	2.52	1.78	7.77	31.10	1.87	**66.66**	**7.97**
1968	**79.24**	11.88	4.65	2.76	2.00	8.78	31.10	1.73	**70.13**	**9.10**
1969	**85.43**	13.44	4.87	2.97	2.24	10.06	33.93	1.67	**75.49**	**9.94**
1970	**91.15**	14.24	5.04	3.25	2.48	11.55	38.59	1.50	**81.02**	**10.13**
1971	**91.99**	14.96	5.19	3.67	2.57	12.12	39.40	1.42	**81.86**	**10.13**
1972	**98.47**	15.90	5.25	3.93	2.93	14.55	41.31	1.71	**87.79**	**10.68**
1973	**99.79**	16.93	5.66	4.20	3.18	14.59	39.45	1.86	**88.20**	**11.59**
1974	**93.41**	16.48	5.52	3.69	2.78	13.11	36.81	1.31	**81.55**	**11.86**
1975	**82.82**	16.13	5.41	3.83	2.63	12.60	30.47	1.34	**73.38**	**9.44**
1976	**81.60**	16.88	5.59	3.99	2.62	12.52	27.83	1.34	**71.47**	**10.11**
1977	**82.76**	17.34	5.71	4.17	2.62	13.37	27.77	1.36	**73.04**	**9.72**
1978	**84.14**	18.35	5.88	4.51	2.65	13.18	28.23	1.4`	**74.74**	**9.40**
1979	**84.55**	18.69	6.06	4.67	2.70	13.48	27.49	1.51	**75.02**	**9.53**
1980	**71.18**	19.15	5.85	4.69	2.10	11.61	19.16	1.35	**64.18**	**7.00**
1981	**66.26**	18.72	5.55	4.50	1.91	10.91	15.65	1.29	**58.71**	**7.55**
1982	**67.25**	19.25	5.73	4.47	1.75	10.48	16.19	1.60	**59.64**	**7.60**
1983	**64.46**	19.57	6.18	4.57	1.66	9.88	12.51	1.92	**56.45**	**8.02**
1984	**81.44**	20.23	6.76	4.83	1.71	9.92	27.84	1.85	**73.26**	**8.18**
1985	**69.78**	20.40	7.11	5.01	1.87	9.71	16.00	1.13	**61.30**	**8.48**
1986	**69.23**	21.47	7.87	5.50	2.02	9.22	12.51	1.18	**59.86**	**9.36**
1987	**67.70**	22.18	8.47	5.82	2.03	8.51	9.94	1.20	**58.30**	**9.40**
1988	**72.32**	23.25	9.37	6.20	1.99	8.39	11.87	1.15	**62.32**	**10.00**
1989	**73.03**	23.92	10.12	6.56	1.94	8.26	11.13	1.13	**63.15**	**9.88**
1990	**73.94**	24.31	10.65	6.59	2.06	8.03	12.00	1.09	**64.77**	**9.17**
1991	**74.51**	24.02	10.69	6.18	2.38	8.02	11.95	1.28	**64.55**	**9.95**
1992	**75.47**	24.04	11.13	6.67	2.47	7.86	11.48	1.15	**64.84**	**10.63**
1993	**75.79**	23.77	11.81	7.11	2.63	7.77	10.77	1.18	**65.07**	**10.73**
1994	**74.96**	22.84	12.91	7.28	2.66	7.49	9.28	1.29	**63.78**	**11.18**
1995	**73.69**	21.95	13.46	7.66	2.77	7.23	7.98	1.30	**62.37**	**11.32**
1996	**75.39**	22.41	14.37	8.05	3.34	7.63	6.85	1.42	**64.10**	**11.29**
1997	**72.50r**	22.25	14.98	8.41	3.34	7.33	3.94	1.27	**61.55**	**10.95**
1998	**71.97**	21.85	15.14	9.24	3.57	7.24	2.93	1.22	**61.24**	**10.72**
1999	**70.99**	21.51	15.19	9.66	3.56	6.81	2.29	1.14	**60.11**	**10.88**

(1) Aggregate monthly and quarterly data on inland deliveries of oil products are published in Tables 16 and 17 of Energy Trends - see Annex F.
(2) Other than DERV fuel.
(3) Includes Orimulsion from 1989. Imports / deliveries of orimulsion ceased in February 1997.
(4) Includes aviation spirit, naphtha (LDF) for gasworks, marine diesel oil and wide cut gasoline.

3.12 Inland deliveries of petroleum, 1960 to 1999[1] (continued)

Million tonnes

	Fuel producers			Final users				
	Electricity generators	Gas works	Refineries	Iron & steel	Other industries	Transport	Domestic	Other final users (5)
1960	5.56	1.10	3.39	2.46	7.61	13.37	1.52	4.16
1961	5.68	1.15	3.57	2.85	8.97	14.83	1.67	3.69
1962	5.92	1.63	3.76	3.13	10.58	15.66	1.98	4.33
1963	5.24	2.13	3.94	3.78	11.94	16.64	2.27	5.00
1964	5.82	2.99	4.08	4.38	13.28	18.05	2.09	5.04
1965	6.43	4.00	4.21	4.88	14.91	19.24	2.23	5.54
1966	7.39	5.08	4.64	4.79	16.51	20.23	2.21	5.92
1967	7.52	6.06	4.75	4.80	17.90	21.59	2.30	6.39
1968	6.52	6.90	5.16	4.98	19.00	22.88	2.55	7.23
1969	8.41	6.85	5.64	5.43	20.22	23.80	2.79	7.97
1970	12.60	4.56	6.03	5.67	21.55	25.00	3.05	8.59
1971	14.68	2.59	6.18	5.29	21.55	26.07	3.01	8.67
1972	18.87	2.21	6.42	5.04	22.14	27.14	3.48	8.91
1973	16.95	2.32	7.05	4.99	22.18	28.96	3.80	8.99
1974	17.21	1.28	6.95	4.02	19.82	27.92	3.38	7.91
1975	12.82	0.59	6.03	3.31	17.89	27.57	3.27	7.93
1976	10.18	0.25	6.34	3.31	18.06	28.60	3.27	8.16
1977	10.60	0.16	6.24	2.94	18.06	29.37	3.31	8.60
1978	11.64	0.35	6.42	2.83	17.55	30.87	3.26	8.25
1979	11.12	0.42	6.49	2.85	17.62	31.58	3.21	8.22
1980	6.52	0.31	6.27	1.59	14.51	31.74	2.55	6.96
1981	4.86	0.25	5.45	1.33	12.67	30.63	2.31	6.65
1982	6.87	0.21	5.55	1.18	11.64	31.31	2.15	6.27
1983	4.65	0.16	5.30	1.02	10.23	32.25	2.14	6.01
1984	20.91	0.16	5.35	0.84	9.39	33.82	2.14	6.00
1985	9.72	0.15	5.18	0.69	8.43	34.46	2.20	5.65
1986	5.66	0.17	5.40	0.67	9.02	36.66	2.32	5.35
1987	5.36	0.09	5.05	0.56	7.36	38.22	2.21	4.66
1988	6.07	0.06	5.29	0.73	8.23	40.62	2.13	4.67
1989	6.17	0.05	5.62	0.75	7.52	42.54	2.11	4.20
1990	7.98	0.05	5.07	0.70	7.03	43.45	2.22	4.12
1991	7.56	0.05	5.26	0.70	7.49	42.86	2.52	4.17
1992	8.32	0.04	4.16	0.68	7.13	43.79	2.58	4.21
1993	6.02	0.04	5.89	0.85	7.17	44.56	2.71	4.16
1994	4.04	0.05	6.04	0.89	7.47	44.82	2.70	4.01
1995	4.37r	0.05	5.99r	0.83r	6.41r	44.81r	2.70	3.74r
1996	3.93r	0.05	6.16r	0.69r	6.39r	46.64r	3.17	3.73r
1997	2.06r	0.05	6.02r	0.68r	5.75r	47.32r	3.06	3.24r
1998	1.66r	0.05	5.92r	0.47r	5.65r	47.81r	3.20	3.02r
1999	1.40	0.05	5.38	0.50	5.58	47.86	2.92	2.55

(5) Mainly agriculture, public administration, commerce and other services.

Chapter 4
Natural gas

Introduction

4.1 This chapter presents figures on the production, transmission and consumption of natural gas.

4.2 The structure of this chapter is unchanged from last year. Four tables are presented with the commodity balances for natural gas and colliery methane forming the first table (Table 4.1). This is followed by a 5 year table showing the supply, transmission and consumption of these gases as a time series (Table 4.2). A more detailed examination of the various stages of natural gas from gross production through to consumption is given in Table 4.3. Table 4.4 is a long term trends table of production and consumption of gas back to 1960.

4.3 Petroleum gases are covered in Chapter 3. Gases manufactured in the coke making and iron and steel making processes (coke oven gas and blast furnace gas) appear in Chapter 2. Biogases (landfill gas and sewage gas) are part of Chapter 7. Details of net selling values of gas for the domestic sector are to be found in Chapter 9.

The gas supply industry

Great Britain

4.4 When British Gas was privatised in 1986, it was given a statutory monopoly over supplies of natural gas (methane) to premises taking less than 732,000 kWh (25,000 therms) a year. Under the Oil and Gas (Enterprise) Act 1982, contract customers taking more than this were able to buy their gas from other suppliers but no other suppliers entered the market until 1990.

4.5 In 1991, the Office of Fair Trading (OFT), followed up an examination of the contract market by the Monopolies and Mergers Commission (MMC) that had taken place in 1988. It reviewed progress towards a competitive market and found that the steps taken in 1988 had been ineffective in encouraging self-sustaining competition. British Gas undertook in March 1992 to allow competitors to take by 1995 at least 60 per cent of the contract market above 732,000 kWh (25,000 therms) a year (subsequently redefined as 45 per cent of the market above 73,200 kWh (2,500 therms)); to release to competitors the gas necessary to

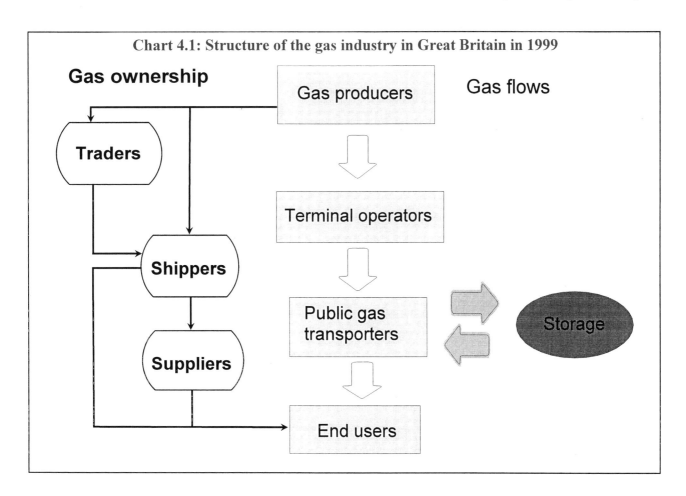

Chart 4.1: Structure of the gas industry in Great Britain in 1999

achieve this; and to establish a separate transport and storage unit with regulated charges. At the same time, the Government took powers in the 1992 Competition and Service (Utilities) Act to reduce or remove the tariff monopoly, and in July 1992 it lowered the tariff threshold from 732,000 kWh (25,000 therms) to 73,200 kWh (2,500 therms).

4.6   Difficulties in implementing the March 1992 undertakings led to further references to the MMC. As a result of the new recommendations made by the MMC in 1993, the President of the Board of Trade decided in December 1993 to require full internal separation of British Gas' supply and transportation activities, but not divestment, and to accelerate removal of the tariff monopoly to April 1996, with a phased opening of the domestic market by the regulator over the following two years.

4.7   In November 1995 the Gas Bill received Royal Assent, clearing the way for the extension of competition into the domestic gas supply market on a phased basis between 1996 and 1998. The first phase, from April 1996, comprised a pilot covering some 500,000 domestic premises in Cornwall, Devon, and Somerset. The scheme was extended (phase 2) to a further 500,000 premises in Dorset and Avon, adjacent to the first pilot area, in February 1997, and to a further million premises in Kent, West Sussex and East Sussex in March 1997. In November 1997, the third phase began with 2.5 million consumers in Scotland and the North East of England able to choose a gas supplier. The remaining 14 million domestic customers were included in the competitive market in the five further stages between February and May 1998. Chapter 9, paragraphs 9.22 to 9.41 give more details of the opening up of the domestic gas and electricity markets to competition and the effects of competition on prices.

4.8   Following the 1995 Act, the business of British Gas was fully separated into two corporate entities. The supply and shipping businesses were devolved to a subsidiary, British Gas Trading Limited, while the transportation business (Transco) remained within British Gas plc. In February 1997, Centrica plc was demerged from British Gas plc (which was itself renamed as BG plc) completing the division of the business into two independent entities. Centrica is now the holding company for British Gas Trading, British Gas Services, the Retail Energy Centres and the company producing gas from the North and South Morecambe fields. BG plc comprises the gas transportation and storage business of Transco, along with British Gas' exploration and production, international downstream, research and technology and property activities. In March 2000 BG plc announced that Transco would become a separately listed company with the rest of the group forming BG International.

4.9   The introduction of effective competition into the contract market has taken both time and heavy regulatory intervention. Since 1990 a number of independent gas suppliers have started to supply natural gas to non-tariff customers, and, where applicable, these supplies are included in the totals. The new supply companies nearly all use the pipeline system owned by Transco to supply gas to their customers. Terms for the use of the pipeline network are set out in Transco's "network code". In some areas low pressure spur networks are being developed by new transporters competing with Transco to bring gas supplies to new customers (mainly domestic). By the end of May 2000 spur networks to over 4,000 housing developments were either operated or due to be operated (when completed) by such new competing transporters. Some very large loads (above 60 GWh) are serviced by pipelines operated independently, some by North Sea producers.

4.10   By the end of 1994, competitors had exceeded the target 45 per cent of the market above 2,500 therms, but virtually all of this was in the firm gas market. From 1995 British Gas' competitors made inroads into the interruptible market and in 1998 Centrica's share of the industrial and commercial market had fallen to around 18 per cent. At the end of 1999, 60 suppliers were active in the contract market. By the end of March 2000, around 5½ million domestic sector customers (27½ per cent) were no longer supplied by British Gas. At the end of 1999, 26 suppliers were licensed to supply gas to domestic customers. The structure of the gas industry in Great Britain as it stood at the end of 1999 is shown in Chart 4.1.

Regional analysis
4.11   Table 4A on the next page, gives the number of consumers with a gas demand below 73,200 kWh per year in 1999. It covers both domestic and small businesses customers receiving gas from the national transmission system. It is this section of the market that was progressively opened up to competition between April 1996 and May 1998. The regions shown are Transco's 13 local distribution zones (LDZs). Table 4B gives the corresponding information for all consumers of gas.

Northern Ireland
4.12   Before 1997 Northern Ireland did not have a public natural gas supply. The construction of a natural gas pipeline from Portpatrick in Scotland to Northern Ireland was completed in 1996, and provided the means of establishing such a system. The primary

Table 4A: Consumption by customers below 73,200 kWh (2,500 therms) annual demand 1999		
Local distribution zones	Number of consumers (thousands)	Gas sales 1999 (GWh)
North Western	2,584	51,980
South Eastern	2,344	45,520
North Thames	2,204	42,913
East Midlands	2,082	41,195
West Midlands	1,878	34,341
Eastern	1,612	33,197
Scotland	1,590	32,259
Southern	1,454	27,877
North Eastern	1,277	24,426
South Western	1,264	22,556
Northern	1,109	23,097
Wales South	759	15.026
Wales North	208	3,857
Great Britain	20,365	398,245

Source: Transco

Table 4B: Consumption by all customers 1999		
Local distribution zones	Number of consumers (thousands)	Gas sales 1999 (GWh)
North Western	2,627	94,410
South Eastern	2,382	66,863
North Thames	2,250	74,599
East Midlands	2,113	76,947
West Midlands	1,908	63,798
Eastern	1,638	52,157
Scotland	1,617	65,148
Southern	1,478	44,540
North Eastern	1,298	44,958
South Western	1,284	38,355
Northern	1,124	42,265
Wales South	769	27,958
Wales North	211	7,512
Great Britain	20,700	699,510

Source: Transco

market is Ballylumford power station, which was purchased by British Gas in 1992 and converted from oil to gas firing (with a heavy fuel oil back up). The onshore line has been extended to serve wider industrial, commercial and domestic markets and this extension is continuing. In 1999, 95 per cent of all gas supplies in Northern Ireland were used to generate electricity.

Competition

4.13   Paragraphs 4.4 to 4.10 above, referred to the developments in recent years in opening up the non-domestic market to competition. About three-quarters of this market (by volume) in the United Kingdom was opened to competition at the end of 1982, and the remainder in August 1992 (with the reduction in the tariff threshold). As mentioned above, however, no other suppliers entered the market until 1990. After 1990 there was a rapid increase in the number of

independent companies supplying gas, although in 1999 there were signs of some consolidation. Chart 4.2 shows how in the mid 1990s sales of gas became less concentrated in the hands of the largest companies. However, for industrial sales in 1999, the trend first seen in 1997 for the largest companies to increase their share of the market has continued. The three largest suppliers now jointly account for 52 per cent of sales to industry compared with 40 per cent in 1996. This has been brought about through larger companies absorbing smaller suppliers and through an expansion of industrial sales by Centrica. For the purpose of this chart industrial sales include sales of gas to autogenerators in the industrial sector, an area that has continued to expand more rapidly than industrial sales for heating and processing. A similar situation is now becoming apparent for commercial sector sales. In 1999 the second to tenth largest firms all gained market share at the expense of the largest and smallest suppliers. Sales of gas for electricity generation have grown the most rapidly in recent years, and are now less concentrated with more suppliers covering the market.

Commodity balances for gas (Table 4.1)

4.14   For the last three years production of natural gas has been greater than supply because exports are larger than imports. However, net exports of natural gas, although growing, were not large in absolute terms, amounting to only 6 per cent of total production in 1999. Imports and exports of natural gas are described in greater detail below (paragraph 4.19).

4.15   Demand for natural gas is traditionally less than supply because of the various measurement differences described in paragraphs 4.46 to 4.49.

4.16   In 1999, 29½ per cent of natural gas demand was for electricity generation (transformation sector). A further 7 per cent was consumed for heating purposes within the energy industries. Less than 1 per cent was accounted for by distribution losses within the gas network. (For an explanation of the items included under losses see paragraphs 4.46 to 4.49.) Of the remaining 63 per cent, 17 per cent was accounted for by the industrial sector with the chemicals industry (excluding natural gas for petrochemical feedstocks), iron and steel and the food industry being the largest consumers. The chemicals sector accounted for over a quarter of the industrial consumption of natural gas.

4.17   Sales of gas to households (domestic sector) produced 33½ per cent of gas demand, while public administration consumed 5 per cent of total demand. Public administration is a larger consumer of gas than any of the individual industrial sectors listed. The

Chart 4.2: Competition in natural gas supplies 1995 to 1999

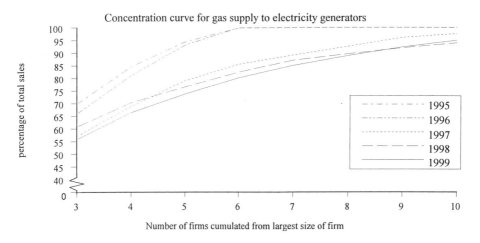

Concentration curve for gas supply to electricity generators

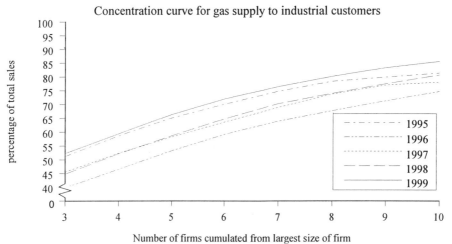

Concentration curve for gas supply to industrial customers

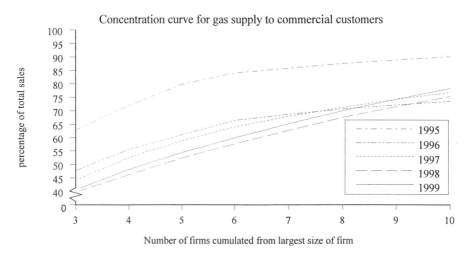

Concentration curve for gas supply to commercial customers

Illustrating increasing competition using concentration curves:

Concentration curves can be used to show the increase (or decrease) in competition within an industry. The sets of curves for the three sectors shown in Chart 4.2 are all constructed in the same way. In any particular year, in a particular sector, the proportion (expressed as a percentage) of total sales of gas (by volume) accounted for by the three firms with the largest sales is calculated. This calculation is repeated for the largest four firms, the largest five firms and so on up to the largest 10 firms. These percentages of total sales are plotted. If each of 10 firms had an equal share of gas sales the plot would form a straight diagonal line from the origin to 100 per cent at the 10 firm point. For a monopoly the curve lies on the vertical axis. When an industry is concentrated in the hands of a few firms, the curve is well above and to the left of the diagonal line and moves towards the diagonal as competition increases. In 1998 the three largest firms in terms of sales to generators accounted for 61 per cent of sales, the six largest 82 per cent, and the ten largest 94 per cent. By 1999 the curve had moved downwards to the right indicating more competition since the largest three firms accounted for only 56 per cent of total sales to industry and the largest six firms 80 per cent, while the proportion for the largest ten was 95 per cent.

commercial, agriculture and miscellaneous sectors together took up 6½ per cent. Non energy use of gas accounted for the remaining 1½ per cent. Non-energy use of natural gas was lower in 1997 than in other recent years because of re-fitting work at one petrochemical plant. As Table 4C, below, shows, non-energy use of gas is small relative to total use (see the technical notes section, paragraph 4.39 for more details on non-energy use of gas).

Table 4C: Non-energy use: share of natural gas demand

	Continental shelf and onshore natural gas
1995	1.8%
1996	1.6%
1997	1.3%
1998	1.3%
1999	1.3%

4.18   Care should be exercised in interpreting the figures for individual industries in these commodity balance tables. As more companies have entered the gas supply market, it has not been possible to ensure consistent classification between and within industry sectors and across years. The breakdown of final consumption includes a substantial amount of estimated data. For about 14 per cent of consumption the allocation to consuming sector is wholly estimated and for a further 6 per cent of consumption the sector figures are partially estimated.

4.19   Imports of natural gas from the Norwegian sector of the North Sea, began to decline in the late 1980s as output from the Frigg field tailed off. In 1999 there was an increase in imports brought about by inflows through the Bacton-Zeebrugge interconnector (although the UK was a net exporter through this interconnector in 1999). Imports added only about 1 per cent to UK production. Exports to mainland Europe, from the United Kingdom's share of the Markham field began in 1992 with Windermere's output being added in 1997. Exports to the Republic of Ireland began in 1995. The interconnector linking the UK's transmission network with Belgium via a Bacton to Zeebrugge pipeline began to operate in October 1998. Since then the interconnector has facilitated both imports and exports of natural gas. Exports accounted for 7½ per cent of production in 1999. Exports of natural gas exceeded imports for the first time in 1997 but grew rapidly (by 46 per cent in 1998 and by over 2½ times in 1999), and as a result were 6½ times the volume of gas imports.

4.20   Chart 4.3 shows the increase in indigenous production and consumption of natural gas over the past five years and how net imports first declined and then became net exports.

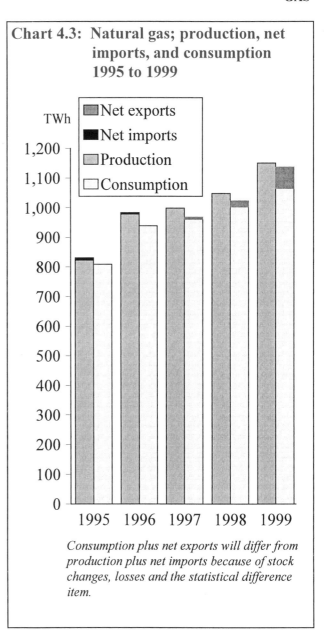

Chart 4.3: Natural gas; production, net imports, and consumption 1995 to 1999

TWh

- Net exports
- Net imports
- Production
- Consumption

Consumption plus net exports will differ from production plus net imports because of stock changes, losses and the statistical difference item.

Supply and consumption of natural gas and colliery methane (Table 4.2)

4.21   These tables summarise the production and consumption of gas from these sources in the United Kingdom over the last 5 years.

4.22   The rapid increase in natural gas use for electricity generation means that the transformation sector now accounts for 29½ per cent of gas demand compared with only 18 per cent in 1995. Although demand from the domestic sector increased by 9 per cent between 1995 and 1999, its share of total demand fell from 40 per cent to 33 per cent.

4.23   Chart 4.4 also shows that the growth in consumption of natural gas between 1991 and 1999 was dominated by the growth in consumption for electricity generation. Most of this gas was used in Combined Cycle Gas Turbine (CCGT) stations, although the use of gas in dual fired conventional steam stations was a growth area in 1997 and 1998. However, growth rates for electricity generation over the last four years first declined from 29½ per cent in 1996, to 25 per cent in 1997 and 6½ per cent in 1998, but then grew again in 1999 to 17½ per cent. Over the four years since 1995, industrial use has grown by 25½ per cent, domestic use by 9 per cent, use by the public administration sector by 11 per cent, use by the commercial sector by 15 per cent, and use in the energy industries other than electricity generation by 34½ per cent.

4.24   Gas use in the domestic sector is particularly dependant on winter temperatures and mild winters in each of the last three years accounted for domestic consumption being lower than in the particularly cold year (by recent standards) of 1996.

4.25   Maximum daily demand for natural gas through the National Transmission System in 1999 was 4,534 GWh on 15 December. On that day natural gas demand in Northern Ireland was 5.1 GWh. This total maximum daily demand was 8 per cent higher than the previous record daily level recorded during the colder winter of 1996 (February).

4.26   It is estimated that sales of gas supplied on an interruptible basis accounted for around 24 per cent of total gas sales in 1999, a similar proportion to 1998.

UK continental shelf and onshore natural gas (Table 4.3)

4.27   This table shows the flows for natural gas from production through transmission to consumption. The footnotes to the table give more information about each table row. This table departs from the standard balance methodology and definitions in order to maintain the link with past data and with data presented quarterly in *Energy Trends (Table 11)*. The relationship between total UK gas consumption shown in this table and total demand for gas given in the balance tables (4.1 and 4.2) is illustrated for 1999 as follows:

	GWh
Total UK consumption (Table 4.3)	994,560
plus Producers own use	64,760
plus Operators own use	5,626
equals	
"Consumption of natural gas" (see paragraph 4.31)	1,064,946
plus Other losses and metering differences (upstream)	3,835
plus Metering differences (transmission)	633
equals	
Total demand (Tables 4.1 and 4.2)	1,069,414

4.28   Gross production increased by 40 per cent between 1995 and 1999. In 1999, the increase was 10 per cent, compared with 5 per cent and 2 per cent respectively in 1998 and 1997. Gas available at UK terminals has increased by a lesser amount (30 per cent) over this four year period mainly because of the increase in exports and decrease in imports described in paragraph 4.19. Producers' and operators' own use of gas have grown in proportion to the volumes of gas produced and transmitted. Output from the transmission system also increased by 30 per cent between 1995 and 1999, while total UK consumption of natural gas increased by 31½ per cent. Consumption increased by more than the output from the transmission system because distribution losses and metering differences have been reduced as a proportion of consumption over these four years.

4.29   For a discussion of the various losses and statistical differences terms in this table see paragraphs 4.46 to 4.49 in the Technical notes and definitions section. Changes to accounting practices within the industry resulted in substantial reductions in the sizes of two of the statistical difference terms since 1997. It is not possible to separate "losses" and "statistical differences" in a completely satisfactory manner. Losses in the distribution system are not separately identifiable and are included under statistical differences. The convention used is set out in paragraph 4.49.

Long term trends

Natural gas and colliery methane production and consumption (Table 4.4)

4.30   Table 4.4 shows data for production, imports, exports, and the consumption of natural gas and colliery methane by major sector in each year from 1960 to 1999. Separate figures are shown for consumption of town gas and methane.

4.31   Total consumption in Table 4.4 is defined to match in with the definition of gas consumption used in the gas tables before the 1999 Digest. This enables a consistent long term series to be presented. Total consumption of natural gas and colliery methane in this table is related to total UK consumption of natural gas in Table 4.3 as follows:

	GWh
Total consumption (Table 4.4)	1,065,427
less Colliery methane	- 481
equals	
Total consumption of natural gas	1,064,946
less Producers own use	- 64,760
less Operators own use	- 5,626
equals	
Total UK consumption (Table 4.3)	994,560

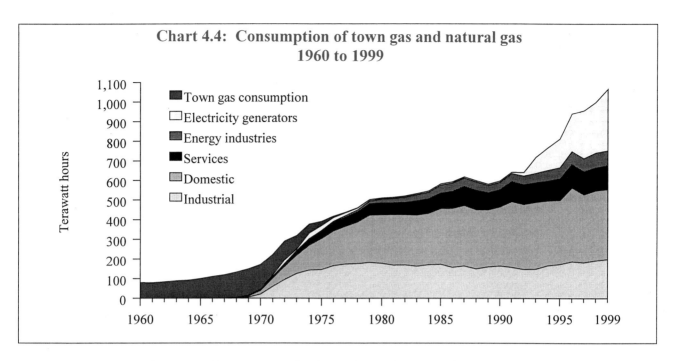

Chart 4.4: Consumption of town gas and natural gas 1960 to 1999

- ■ Town gas consumption
- □ Electricity generators
- ▨ Energy industries
- ■ Services
- ▨ Domestic
- ▢ Industrial

Paragraph 4.27 shows how natural gas consumption in Table 4.3 relates to total demand in the balances Tables 4.1 and 4.2.

4.32   Chart 4.4 illustrates the data in Table 4.4 and shows that the supply of town gas was on a rising path during the 1960s and did not decline until after natural gas supply had become established at the end of that decade. The demise of town gas was then mirrored by the surge of natural gas supply to the mid-1970s. Thereafter, the supply of natural gas continued to grow less rapidly, with indigenous production bolstered from 1977 by increasing imports from the Norwegian sector of the North Sea. In 1998 these imports fell to only 7 per cent of their peak in the mid-1980s, but rose slightly in 1999, because of inflows through the Bacton-Zeebrugge interconnector. In 1999 imports added only 1 per cent to indigenous production compared with about one-third in the mid-1980s. This is largely because of the depletion of the (mainly Norwegian) Frigg field, but also due to the resurgence of UK production which has achieved a new record each year since 1989. Production of gas in 1999 was nearly 2½ times that of 10 years earlier. 1992 saw the first exports of natural gas from the United Kingdom's

share of the Markham gas field. In 1995 these were supplemented by the first exports to the Republic of Ireland, followed by the start of gas exports from the Windermere field via the Markham field during 1997, and exports by the UK-Belgium interconnectors during 1998. By 1999 exports were 6½ times the volume of imports.

4.33   Table 4.4 also shows that the bulk of the rapid growth in the 1970s in consumption of natural gas was in the domestic and industrial sectors. In the 1980s and early 1990s there was a fall in industrial use, but gas consumption by industry has been on an upward trend since 1992 and in 1996 it exceeded the previous peak of 1985. Since 1980 there has been a doubling of gas consumption by the service sector (which for this table is defined as including public administration, commercial activities and agriculture) while domestic sector consumption has increased by 44 per cent over the same period. The increase in total consumption accelerated in the early 1990s because of the large increase in consumption by electricity generators. Even if consumption by electricity generators is excluded consumption in 1999 was around 1½ times its level at the end of the 1970s.

Technical notes and definitions

4.34 These notes and definitions are in addition to the technical notes and definitions covering all fuels and energy as a whole in Chapter 1, paragraphs 1.45 to 1.78. For notes on the commodity balances and definitions of the terms used in the row headings see the Annex A, paragraphs A.7 to A.41.

Definitions used for production and consumption

4.35 **Natural gas** production in Tables 4.1 and 4.2 relates to the output of indigenous methane at land terminals and gas separation plants (includes producers' and processors' own use). For further explanation, see the Annex C, paragraph C.18 under 'Production of oil and gas'. Output of the Norwegian share of the Frigg and Murchison fields from these installations is included under imports. A small quantity of onshore produced methane (other than colliery methane) is also included.

4.36 Table 4.3 shows production, transmission and consumption figures for UK continental shelf and onshore natural gas. Production includes waste and own use for drilling, production and pumping operations, but excludes gas flared. Gas available in the United Kingdom excludes waste, own use for drilling etc, stock change, and includes imports net of exports. Gas transmitted (input into inland transmission systems) is after stock change, own use, and losses at inland terminals. The amount consumed in the United Kingdom differs from the total gas transmitted by the gas supply industry, because of losses in transmission, differences in temperature and pressure between the points at which the gas is measured, delays in reading meters and consumption in the works, offices, shops, etc of the undertakings. The figures include an adjustment to the quantities billed to consumers to allow for the estimated consumption remaining unread at the end of the year.

4.37 **Colliery methane** production is colliery methane piped to the surface and consumed at collieries or transmitted by pipeline to consumers. As the output of deep-mined coal declines so does the production of colliery methane, unless a use can be found for gas that was previously vented.

4.38 **Transfers** of natural gas include natural gas use within the iron and steel industry for mixing with blast furnace gas to form a synthetic coke oven gas. For further details see paragraph 2.52 in Chapter 2.

4.39 **Non-energy gas**: Non-energy use is gas used as feedstock for petrochemical plants in the chemical industry as raw material for the production of

ammonia (an essential intermediate chemical in the production of nitrogen fertilisers) and methanol. The contribution of liquefied petroleum gases (propane and butane) and other petroleum gases are shown in Tables 3.4 to 3.6 of Chapter 3. Firm data for natural gas are not available, but estimates for 1995 to 1999 are shown in Table 4.2. and estimates for 1997 to 1999 in Table 4.1 Estimates for 1995 to 1998 have been obtained from the National Atmospheric Emissions Inventory (NAEI); 1999 data are DTI extrapolations.

Sectors used for sales/consumption

4.40 For definitions of the various sectors used for sales and consumption analyses see the Chapter 1 paragraphs 1.74 to 1.78 and Annex A paragraphs A.30 to A.41. However, **miscellaneous** has a wider coverage than in the commodity balances of other fuels. This is because some gas supply companies are currently unable to provide a full breakdown of the services sector and the gas they supply to consumers is allocated to miscellaneous when there is no reliable basis for allocating it elsewhere.

Data collection

4.41 Production figures are generally obtained from returns made under the Department of Trade and Industry's Petroleum Production Reporting System and from other sources. DTI obtain data on the transmission of natural gas from BG Transco (who operate the National Transmission System) and from other pipeline operators. Data on consumption are based on returns from gas suppliers and UKCS producers who supply gas directly to customers.

4.42 The production data are for the United Kingdom (including natural gas from the UKCS - offshore and onshore). The restoration of a public gas supply to parts of Northern Ireland in 1997 (see paragraph 4.12 means that all tables in this chapter (except 4A and 4B) cover the United Kingdom.

4.43 DTI carry out an annual survey of gas suppliers to obtain details of gas sales to the various categories of consumer. Estimates are included for the suppliers with the smallest market share since the DTI inquiry covers only the largest suppliers (i.e. those with more than about a ½ per cent share of the UK market up to 1997 and those known to supply more than 1,750 GWh per year for 1998 onwards).

Period covered

4.44 Figures generally relate to years ended 31 December. However, data for natural gas for electricity generation relate to periods of 52 weeks as set out in Chapter 5, paragraphs 5.66 and 5.67.

Monthly and quarterly data

4.45 Monthly data on natural gas production and supply, and quarterly data on natural gas consumption, are provided in Tables 11 and 12 of DTI's monthly statistical bulletin *Energy Trends*. See Annex F for more information about *Energy Trends*.

Statistical and metering differences

4.46 In Table 4.3 there are several headings that refer to statistical or metering differences. These arise because measurement of gas flows, in volume and energy terms, takes place at several points along the supply chain. The main sub-headings in the table represent the instances in the supply chain where accurate reports are made of the gas flows at that particular key point in the supply process. It is possible to derive alternative estimates of the flow of gas at any particular point by taking the estimate for the previous point in the supply chain and then applying the known losses and gains in the subsequent part of the supply chain. The differences seen when the actual reported flow of gas at any point and the derived estimate are compared are separately identified in the table wherever possible, under the headings statistical or metering differences.

4.47 The differences arise from several factors:-

- Limitations in the accuracy of meters used at various points of the supply chain. While standards are in place on the accuracy of meters, there is a degree of error allowed which, when large flows of gas are being recorded, can become significant.

- Differences in the methods used to calculate the flow of gas in energy terms. For example, at the production end, rougher estimates of the calorific value of the gas produced are used which may only be revised periodically, rather than the more accurate and more frequent analyses carried out further down the supply chain. At the supply end, although the calorific value of gas shows day-to-day variations, for the purposes of recording the gas supplied to customers a single calorific value is used. Until 1997 this was the lowest of the range of calorific values for the actual gas being supplied within each LDZ, resulting in a "loss" of gas in energy terms. In 1997 there was a change to a "capped flow-weighted average" algorithm for calculating calorific values resulting in a reduction in the losses shown in the penultimate row of Table 4.3. This change in algorithm, along with improved meter validation and auditing procedures, has also reduced the level of "metering differences" row within the downstream part of Table 4.3.

- Differences in temperature and pressure between the various points at which gas is measured. Until February 1997 British Gas used "uncorrected therms" on their billing system for tariff customers when converting from a volume measure of the gas used to an energy measure. This made their supply figure too small by a factor of 2.2 per cent, equivalent to about 1 per cent of the wholesale market.

- Differences in the timing of reading meters. While National Transmission System meters are read daily, customers' meters are read less frequently (perhaps only annually for some domestic customers) and profiling is used to estimate consumption. Profiling will tend to under estimate consumption in a strongly rising market.

- Other losses from the system, for example, leakage from offshore gas pipelines, leakage from the local distribution systems, theft through meter tampering.

4.48 The headings in Table 4.3 show where, in the various stages of the supply process, it has been possible to identify these metering differences as having an effect. Usually they are aggregated with other net losses as the two factors cannot be separated. Whilst the factors listed above can give rise to either losses or gains, losses are more common.

4.49 The box below shows how in 1999 the wastage, losses and metering differences figures in Table 4.3 are related to the losses row in the balance Tables 4.1 and 4.2:

Table 4.3	GWh
Upstream gas industry: Other losses and metering differences	3,835
Downstream gas industry: Transmission system metering differences	633
Tables 4.1 and 4.2 Losses	4,468

Similarly the statistical difference row in Tables 4.1 and 4.2 is made up of the following components in 1999:

Table 4.3		GWh
	Statistical difference between gas available from upstream and gas input to downstream	- 461
plus	Downstream gas industry: Distribution losses, metering differences and transfers	17,410
less	Transfers (Tables 4.1 and 4.2)	-506
Tables 4.1 and 4.2 Statistical difference		16,443

Contact: Mike Janes (Statistician)
mike.janes@dti.gsi.gov.uk
020-7215 5186

Natasha Chopra
natasha.chopra@dti.gsi.gov.uk
020-7215 2718

4.1 Commodity balances 1997-1999

Natural gas

GWh

	1997			1998			1999		
	Natural gas	Colliery methane	Total natural gas	Natural gas	Colliery methane	Total natural gas	Natural gas	Colliery methane	Total natural gas
Supply									
Production	998,343	528	998,871	1,048,353	474	1,048,827	1,150,133	481	1,150,614
Other sources	-	-	-	-	-	-	-	-	-
Imports	14,062	-	14,062	10,582	-	10,582	12,862	-	12,862
Exports	-21,666	-	-21,666	-31,604	-	-31,604	-84,433	-	-84,433
Marine bunkers	-	-	-	-	-	-	-	-	-
Stock change (1)	-4,119	-	-4,119	-374r	-	-374r	+7,801	-	+7,801
Transfers (3)	-496	-	-496	-608	-	-608	-506	-	-506
Total supply	**986,124**	**528**	**986,652**	**1,026,349r**	**474**	**1,026,823r**	**1,085,857**	**481**	**1,086,338**
Statistical difference (2)	**+12,947r**	**-**	**+12,947r**	**+16,863r**	**-**	**+16,863r**	**+16,443**	**-**	**+16,443**
Total demand	**973,177r**	**528**	**973,705r**	**1,009,486r**	**474**	**1,009,960r**	**1,069,414**	**481**	**1,069,895**
Transformation	**250,155r**	**35**	**250,190r**	**266,600r**	**30**	**266,630r**	**313,319**	**93**	**313,412**
Electricity generation	250,155r	35	250,190r	266,600r	30	266,630r	313,319	93	313,412
Major power producers	223,691r	-	223,691r	236,994r	-	236,994r	281,306	-	281,306
Autogenerators	26,464r	35	26,499r	29,606r	30	29,636r	32,013	93	32,106
Petroleum refineries	-	-	-	-	-	-	-	-	-
Coke manufacture	-	-	-	-	-	-	-	-	-
Blast furnaces	-	-	-	-	-	-	-	-	-
Patent fuel manufacture	-	-	-	-	-	-	-	-	-
Other	-	-	-	-	-	-	-	-	-
Energy industry use	**67,429r**	**293**	**67,722r**	**75,716r**	**264**	**75,980r**	**76,361**	**238**	**76,599**
Electricity generation	-	-	-	-	-	-	-	-	-
Oil and gas extraction	58,281	-	58,281	65,468	-	65,468	64,760	-	64,760
Petroleum refineries	3,122r	-	3,122r	3,772r	-	3,772r	3,509	-	3,509
Coal extraction	193	293	486	67	264	331	160	238	398
Coke manufacture	15	-	15	7	-	7	13	-	13
Blast furnaces	342	-	342	527	-	527	643	-	643
Patent fuel manufacture	-	-	-	-	-	-	-	-	-
Pumped storage	-	-	-	-	-	-	-	-	-
Other	5,476	-	5,476	5,875r	-	5,875r	7,276	-	7,276
Losses (4)	**13,462r**	**-**	**13,462r**	**7,928r**	**-**	**7,928r**	**4,468**	**-**	**4,468**
Final consumption	**642,131r**	**200**	**642,331r**	**659,242r**	**180**	**659,422r**	**675,266**	**150**	**675,416**
Industry	**170,651r**	**200**	**170,851r**	**171,882r**	**180**	**172,062r**	**184,963**	**150**	**185,113**
Unclassified	-	200	200	-	180	180	-	150	150
Iron and steel	20,577	-	20,577	21,020r	-	21,020r	21,936	-	21,936
Non-ferrous metals	4,622r	-	4,622r	5,026r	-	5,026r	5,542	-	5,542
Mineral products	14,618r	-	14,618r	14,168r	-	14,168r	14,371	-	14,371
Chemicals	46,185r	-	46,185r	45,069r	-	45,069r	51,185	-	51,185
Mechanical Engineering etc	9,621r	-	9,621r	9,832r	-	9,832r	10,148	-	10,148
Electrical engineering etc	2,916r	-	2,916r	3,407r	-	3,407r	3,941	-	3,941
Vehicles	9,240r	-	9,240r	9,776r	-	9,776r	10,591	-	10,591
Food, beverages etc	26,690r	-	26,690r	27,506r	-	27,506r	29,001	-	29,001
Textiles, leather, etc	6,746r	-	6,746r	6,768r	-	6,768r	6,966	-	6,966
Paper, printing etc	13,828r	-	13,828r	14,350r	-	14,350r	15,365	-	15,365
Other industries	13,888r	-	13,888r	13,264r	-	13,264r	14,282	-	14,282
Construction	1,720	-	1,720	1,696	-	1,696	1,635	-	1,635
Transport	**-**	**-**	**-**	**-**	**-**	**-**	**-**	**-**	**-**
Air	-	-	-	-	-	-	-	-	-
Rail	-	-	-	-	-	-	-	-	-
Road (5)	-	-	-	-	-	-	-	-	-
National navigation	-	-	-	-	-	-	-	-	-
Pipelines	-	-	-	-	-	-	-	-	-
Other	**459,200r**	**-**	**459,200r**	**474,025r**	**-**	**474,025r**	**476,968**	**-**	**476,968**
Domestic	345,532	-	345,532	355,895	-	355,895	356,067	-	356,067
Public administration	53,203r	-	53,203r	52,345r	-	52,345r	51,365	-	51,365
Commercial	36,373r	-	36,373r	40,722r	-	40,722r	41,122	-	41,122
Agriculture	1,443	-	1,443	1,344	-	1,344	1,486	-	1,486
Miscellaneous	22,649	-	22,649	23,719r	-	23,719r	26,928	-	26,928
Non energy use	**12,280r**	**-**	**12,280r**	**13,335r**	**-**	**13,335r**	**13,335**	**-**	**13,335**

(1) Stock fall (+), stock rise (-).
(2) Total supply minus total demand.

(3) Natural gas used in the manufacture of synthetic coke oven gas.

(4) See paragraph 4.49.
(5) See footnote 5 to Table 4.2.

4.2 Supply and consumption of natural gas and colliery methane[1]

GWh

	1995	1996	1997	1998	1999
Supply					
Production	823,336	979,019	998,871	1,048,827	1,150,614
Imports	19,457	19,804	14,062	10,582	12,862
Exports	-11,232	-15,203	-21,666	-31,604	-84,433
Stock change (2)	+9,537	-2,749	-4,119	-374r	+7,801
Transfers	-	-270	-496	-608	-506
Total supply	**841,098**	**980,601**	**986,652**	**1,026,823r**	**1,086,338**
Statistical difference (3)	+20,890	+25,611	+12,947r	+16,863r	+16,443
Total demand	**820,208**	**954,990**	**973,705r**	**1,009,960r**	**1,069,895**
Transformation	**154,393r**	**199,932r**	**250,190r**	**266,630r**	**313,412**
Electricity generation	154,393r	199,932r	250,190r	266,630r	313,412
Major power producers	133,065	176,702	223,691r	236,994r	281,306
Autogenerators	21,328r	23,230r	26,499r	29,636r	32,106
Other	-	-	-	-	-
Energy industry use	**57,010r**	**65,905r**	**67,722r**	**75,980r**	**76,599**
Electricity generation	-	-	-	-	-
Oil and gas extraction	49,249	55,871	58,281	65,468	64,760
Petroleum refineries	2,457r	2,975r	3,122r	3,772r	3,509
Coal extraction	729	670	486	331	398
Coke manufacture	11	23	15	7	13
Blast furnaces	434	478	342	527	643
Other	4,130	5,888	5,476	5,875r	7,276
Losses (4)	**11,423**	**16,256**	**13,462r**	**7,928r**	**4,468**
Final consumption	**597,382r**	**672,897r**	**642,331r**	**659,422r**	**675,416**
Industry	**147,464r**	**162,832r**	**170,851r**	**172,062r**	**185,113**
Unclassified	203	200	200	180	150
Iron and steel	19,988r	21,159r	20,577	21,020r	21,936
Non-ferrous metals	3,993r	4,848r	4,622r	5,026r	5,542
Mineral products	13,828r	14,567r	14,618r	14,168r	14,371
Chemicals	33,394r	36,465r	46,185r	45,069r	51,185
Mechanical engineering etc	8,413r	9,943r	9,621r	9,832r	10,148
Electrical engineering etc	2,468r	2,955	2,916r	3,407r	3,941
Vehicles	8,472r	9,437r	9,240r	9,776r	10,591
Food, beverages etc	23,631r	27,135r	26,690r	27,506r	29,001
Textiles, leather, etc	5,493r	6,897	6,746r	6,768r	6,966
Paper, printing etc	14,231r	14,991r	13,828r	14,350r	15,365
Other industries	11,628r	12,449r	13,888r	13,264r	14,282
Construction	1,722	1,786	1,720	1,696	1,635
Transport	**-**	**-**	**-**	**-**	**-**
Road (5)	-	-	-	-	-
Other	**435,030r**	**495,177r**	**459,200r**	**474,025r**	**476,968**
Domestic	326,010	375,841	345,532	355,895	356,067
Public administration	46,308r	51,411r	53,203r	52,345r	51,365
Commercial	35,729	39,156	36,373r	40,722r	41,122
Agriculture	1,210	1,420	1,443	1,344	1,486
Miscellaneous	25,773	27,348	22,649	23,719r	26,928
Non energy use	**14,888**	**14,888**	**12,280r**	**13,335r**	**13,335**

(1) Colliery methane figures included within these totals are as follows:

	1995	1996	1997	1998	1999
Total production	**610**	**566**	**528**	**474**	**481**
Electricity generation	46	40	35	30	93
Coal extraction	361	326	293	264	238
Other industries	203	200	200	180	150
Total consumption	**610**	**566**	**528**	**474**	**481**

(2) Stock fall (+), stock rise (-).
(3) Total supply minus total demand.
(4) For an explanation of what is included under losses, see paragraphs 4.49.

(5) A small amount of natural gas is consumed by road transport, but gas use in this sector is predominantly of petroleum gases, hence road use of gas is reported in the petroleum products balances in Chapter 3.

4.3 UK continental shelf and onshore natural gas production and supply[1]

GWh

	1995	1996	1997	1998	1999
Upstream gas industry:					
Gross production [2]	822,726	978,453	998,343	1,048,353	1,150,133
Minus Producers' own use [3]	49,249	55,871	58,281	65,468	64,760
Exports [4]	11,232	15,203	21,666	31,604	84,433
Stock change (pipelines) [5]	+390	-883	-2,220	-1,721r	-856
Waste [6]	481	114	94	89r	-
Other losses and metering differences [7][8]	3,407	5,623	6,794	7,419r	3,835
Plus Imports of gas	19,457	19,804	14,062	10,582	12,862
Gas available at terminals [9]	777,424	922,329	927,790	956,076	1,010,823
Minus Statistical difference [8]	-1,450	-1,469	-1,081	+734	-461
Downstream gas industry:					
Gas input into the national transmission system [10]	778,874	923,798	928,871	955,342	1,011,284
Minus Operators' own use [11]	3,311	4,576	4,066	4,337	5,626
Stock change (storage sites) [12]	-9,927	+3,632	+6,339	+2,095	-6,945
Metering differences [8]	7,535	10,519	6,668	509	633
Gas output from the national transmission system [13]	777,955	905,071	911,798	948,401	1,011,970
Minus Distribution losses, metering differences and transfers [8][14]	22,340	27,350	14,430r	16,648r	17,410
Total UK consumption [15]	**755,615**	**877,721**	**897,368r**	**931,753r**	**994,560**

(1) A monthly update of natural gas production and supply is given in Table 11 of Energy Trends.

(2) Includes waste and producers own use, but excludes gas flared.

(3) Gas used for drilling, production and pumping operations.

(4) Due to the confidential nature of some of the data, it is not possible to separately identify exports that are direct from the UKCS from those that are, under a strict interpretation, exports by the downstream gas industry, flowing as they do from part of the national transmission system. As such, both types of export are included under this heading.

(5) Gas held within the UKCS pipeline system. As sections are opened and closed between fields, gas moves in and out of the system, hence it is regarded as a change in stocks.

(6) Gas vented from oil and gas platforms as part of the production process. With effect from 1999 gas vented is deducted from the Gross Production figure.

(7) Losses due to pipeline leakage.

(8) Measurement of gas flows, in volume and energy terms, occurs at several points along the supply chain. As such, differences are seen between the actual recorded flow through any one point and estimates calculated for the flow of gas at that point. More detail on the reasons for these differences is given in the technical notes and definitions section of this chapter, paragraphs 4.46 to 4.49.

(9) The volume of gas available at terminals for consumption in the UK as recorded by the terminal operators. The percentage of gas available for consumption in the UK from indigenous sources in 1998 was 98.9 per cent, compared with 98.5 per cent in 1997.

(10) Gas received as reported by the pipeline operators. The pipeline operators include Transco, who run the national pipeline network, and other pipelines that take North Sea gas supplies direct to consumers.

(11) Gas consumed by pipeline operators in pumping operations and on their own sites, office, etc.

(12) Stocks of gas held in specific storage sites, either as liquefied natural gas, pumped into salt cavities or stored by pumping the gas back into an offshore field. Stock rise (+), stock fall (-).

(13) Including public gas supply, direct supplies by North Sea producers, third party supplies and stock changes.

(14) Includes losses due to leakage through the local distribution system, theft or the effect of accounting procedures (estimated as roughly 1 per cent of demand each for 1995 and 1996). Changes to accounting procedures in 1997 led to a substantial reduction in this element in 1997 which has been maintained in 1998 and 1999. Transfers are the use within the iron and steel industry for use in the manufacture of synthetic coke oven gas.

(15) See paragraph 4.27 for an explanation of the relationship between these "Total UK consumption" figures and "Total demand" shown within the balance tables.

4.4 Natural gas and colliery methane production and consumption 1960 to 1999

	Production		Imports	Exports	Total for consumption			Domestic	
	Town gas *(1)*	Methane *(2)*	Methane *(3)*	Methane	**Total**	Town gas	Methane *(2)*	Town gas	Methane
1960	66,615	791	-	-	**79,452**	79,012	440	38,041	-
1961	65,267	821	-	-	**78,602**	78,162	440	38,099	-
1962	69,048	1,231	-	-	**83,174**	82,441	733	41,059	-
1963	70,982	1,641	-	-	**88,508**	87,482	1,026	45,133	-
1964	68,110	1,846	1,172	-	**91,468**	90,442	1,026	47,302	-
1965	69,370	1,934	7,620	-	**99,791**	98,648	1,143	54,775	
1966	75,407	1,876	7,444	-	**110,107**	108,905	1,202	63,802	
1967	78,865	6,741	8,939	-	**118,635**	117,316	1,319	72,447	29
1968	82,499	23,768	11,606	-	**132,322**	128,336	3,986	82,089	821
1969	76,462	56,797	12,192	-	**148,909**	133,816	15,093	88,683	5,422
1970	49,617	121,712	9,759	-	**171,564**	125,933	45,631	85,430	18,376
1971	24,882	201,721	9,730	-	**222,616**	104,245	118,371	73,502	41,675
1972	17,848	291,078	8,968	-	**290,287**	95,834	194,453	64,974	67,172
1973	21,336	317,132	8,587	-	**319,917**	68,286	251,631	46,598	94,515
1974	12,221	382,253	7,122	-	**377,388**	44,840	332,548	30,450	127,339
1975	5,393	397,932	9,818	-	**391,250**	21,013	370,237	14,507	158,141
1976	1,700	421,700	11,254	-	**417,655**	6,535	411,120	4,250	177,279
1977	762	440,544	19,548	-	**436,793**	2,051	434,742	1,290	191,844
1978	615	422,257	55,361	-	**460,297**	938	459,359	557	212,242
1979	674	425,832	95,424	-	**502,382**	1,055	501,327	586	240,465
1980	586	404,760	116,291	-	**508,684**	909	507,775	557	246,766
1981	557	401,742	124,262	-	**512,112**	791	511,321	469	256,379
1982	557	405,815	115,001	-	**518,149**	674	517,475	410	255,118
1983	586	416,454	124,497	-	**528,642**	528	528,114	322	259,661
1984	557	414,314	147,415	-	**544,584**	498	544,086	293	261,507
1985	498	461,851	147,122	-	**581,717**	469	581,248	293	283,517
1986	440	483,040	137,099	-	**588,691**	410	588,281	234	299,929
1987 *(4)*	322	508,126	128,893	-	**614,247**	322	613,925	147	307,578
1988	88	489,133	115,441	-	**594,766**	88	594,678	29	300,515
1989	-	478,931	113,770	-	**580,522**	-	580,522	-	290,557
1990	-	528,843	79,833	-	**597,046**	-	597,046	-	300,410
1991	-	588,822	72,007	-	**641,763**	-	641,763	-	333,963
1992	-	598,761	61,255	620	**640,818**	-	640,818	-	330,101
1993	-	703,971	48,528	6,824	**717,357**	-	717,357	-	340,162
1994	-	751,588	33,053	9,557	**764,667**	-	764,667	-	329,710
1995	-	823,336	19,457	11,232	**808,786**	-	808,786	-	326,010
1996	-	979,019	19,804	15,203	**938,734**	-	938,734	-	375,841
1997	-	998,871	14,062	21,666	**960,243r**	-	960,243r	-	345,532
1998	-	1,048,827	10,582	31,604	**1,002,032r**	-	1,002,032r	-	355,895
1999	-	1,150,614	12,862	84,433	**1,065,427**	-	1,065,427	-	356,067

(1) In most years production of town gas is less than consumption because of transfers into town gas of north sea and imported methane.
(2) Includes colliery methane.
(3) Before 1977 imports were of liquefied natural gas. These imports continued until the early 1980s.
(4) From 1987 data for industrial use of gas exclude gas used for electricity generation within industry (see paragraph 1.52).

4.4 Natural gas and colliery methane production and consumption 1960 to 1999 (continued)

GWh

Industrial (5)		Electricity generators	Other energy industries (6)		Services (7)		
Town gas	Methane (2)	Methane (2)	Town gas (8)	Methane (2)	Town gas	Methane	
25,263	-	-	-	-	15,709	440	1960
24,882	-	-	-	-	15,533	440	1961
24,823	-	-	-	-	16,559	733	1962
25,233	-	-	-	-	17,115	1,026	1963
25,907	-	-	-	-	16,881	1,026	1964
26,933	-	-	-	-	16,940	1,143	1965
27,314	-	-	-	-	17,789	1,202	1966
26,435	-	-	-	-	18,434	1,290	1967
26,376	1,495	-	-	-	19,870	1,671	1968
24,911	5,773	1,279	-	-	20,222	2,619	1969
20,691	20,808	1,858	-	1,160	19,812	3,428	1970
12,075	60,431	7,808	-	926	18,669	7,531	1971
13,423	94,662	18,563	-	633	17,438	13,423	1972
9,173	125,552	8,453	-	2,743	12,514	20,369	1973
5,744	143,341	28,967	-	3,094	8,646	29,806	1974
2,579	146,067	25,245	-	3,241	3,898	37,542	1975
791	165,644	19,501	-	3,563	1,231	45,132	1976
352	173,820	15,310	-	7,637	410	46,131	1977
176	176,253	10,006	-	9,952	205	50,906	1978
205	182,232	7,104	-	14,143	264	57,382	1979
147	177,513	4,027	-	19,096	205	60,373	1980
147	168,574	4,174	-	22,320	176	59,874	1981
88	169,717	3,793	-	26,657	176	62,190	1982
59	163,123	2,357	-	30,819	147	72,154	1983
59	170,831	5,317	-	33,193	147	73,238	1984
29	172,941	5,873	-	41,135	147	77,781	1985
29	157,496	2,269	-	43,421	147	85,166	1986
29	164,442	2,415	-	43,743	147	95,746	1987 (4)
-	149,935	2,407	-	44,109	59	97,712	1988
-	159,701	6,210	-	37,850	-	86,204	1989
-	164,595	6,513	-	39,159	-	86,369	1990
-	157,932	6,650	-	41,472	-	101,746	1991
-	147,218	17,969	-	45,660	-	99,871	1992
-	148,522	81,848	-	47,006	-	99,819	1993
-	161,815r	117,606r	-	54,700r	-	100,836r	1994
-	162,797r	154,393r	-	56,565r	-	109,020r	1995
-	178,221r	199,932r	-	65,404r	-	119,336r	1996
-	183,488r	250,190r	-	67,365r	-	113,668r	1997
-	185,931r	266,630r	-	75,446	-	118,130	1998
-	199,104	313,412	-	75,943	-	120,901	1999

(5) Industrial consumption in Tables 4.1 and 4.2 plus use in coke manufacture and blast furnaces and non energy gas use.
(6) Energy industry use in Tables 4.1 and 4.2 less use in coke manufacture and blast furnaces.
(7) Public administration, commercial, agriculture and miscellaneous in Tables 4.1 and 4.2.
(8) Town gas consumption by the energy industries is included with the industrial sector.

Chapter 5
Electricity

Introduction

5.1   This Chapter presents statistics on electricity from generation through to sales. Also shown are data for the capacity of plant, for fuel use and for load factors and efficiencies.

5.2   The structure of this Chapter has been changed from last year following comments received. As before, commodity balances for electricity for each of the last three years form the introductory table (Table 5.1). The supply and consumption of elements of the electricity balance are presented as 5-year time series in Table 5.2. Table 5.3 is a new table separating out the public distribution system for electricity from electricity generated and consumed by autogenerators and uses a commodity balance format. In effect this re-instates data last published in the 1998 Digest[1]. Fuels used to generate electricity in the United Kingdom in each of the last five years are covered in Table 5.4, while Table 5.5 adopts a changed and simplified format to show the relationship between fuels used, generation and supply in each of the latest three years. Table 5.6 is also new and shows the relationship between the commodity balance definitions and traditional Digest definitions for electricity so that the most recent data can be linked to the long term trends data presented in Tables 5.11 and 5.12. As in previous years, tables on plant capacity (Tables 5.7 and 5.8) and on plant loads and efficiency (Table 5.9) have been included, as has a long term trends table on fuel use (Table 5.10).

5.3   A large number of revisions appear in the tables this year and are marked with an 'r' suffix. These arise because of changes to the methodology for Combined Heat and Power (CHP), outlined in Chapter 6, affecting electricity generation, supply, capacity and fuel use in each of the last five years.

Structure of the industry

5.4   The structure of the electricity industry in Great Britain in 1999 is illustrated in Chart 5.1. In England and Wales generators and suppliers trade electricity through the Electricity Pool. The Pool is regulated by its members and operated by the National Grid Company which also owns the transmission network. Commercial contracts between generators and suppliers are used to hedge against the uncertainty of future prices in the Pool. Électricité de France (EdF),

together with the generation businesses of Scottish Power and Scottish and Southern Energy, are external members of the England and Wales Pool. Each of these has a number of commercially negotiated contracts to sell electricity through the interconnectors to suppliers in England and Wales.

5.5   In addition to their distribution activities the Regional Electricity Companies (RECs) retain the obligation to supply electricity to customers in their own areas in competition with other suppliers. In turn they may also supply customers in the competitive market nation-wide.

5.6   A number of the major generators also operate as suppliers in the competitive market. In recent years there has been a move towards vertical integration with some generators acquiring supply businesses, and some REC owners acquiring generation businesses.

5.7   In 1999 OFGEM consulted on proposals for new electricity trading arrangements in England and Wales. The proposals are based on bi-lateral trading between generators, suppliers, traders and customers and are designed to be more efficient and provide greater choice for market participants whilst maintaining the operation of a secure and reliable electricity system. The proposals include forwards and futures markets, a balancing mechanism to enable the National Grid Company, as systems operator, to balance the system, and a settlement process. The Government expects the new arrangements to be in place by the autumn of 2000.

5.8   In Scotland, the two main companies, Scottish Power and Scottish and Southern Energy, cover the full range of electricity provision. They operate generation, transmission, distribution and supply businesses. Like the RECs in England and Wales, they retain the obligation to supply customers in their own areas in competition with other suppliers. The entire output of the two nuclear power stations in Scotland, which are owned by British Energy plc, is sold to these two Public Electricity Suppliers (PESs) under long-term contracts. In addition, there are about 25 small independent hydro stations and some independent generators operating fossil-fuelled stations which sell their output to the PESs.

---

[1] *Table 6.3 in the 1998 Digest of UK Energy Statistics. .*

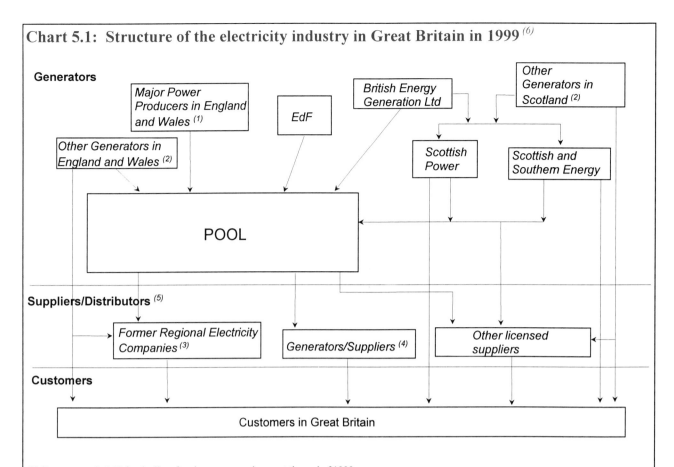

Chart 5.1: Structure of the electricity industry in Great Britain in 1999 [6]

Generators

Major Power Producers in England and Wales [1]

Other Generators in England and Wales [2]

EdF

British Energy Generation Ltd

Other Generators in Scotland [2]

Scottish Power

Scottish and Southern Energy

POOL

Suppliers/Distributors [5]

Former Regional Electricity Companies [3]

Generators/Suppliers [4]

Other licensed suppliers

Customers

Customers in Great Britain

(1) See paragraph 5.58 for the list of major power producers at the end of 1999.

(2) Generators other than major power producers, some of which, as licence exempt suppliers, meet their own electricity needs and sell electricity directly to other local customers.

(3) Public electricity supply licence holders in England and Wales, namely: East Midlands Electricity (now part of PowerGen), NORWEB, Eastern Electricity, SEEBOARD, London Electricity (now part of EdF), SWALEC (supply licence held by British Energy), MANWEB (now part of Scottish Power), SWEB (supply licence held by London Electricity), Midlands (now part of National Power), Southern Electric (now part of Scottish and Southern Energy), Yorkshire Electricity, Northern Electric.

(4) Generators holding Second Tier supply licences are: National Power, PowerGen, Scottish Power, Scottish and Southern Energy, BNFL Magnox, British Energy Generation.

(5) Each public electricity supplier owns and operates the distribution network within its own authorised area. Distributors are therefore the same companies as those listed in note 3 plus Scottish Power and Scottish and Southern Energy with authorised areas in Scotland. The distribution business of SWALEC was part of Hyder in 1999, and the distribution business of SWEB is a separate company (Western Power Distribution).

(6) Note that in 2000 under the New Electricity Trading Arrangements the structure of the electricity industry will change, see paragraph 5.7.

5.9 The electricity supply industry in Northern Ireland is also in private hands. Northern Ireland Electricity plc (NIE) (part of the Viridian Group) is responsible for power procurement, transmission, distribution and supply in the Province. Generation is in the hands of three private sector companies who own the four major power stations. There is a link (re-established in 1996) between the Northern Ireland grid and that of the Irish Republic along which electricity is both imported and exported.

5.10 In Great Britain competition in generation and supply has developed as follows:

(a) Since 1 April 1990, customers with peak loads of more than 1 MW (about 45 per cent of the non-domestic market) have been able to choose their supplier. The Office of Electricity Regulation estimated that in 1990/91 customers accounting for 43 per cent of the output in the 1 MW market in England and Wales chose to take their supply from a company other than their local REC. By 1998/99 this had increased to 80 per cent.

(b) Since 1 April 1994 customers with peak loads of more than 100 kW have been able to choose their supplier. The Office of Gas and Electricity Markets (OFGEM) estimates that in 1998/99 customers accounting for 61 per cent of the output in the 100 kW to 1 MW market in England and Wales chose to take their supply from a company other than their local REC.

(c) Between September 1998 and May 1999 the remaining part of the electricity market (i.e. below 100 kW peak load) was opened up to competition. Chapter 9, paragraphs 9.22 to 9.41 give more details of the opening up of the domestic gas and electricity markets to competition and the effects of competition on prices.

(d) Since vesting, a number of new companies have entered the generation market. In 1999 these companies produced over a third of the electricity generated by major power producers in the United Kingdom.

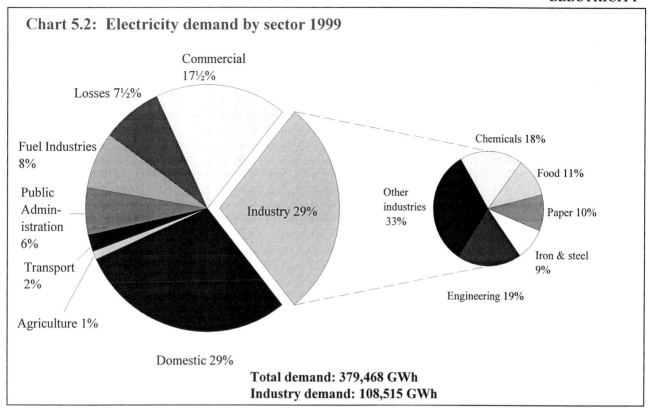

Chart 5.2: Electricity demand by sector 1999

Commercial 17½%

Losses 7½%

Fuel Industries 8%

Public Admin-istration 6%

Transport 2%

Agriculture 1%

Domestic 29%

Industry 29%

Other industries 33%

Chemicals 18%

Food 11%

Paper 10%

Iron & steel 9%

Engineering 19%

Total demand: 379,468 GWh
Industry demand: 108,515 GWh

Commodity balances for electricity (Table 5.1)

5.11 The top section of the balance table shows that 96 per cent of UK electricity supply in 1999 was home produced and 4 per cent was from imports. Of the 367 TWh produced (including pumped storage production) 92 per cent was from major power producers and 8 per cent from autoproducers, 28 per cent was from primary sources and 72 per cent from secondary sources.

5.12 Electricity generated by each type of fuel is shown in the final section of the commodity balance table. The link between electricity generated and electricity supplied is made in Table 5.5 and electricity supplied by each type of fuel is illustrated in Chart 5.3. Paragraph 5.28 examines further the ways of presenting each fuel's contribution to electricity production.

5.13 Demand for electricity is predominantly from final consumers who accounted for 84½ per cent in 1999. The remaining 15½ per cent is split 8 per cent to energy industry use and 7½ per cent to losses. 57 per cent of the energy industries' use of electricity is by the electricity industry itself, with petroleum refineries being the next most significant consumer. The losses item represents transmissions losses from the high voltage transmission system (about 20 per cent of the figure in 1999), distribution losses which occur between the gateways to the public supply system's network and the customers meters (about 75 per cent) plus a small amount lost through theft or meter fraud (5 per cent) (see also paragraphs 5.74 to 5.75).

5.14 Industrial consumption was 34 per cent of final consumption in 1999, slightly less than consumption by households (34½ per cent), with transport and the services sector accounting for the remaining 31½ per cent. Within the industrial sector the four largest consuming sectors are chemicals, food, paper and iron and steel, which together account for nearly 49 per cent of industrial consumption. The iron and steel sector excludes blast furnace and coke oven uses of electricity which are included under energy industry uses. This is because electricity is used by coke ovens and blast furnaces in the transformation of solid fuels into coke, coke oven gas and blast furnace gas. Taken together the engineering industries accounted for a further 19 per cent. A note on the estimates included within these figures is to be found at paragraph 5.76. Chart 5.2 shows diagrammatically the demand for electricity in 1999.

5.15 The transport sector covers electricity consumed by companies involved in transport, storage and communications. Within the overall total of nearly 8,500 GWh it is known that national railways consume about 2,700 GWh each year for traction purposes, and this figure has been included in the balances.

Supply and consumption of electricity (Table 5.2)

5.16 There was a 1¾ per cent increase in the supply of electricity in 1999. Production (including pumped storage production) increased by 1¼ per cent but there was a 15 per cent increase in imports of electricity. The main cause of this increase in imports was the low level of imports from France in the second half of

Table 5A: Electricity distributed by public electricity suppliers, 1999[1][2]

	Area (sq. km)	Number of customers (thousand)	Customer density (No per sq. km)	Electricity distributed (GWh)
Eastern	20,300	3,257	160	33,120
South	16,900	2,647	157	30,333
East Midlands	16,000	2,300	144	26,676
Midlands	13,300	2,250	169	26,007
North West	12,500	2,202	176	24,088
South Scotland	22,950	1,860	81	22,710
Yorkshire	10,700	2,072	194	21,979
South East	8,200	2,108	257	20,050
London	665	1,982	2,980	17,868
Merseyside and North Wales	12,200	1,382	113	17,025
North East	14,400	1,441	100	16,012
South West	14,400	1,323	92	14,389
South Wales	11,800	980	83	12,294
North Scotland	54,390	640	12	8,189
Northern Ireland	13,506	681	50	7,444
Total	242,211	27,125	112	298,184

1. The figures for the area and number of customers are taken from OFFER's Reviews of Public Electricity Suppliers 1998 to 2000 (PES Business Plans Consultation Paper) December 1998 and relate to 1997/98. Electricity distributed is taken from returns made to DTI for 1999.

2. The figures for electricity distributed exclude electricity sold directly to customers over high voltage lines.

1998 following problems at three French nuclear stations in mid summer. While the volume of imports from France rose in 1999, it was still over 10 per cent lower than the average annual volume recorded in the mid 1990s. Exports of electricity in 1999 rose because of the increased use of the interconnector between the Irish Republic and Northern Ireland.

5.17 Energy industry use of electricity as a proportion of electricity demand rose slightly in 1999, but it was still below the proportions recorded in the mid 1990s as might be expected with gas fired stations replacing coal fired. While losses represent 7½ per cent of total demand, this proportion is the same as in 1998 and was lower than in any of the previous five years. Industrial consumption of electricity grew by 3 per cent in 1999, reflecting a continuing recovery in industrial output, while consumption in the services sector rose by 1 per cent. Consumption by transport,

storage and communications also grew by 1 per cent. Domestic sector sales increased by just under 1 per cent in 1999, and were 2½ per cent higher than the previous peak for this sector in 1996. The actual level in any one year is influenced by temperatures in the winter months, as customers adjust heating levels in their homes. On average, temperatures in the winter months were lower in 1996 and 1999 than in the milder years of 1995, 1997 and 1998.

Regional electricity data

5.18 The restructuring of the electricity industry in 1990 and the privatisation of the electricity companies meant that it was no longer possible for this Digest to present regional data on the supply of electricity, as it would disclose information about individual businesses which were in competition with each other. However, distribution, the physical delivery of electricity, is a monopoly activity for each public electricity supplier inside its own geographic area. These areas vary in terms of square kilometres covered, number of customers and electricity distributed within the area as Table 5A shows.

5.19 The difference between total electricity distributed, shown in this table and total consumption is accounted for by electricity, sold directly to large users by generators via the high voltage grid without passing through the distribution network of the regional companies, electricity consumed by the company generating the electricity (autogeneration plus electricity industry own use), and transmission losses.

Numbers of consumers and consumption bands

5.20 An estimate of the total number of consumers of electricity in terms of sites supplied is also available from information provided to DTI by the electricity companies. This information can be broken down into broad sectors and into consumption bands as Table 5B shows. The figures in this table relate to 1998 and have been adjusted within DTI to remove the double counting of sites that changed supplier during the year. For these reasons the numbers are not directly comparable with the customer numbers in Table 5A.

Table 5B: Number of sites by sector and size band, 1998, adjusted for double counting of customers changing supplier in mid-year

	Industry	Transport storage and communications	Commercial, agriculture, public lighting, public administration and other services	Domestic	Total
Less than 9 GWh	174,408	37,225	2,512,427	25,224,122	27,588,182
9 GWh but less than 20 GWh	928	54	1,030	-	2,012
20 GWh but less than 40 GWh	264	33	96	-	393
40 GWh or more	275	26	36	-	337
Total	175,875	37,338	2,153,589	25,224,122	27,590,924

Commodity balances for the public distribution system and for other generators (Table 5.3)

5.21 Table 5.3 is a new table this year. In last year's re-organisation of the electricity chapter, the table that showed consumption divided between electricity distributed over the public distribution system and electricity provided by other generators (autogeneration) was dropped. This year it has been re-instated in commodity balance format with the domestic sector expanded to show consumption by payment type and the commercial sector expanded to show detailed data beyond that presented in Tables 5.1 and 5.2. Autogeneration is the generation of electricity wholly or partly for a company's own use as an activity which supplements the primary activity.

5.22 The proportion of electricity supplied by generators other than major power producers continues to increase, growing from 6.7 per cent in 1997 to 7.2 per cent in 1998 and 7.8 per cent in 1999. In each of these years between 22 and 23 per cent of this electricity was transferred to the public distribution system.

5.23 In 1999, 5.2 per cent of final consumption of electricity was by other generators and did not pass over the public distribution system. This compares with 4.7 per cent in 1998 and 4.3 per cent in 1997. A greater proportion of electricity is self generated in the energy industries (particularly at petroleum refineries where in 1999 the proportion was over three quarters) with the proportion close to 20 per cent in all three years shown in the table.

5.24 About 13 per cent of the industrial demand for electricity was met by autogeneration. There was also a lesser proportion (about 2 per cent) from autogeneration within the commercial and transport sectors. Table 1.8 in Chapter 1 shows the fuels used by autogenerators to generate this electricity within each major sector and also the quantities of electricity generated and consumed.

5.25 Within the domestic sector, about 40 per cent of the electricity consumed was purchased under some form of off-peak pricing structure. About 15 per cent of consumption was through prepayment systems. In both cases these proportions are little changed from 1998.

Fuel used in generation (Table 5.4)

5.26 In this table fuel used by electricity generators is measured in both original units and for comparative purposes, in the common unit of million tonnes of oil equivalent. In Table 5.5 figures are quoted in a third unit, namely GWh, in order to show the link between fuel use and electricity generated.

5.27 The energy supplied basis defines the primary input (in million tonnes of oil equivalent) needed to produce 1 TWh of hydro, wind, or imported electricity as:

$$\text{Electricity generated (TWh)} \times 0.08598 \,.$$

The primary input needed to produce 1 TWh of nuclear electricity is similarly

$$\frac{\text{Electricity generated (TWh)} \times 0.08598}{\text{Thermal efficiency of nuclear stations}}$$

In the United Kingdom the thermal efficiency of nuclear stations has risen in stages from 32 per cent in 1982 to 37 per cent in 1999 (see Table 5.9 and paragraph 5.65 for the definition)[2]. The factor of 0.08598 is the energy content of one TWh divided by the energy content of one million tonnes of oil equivalent (see page 247 and inside back cover flap).

5.28 Figures on fuel use for electricity generation can be measured in two ways. Table 5.4 illustrates one method by using the volumes of **fuel input** to power stations (after conversion of inputs to an oil equivalent basis), but this takes no account of how efficiently that fuel is converted into electricity. The fuel input basis is the most appropriate to use for analysis of the quantities of particular fuels used in electricity generation (eg to determine the amount of coal at risk from displacement by gas or other fuels). A second method uses the amount of electricity generated and supplied by each fuel. This **output** basis is appropriate for comparing how much, and what percentage, of electricity generation comes from a particular fuel. It is the most appropriate method to use to examine the dominance of any fuel, and diversity issues. Percentage shares based on fuel outputs reduce the contribution of coal and nuclear, and increase the contribution of gas (by about 5 percentage points in 1999) compared with the fuel inputs basis, because of the higher conversion efficiency of the latter. This output basis is used in Chart 5.3, taking electricity supplied (gross) figures from Table 5.5. Trends in fuel used on this electricity supplied basis are described in the section on Table 5.5, in paragraphs 5.30 to 5.33, below.

5.29 Table 5.10 gives an historical series of fuel used in generation on a consistent, energy supplied, fuel input basis.

---

[2] *Note that the International Energy Agency uses 0.33 in its calculations, which is the European average thermal efficiency of nuclear stations in 1989, measured in net terms rather than the UK's gross terms.*

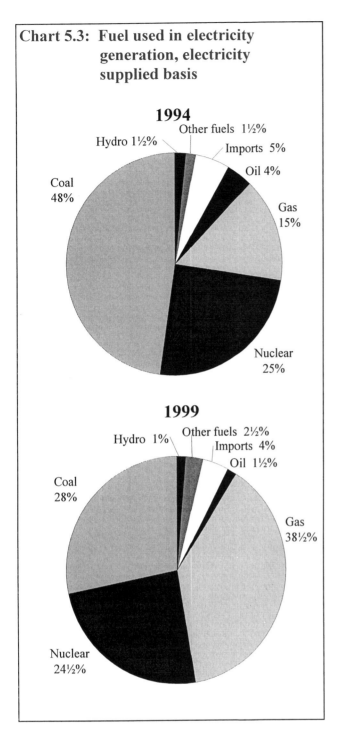

Chart 5.3: Fuel used in electricity generation, electricity supplied basis

1994

Other fuels 1½%
Hydro 1½%
Imports 5%
Oil 4%
Coal 48%
Gas 15%
Nuclear 25%

1999

Hydro 1%
Other fuels 2½%
Imports 4%
Oil 1½%
Coal 28%
Gas 38½%
Nuclear 24½%

92 per cent of electricity generation in 1999. Generation by other generators was 10 per cent higher than a year earlier.

5.32 The mix of plant used to generate the electricity within the UK continues to evolve. Generation from coal fired stations fell by 13½ per cent during 1999. Even allowing for the fact that coal stations were called on to make up for the reduced level of electricity imports from France during the latter part of 1998, the reduction was still substantial, representing a 11½ per cent fall on 1997. Generation from gas increased by 21 per cent in 1999 with more new stations contributing than in 1998 when the increase was 7 per cent.

5.33 Table 5.5 also shows electricity supplied data. These data take into account the fact that some stations use relatively more electricity in the generation process itself. In total, electricity supplied (gross) was 5 per cent less than the volume generated in 1999, but for nuclear stations it was 9 per cent less while for gas fired stations it was only 2 per cent less. Chart 5.3 shows how shares of the generation market in terms of electricity output have changed over last five years. Gas' share of electricity supplied (net) has moved up sharply from 15 per cent in 1994 to 38½ per cent in 1999 while coal's share has fallen from 48 per cent to 28 per cent. Nuclear's share rose to a peak in 1997 but then fell back, and in 1999 recorded the same share as in 1994. Oil's share has fallen by 2½ percentage points over the five years shown.

Relating measurements of supply, consumption and availability (Table 5.6)

5.34 Table 5.6 appeared in the 1999 Digest as Table 5C. The balance methodology has introduced some new terms that cannot be readily employed for earlier years' data because statistics were not available in sufficient detail. Table 5.6 shows the relationship between these terms for the latest five years. For the full definitions of the terms used in the commodity balances see the Annex A, paragraphs A.7 to A.41.

Electricity generated, and supplied (Table 5.5)

5.30 Until 1995 data on electricity generation and supply were collected by type of station. From 1996 onwards data on generation and supply have been collected by fuel, and figures in Table 5.5 are presented on this basis. The table retains figures for generation from conventional steam stations and from combined cycle gas turbine stations.

5.31 Total electricity generated in the United Kingdom in 1999 was 1¼ per cent higher than the level of 1998. Over the period 1995 to 1999 the average rate of growth of electricity generation has been just over 2¼ per cent a year. Major power producers (as defined in paragraph 5.58) accounted for

Plant capacity (Tables 5.7 and 5.8)

5.35 Table 5.7 shows capacity, i.e. the maximum power available at any one time, for major power producers and other generators by type of plant. A change was made in 1996 to record capacities for major power producers at the end of December rather than the previous position of recording capacities at the end of March of the following year. Because generators had been closing surplus capacity during March (at the end of the peak winter season) the end March figure did not always show the capacity that had been available to meet the previous winter's peak.

5.36 In 1999 there was an increase of 1,700 MW (2½ per cent) in the capacity of major power producers. Over 250 MW of coal fired capacity (net of closures) was added because one previously mothballed set on a station that changed ownership was re-included in the total. 1,500 MW of new CCGT capacity is included in 1999 with new stations coming on stream at Cottam and Sutton Bridge. In addition that part of Seabank which did not begin to operate until 1999 is included in the 1999 total. In December 1999 major power producers accounted for 93 per cent of the total generating capacity, which is a slightly lower proportion than at the end of 1998 because of the 9 per cent increase in the capacity of other generators.

5.37 A breakdown of the capacity of the major power producers' plant at the end of March each year from 1993 to 1996 and at the end of December for 1996 to 1999 is shown in Chart 5.4.

5.38 In Table 5.8 the data for industrial, commercial and transport undertakings are shown according to the industrial classification of the generator. Over a fifth of the capacity is in the chemicals sector. Petroleum refineries have nearly 17 per cent of capacity, engineering and other metal trades, have ten per cent and paper, printing and publishing 9 per cent.

Plant loads, demand and efficiency (Table 5.9)

5.39 Table 5.9 shows the maximum load met each year, load factors (by type of plant and for the system in total) and indicators of thermal efficiency. Prior to 1996, figures were collected on a financial year basis. Maximum demand figures cover the winter period ending the following March.

5.40 Maximum demand during the winter of 1999/2000 occurred in December 1999. This was 1.6 per cent above the previous maximum achieved in December 1997 and 2.7 per cent higher than the maximum demand in the previous winter (1998/99). Maximum demand in 1999/2000 was 82½ per cent of the capacity of major power producers (Table 5.7) as measured at the end of December 1999, compared with just under 82½ per cent in 1998/99 and 83½ per cent in 1997/98.

5.41 Plant load factors measure how intensively each type of plant has been used. Over the last five years conventional steam plant has been used less intensively and CCGT stations more intensively. However, in 1998 the increased use of coal-fired stations (see paragraph 5.32) combined with some extended maintenance of CCGTs created a small departure from this trend. The reduced load factor for CCGTs in 1998 enabled nuclear stations to record the highest load factor as they were in use for over 80 per cent of the time, but in 1999 the nuclear load factor fell back to 77 per cent. Ample rainfall allowed hydro and pumped storage to be used more than in each of the four previous years.

5.42 Thermal efficiency measures the efficiency with which the heat energy in fuel is converted into electrical energy. Efficiencies of conventional steam stations have remained fairly static over the last five years. CCGT efficiencies in Table 5.9 understate their potential operating efficiencies because almost all of these years include a build up in CCGT capacity and during the start up phase new stations operate at much lower efficiencies. Overall, an increasing trend in efficiency is shown. The efficiencies presented in this table are calculated using **gross** calorific values to obtain the energy content of the fuel inputs. If **net** calorific values are used efficiencies are higher, for example CCGT efficiencies rise by about 5 percentage points.

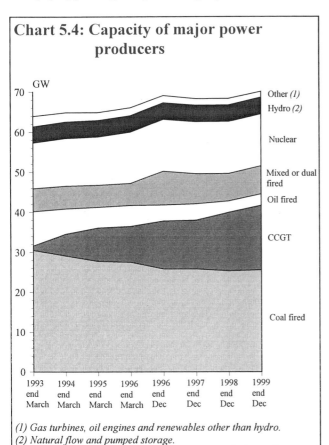

Chart 5.4: Capacity of major power producers

GW

Other (1)
Hydro (2)
Nuclear
Mixed or dual fired
Oil fired
CCGT
Coal fired

1993 end March, 1994 end March, 1995 end March, 1996 end March, 1996 end Dec, 1997 end Dec, 1998 end Dec, 1999 end Dec

(1) Gas turbines, oil engines and renewables other than hydro.
(2) Natural flow and pumped storage.

Long term trends

Fuel input for electricity generation (Table 5.10)

5.43   This table extends the series shown in Table 5.4 back to 1965; data for 1960 are also shown. For the period up to 1987, only fuel inputs for electricity generation at stations owned by the major power producers, transport undertakings, and industrial hydro-electric and nuclear power stations are given; data for conventional thermal electricity generated by industrial producers are not available for this period. From 1987 onwards the table covers **all** generating companies.

5.44   The unit of measurement used in this table is the tonne of oil equivalent. An outline of the method used for converting both fossil and non-fossil fuel energy sources to this unit is given in paragraph 5.27, above.

5.45   Trends in fuel input for electricity generation are shown in Chart 5.5.

5.46   In 1960, coal provided over 80 per cent of the fuel input for electricity generation, with oil making up virtually all the rest. Between 1968 and 1972, oil use increased strongly, rising to account for 29 per cent of fuel input in 1972, but after the oil supply crisis in the following year, its use declined, apart from a temporary increase during the 1984/85 miners' dispute. By 1999, the use of oil for electricity generation had fallen to 1½ per cent. Nuclear generation has grown steadily from virtually zero in 1960 until in 1998 it reached a peak when its oil equivalent input amounted to 29 per cent of total fuel input. Between 1975 and 1990 a European Community directive limited the use of natural gas in public supply power stations. Since 1991 the role of gas in electricity generation has grown rapidly, its share rising from 2 per cent in 1992 to 13½ per cent in 1994, 21½ per cent in 1996, and 28½ per cent in 1998. Then in 1999 its share exceeded that of both coal and nuclear and reached 34 per cent. Coal still provided a substantial input, but in 1999 its share had fallen to 32 per cent, having been 50 per cent as recently as 5 years earlier, and 65 per cent 10 years earlier.

Electricity supply, availability and consumption (Table 5.11)

5.47   Figures for the supply, availability and consumption of electricity are given in Table 5.11, This table retains the nomenclature of previous electricity chapters in the Digest whereas the balance methodology has introduced new nomenclature (see paragraph 5.34, above and Table 5.6). The series

extended back to 1965 and data for 1960 are also shown.

5.48   For the period up to 1986 the data for electricity supplied cover major power producers, transport undertakings and industrial hydro and nuclear stations only. Purchases from other electricity producers are also included, along with net imports, to give electricity available.   Losses are deducted from electricity available to give consumption, which is shown by type of consumer.   Availability and consumption before 1986 exclude electricity consumed or sold by other generators without passing through the public distribution system.

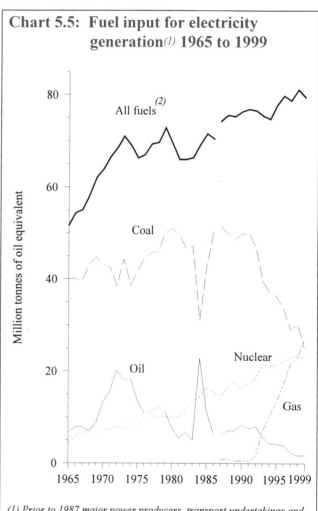

Chart 5.5: Fuel input for electricity generation[1] 1965 to 1999

(1) Prior to 1987 major power producers, transport undertakings and industrial hydro and nuclear stations only.
From 1987 all generators are covered, hence there is a break in the series for all fuels other than nuclear.
(2)Including hydro, other renewables, coke and other fuels, but excluding electricity imports.

5.49   The table shows that virtually all electricity available came from home supply until 1986 when the interconnector between France and England commenced operations. In 1999, net imports from France, combined with net imports into Northern Ireland from the Irish Republic over the re-instated interconnector, amounted to 4 per cent of total electricity available, although this was less than the peak contribution of imports of 5¼ per cent in 1994.

Losses in transmission and distribution have fallen from 10 per cent of electricity available in 1960 to 8 per cent in 1999. Consumption of electricity by fuel industries as a proportion of total electricity consumption has fallen from 6 per cent in 1960 to 2½ per cent in 1999. Thus while electricity available increased 3-fold between 1960 and 1998, the proportion reaching final users rose from 84 per cent in 1960 to nearly 90 per cent in 1999.

Electricity generated and supplied (Table 5.12)

5.50 Figures for the generation and supply of electricity are given in Table 5.12. This table retains the nomenclature of previous electricity chapters in the Digest whereas the balance methodology has introduced new nomenclature (see paragraph 5.34, above and Table 5.6.) Data are given for major power producers, for other generators and for all generators in total, with separate series for the different types of power station.

5.51 Over the period 1960 to 1999, total gross electricity supplied by all generating companies increased at an average annual rate of 2½ per cent.

5.52 In the period between 1960 and 1994 electricity output by generators other than the major producers fluctuated between 13,000 and 18,000 GWh, but moved up to over 20,000 GWh in 1995. Subsequently it has increased every year to reach 28,400 GWh in 1999, mainly as a result of the greater capacity of combined heat and power systems now in use (see Chapter 6). As generation by the major producers increased, so the contribution of other generators to total supply fell from 10½ per cent in 1960 to under 6 per cent in the early 1990s but it has since increased again to reach 8 per cent in 1999. Trends in electricity supplied by all generators by type of plant are illustrated in Chart 5.6.

5.53 In 1960, conventional thermal power stations produced 96 per cent of the gross electricity supplied and output from these stations rose, peaking in 1990. Two factors have led to the subsequent decline. Firstly, it was the development of nuclear generation, which by 1997 accounted for 27 per cent of the electricity supplied by generating companies in the United Kingdom, but has since fallen back slightly. Secondly, since the early 1990s it was the growth of combined cycle gas turbine stations (CCGTs) which overtook nuclear in 1997 and in 1999 supplied 34 per cent. Gross supply from non-CCGT thermal stations has fallen by 43 per cent from the 1990 peak and by 1999 they accounted for 39 per cent of electricity supplied.

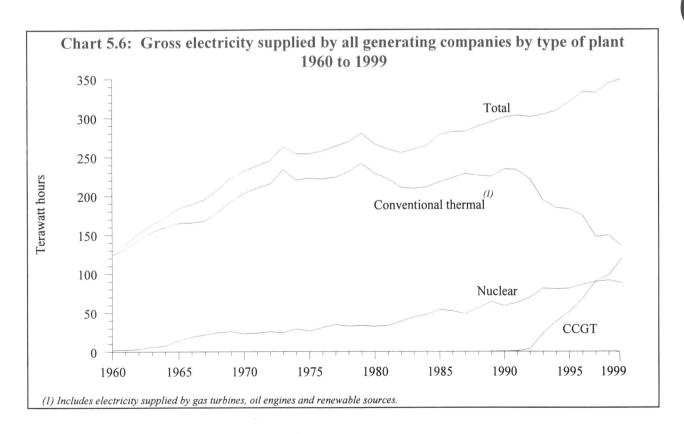

Chart 5.6: Gross electricity supplied by all generating companies by type of plant 1960 to 1999

(1) Includes electricity supplied by gas turbines, oil engines and renewable sources.

Technical notes and definitions

5.54 These notes and definitions are in addition to the technical notes and definitions covering all fuels and energy as a whole in Chapter 1, paragraphs 1.45 to 1.78. For notes on the commodity balances and definitions of the terms used in the row headings see the Annex A, paragraphs A.7 to A.41.

Electricity generation from renewable sources

5.55 Figures on electricity generation from renewable energy sources are included in the tables in this section. Further detailed information on the use of renewable energy sources and, in particular, the capacity, fuel use and the amount of electricity generated from such sources are included in Chapter 7 on Renewable energy sources.

Combined heat and power

5.56 Electricity generated from combined heat and power (CHP) plants is also included in the tables in this section. However, more detailed analyses of CHP plant are set out in Chapter 6.

Generating companies

5.57 Following the restructuring of the electricity supply industry in 1990, the term "Major generating companies" was introduced into the electricity tables to describe the activities of the former nationalised industries and distinguish them from those of autogenerators and new independent companies set up to generate electricity. The activities of the autogenerators and the independent companies were classified under the heading "Other generating companies". In the 1994 Digest a new terminology was adopted. The new independent producers now make a significant contribution to electricity supply and all companies whose prime purpose is the generation of electricity are included under the heading "Major power producers" (or MPPs). The term "Other generators" ("Autogenerators" in the balance tables) is restricted to companies who produce electricity as part of their manufacturing or other commercial activities but whose main business is not electricity generation. "Other generators" also covers generation by energy services companies at power stations on an industrial or commercial site where the main purpose is the supply of electricity to that site, even if the energy service company is a subsidiary of a major power producer.

5.58 **Major power producers at the end of 1999 were:-**
AES Electric Ltd, Anglian Power Generation, Barking Power Ltd, BNFL Magnox, Coolkeeragh Power Ltd, Corby Power Ltd, Derwent Co-generation Ltd, Edison Mission Energy Ltd, Fellside Heat and Power Ltd, Fibrogen Ltd, Fibropower Ltd, Fibrothetford Ltd, Fife Power Ltd, Humber Power Ltd, Lakeland Power Ltd, Medway Power Ltd., Midlands Power (UK) Ltd, National Power plc, NIGEN, Nuclear Electric, Peterborough Power Ltd, PowerGen plc, Premier Power Ltd, Regional Power Generators Ltd, Rocksavage Power Company Ltd, Sita Tyre Recycling Ltd, Scottish Power plc, Scottish and Southern Energy plc, Seabank Power Ltd, SELCHP Ltd, South Western Electricity, Sutton Bridge Power Ltd, Teesside Power Ltd, TXU Europe Power Ltd.

Types of station

5.59 The various types of station identified in the tables of this chapter are as follows:

Conventional steam stations are stations which generate electricity by burning fossil fuels to convert water into steam, which then powers steam turbines.

Nuclear stations are also steam stations but the heat needed to produce the steam comes from nuclear fission.

Gas turbines use pressurised combustion gases from fuel burned in one or more combustion chambers to turn a series of bladed fan wheels and rotate the shaft on which they are mounted. This then drives the generator. The fuel burnt is usually natural gas or gas oil.

Combined cycle gas turbine (CCGT) stations combine gas turbines and steam turbines connected to one or more electrical generators in the same plant. This enables electricity to be produced at higher efficiencies than is otherwise possible when either gas or steam turbines are used in isolation. The gas turbine (usually fuelled by natural gas or oil) produces mechanical power (to drive the generator) and waste heat. The hot exhaust gases (waste heat) are fed to a boiler, where steam is raised at pressure to drive a conventional steam turbine which is also connected to an electrical generator.

Natural flow hydro-electric stations use natural water flows to turn turbines.

Pumped storage hydro-electric stations use electricity to pump water into a high level reservoir. This water is then released to generate electricity at peak times. Where the reservoir is open, some natural flow electricity is also generated by the stations; this is included with natural flow generation. As electricity is used in the pumping process, pumped storage stations are net consumers of electricity.

Other stations include wind turbines and stations burning fuels such as landfill gas, sewage sludge and waste.

Public distribution system

5.60 This comprises the grids in England and Wales, Scotland and Northern Ireland.

Sectors used for sales/consumption

5.61 The various sectors used for sales and consumption analyses are standardised across all chapters of the 2000 Digest. For definitions of the sectors see the Chapter 1 paragraphs 1.74 to 1.78 and Annex A paragraphs A.30 to A.41.

Declared net capability and declared net capacity

5.62 Declared net capability is the maximum power available for export from a power station on a continuous basis minus any power imported by the station from the network to run its own plant. It represents the nominal maximum capability of a generating set to supply electricity to consumers. The registered capacity of a generating set differs from declared net capability in that, for registered capacity, not all power consumed by the plant is subtracted from the normal full load capacity, only the MW consumed by the generating set through its transformer when generating at its normal full load capacity.

5.63 Declared net capacity is used to measure the maximum power available from generating stations that use renewable resources. For wind and tidal power a factor is applied to declared net capability to take account of the intermittent nature of the energy source (eg 0.43 for wind and 0.33 for tidal).

Load factors

5.64 The following definitions are used in Table 5.9:

Maximum load - Twice the largest number of units supplied in any consecutive thirty minutes commencing or terminating at the hour.

Simultaneous maximum load met - The maximum load on the grid at any one time. It is measured by the sum of the maximum load met in England and Wales and the loads met at the same time by companies in other parts of the United Kingdom.

Plant load factor - The average hourly quantity of electricity supplied during the year, expressed as a percentage of the average output capability at the beginning and the end of year.

System load factor - The average hourly quantity of electricity available during the year expressed as a percentage of the maximum demand nearest the end of the year or early the following year.

Thermal efficiency

5.65 Thermal efficiency is the efficiency with which heat energy contained in fuel is converted into electrical energy. It is calculated for fossil fuel burning stations by expressing electricity generated as a percentage of the total energy content of the fuel consumed (based on average gross calorific values). For nuclear stations it is calculated using the quantity of heat released as a result of fission of the nuclear fuel inside the reactor. The efficiency of CHP systems is discussed separately in Chapter 6, paragraphs 6.25 and 6.36. Efficiencies based on gross calorific value of the fuel (sometimes referred to as higher heating values or HHV) are lower than the efficiencies based on net calorific value (or lower heating value LHV). The difference between HHV and LHV is due to the energy associated with the latent heat of the evaporation of water products from the steam cycle which cannot be recovered and put to economic use.

Period covered

5.66 Figures for the major power producers relate to periods of 52 weeks as follows:-

Year	52 weeks ended
1993	2 January 1994
1994	1 January 1995
1995	31 December 1995
1996	29 December 1996
1997	28 December 1997
	53 weeks ended
1998	3 January 1999
	52 weeks ended
1999	2 January 2000

5.67 Figures for industrial and transport undertakings relate to years ended 31 December, except for the iron and steel industry where figures relate to the following 52 week periods:-

Year	52 weeks ended
1993	1 January 1994
1994	31 December 1994
1995	30 December 1995
1996	28 December 1996
1997	27 December 1997
	53 weeks ended
1998	2 January 1999
	52 weeks ended
1999	1 January 2000

5.68 Statistical years that contain 53 weeks are adjusted to 52 week equivalents by taking 5/6ths of the 6-week December period values.

5.69 Figures for financial year 1995/96 in Table 5.9 relate to year ended 31 March 1996.

Monthly and quarterly data

5.70 Monthly and quarterly data on fuel use, electricity generation and supply and electricity availability and consumption are provided in DTI's monthly statistical bulletin *Energy Trends*. Monthly data on fuel used in electricity generation by major power producers are given in Table 19 of the bulletin, and quarterly data on fuel used by all generating companies in Table 18. Quarterly data on electricity generation and supply by all generators are given in Table 20 and monthly data on supplies by type of plant and type of fuel are given in Table 22. Monthly data on availability and consumption of electricity by the main sectors of the economy are given in Table 23, while electricity supplied by other generating companies is provided quarterly in Table 21. See Annex F for more information about *Energy Trends*.

Data collection

5.71 For Major Power Producers, as defined in paragraph 5.58 the data in these tables are obtained from the results of an annual DTI inquiry sent to each company covering generating capacity, fuel use, generation, sales and distribution of electricity.

5.72 Another annual inquiry is sent to regional electricity companies, to Northern Ireland Electricity and to other licensed suppliers of electricity to establish electricity sales and electricity distributed by these companies.

5.73 Companies that generate electricity mainly for their own use (known as autogenerators or autoproducers - see paragraph 5.57, above) are covered by an annual inquiry commissioned by DTI but carried out by the Office for National Statistics from their Newport offices. However, this inquiry covers only generators with capacities greater than 250 kWe and DTI estimates fuel use and electricity generation for smaller companies, unless they have combined heat and power plants in which case they are included in the estimates made by ETSU described in Chapter 6. There are two areas of autogeneration that are covered by direct data collection by DTI, mainly because the return contains additional energy information needed by the Department. These are the Iron and Steel industry, and generation on behalf of London Underground.

Losses and statistical differences

5.74 Statistical differences are included in Tables 5.1, 5.2 and 5.3. These arise because data collected on production and supply do not match exactly with data collected on sales or consumption. One of the reasons for this is that some of the data are based on different calendars as described in paragraphs 5.66 and 5.67, above. Sales data based on calendar years will always include more electricity consumption than the slightly shorter statistical year of exactly 52 weeks.

5.75 Of the losses shown in the commodity balance for electricity of 28,300 GWh in 1999, it is estimated that about 5,300 GWh (1½ per cent of electricity available) were lost from the high voltage transmission system of the National Grid and 21,500 GWh (6 per cent) between the grid supply points (the gateways to the public supply system's distribution network) and customers' meters. The balance is accounted for by theft and meter fraud, accounting differences and calendar differences (as described in paragraph 5.74, above).

5.76 Care should be exercised in interpreting the figures for individual industries in the commodity balance tables. As new suppliers have entered the market and companies have moved between suppliers, it has not been possible to ensure consistent classification between and within industry sectors and across years. The breakdown of final consumption includes some estimated data. For about 5 per cent of consumption of electricity supplied by the public distribution system the sector figures are partially estimated.

Contact: *Mike Janes (Statistician)*
                mike.janes@dti.gsi.gov.uk
                020-7215 5186

                Joe Ewins
                joe.ewins@dti.gsi.gov.uk
                020-7215 5190

5.1 Commodity balances 1997 to 1999

Electricity

GWh

	1997	1998	1999
Total electricity			
Supply			
Production	346,965r	360,390r	363,896
Other sources(1)	1,486	1,624	2,902
Imports	16,615	12,630	14,507
Exports	-41	-162	-263
Marine bunkers	-	-	-
Stock change (2)	-	-	-
Transfers	-	-	-
Total supply	**365,025r**	**374,482r**	**381,042**
Statistical difference (3)	-1,721r	+1,192r	+1,574
Total demand	**366,746r**	**373,290r**	**379,468**
Transformation	-	-	-
Electricity generation	-	-	-
Major power producers	-	-	-
Autogenerators	-	-	-
Petroleum refineries	-	-	-
Coke manufacture	-	-	-
Blast furnaces	-	-	-
Patent fuel manufacture	-	-	-
Other	-	-	-
Energy industry use	**28,519r**	**29,677r**	**30,817**
Electricity generation	16,503r	17,686r	17,456
Oil and gas extraction	674	537	408
Petroleum refineries	5,284r	5,192r	5,351
Coal extraction	-	-	-
Coke manufacture	1,528	1,342	1,358
Blast furnaces	904	921	948
Patent fuel manufacture	-	-	-
Pumped storage	2,477	2,594	3,774
Other	1,149	1,405	1,522
Losses	**28,670**	**27,969**	**28,298**
Final consumption	**309,557**	**315,644r**	**320,353**
Industry	**104,914r**	**105,547r**	**108,515**
Unclassified	-	-	-
Iron and steel	9,649r	9,699r	9,899
Non-ferrous metals	5,261r	5,698r	5,895
Mineral products	7,097r	7,051r	7,217
Chemicals	19,384r	19,493r	19,851
Mechanical engineering etc	8,333r	8,552r	8,845
Electrical engineering etc	6,113r	5,996r	6,006
Vehicles	5,642r	5,595r	5,616
Food, beverages etc	11,568r	11,765r	12,003
Textiles, leather, etc	3,736r	3,666r	3,751
Paper, printing etc	10,781r	10,651r	10,909
Other industries	15,803r	15,842r	17,095
Construction	1,547	1,539	1,428
Transport	**8,406r**	**8,372r**	**8,459**
Air	-	-	-
Rail (4)	2,700	2,700	2,700
Road	-	-	-
National navigation	-	-	-
Pipelines	-	-	-
Other	**196,237r**	**201,725r**	**203,379**
Domestic	104,455	109,610	110,408
Public administration	21,688r	21,891r	22,686
Commercial	66,286	66,352	66,448
Agriculture	3,808r	3,872r	3,837
Miscellaneous	-	-	-
Non energy use	-	-	-

Electricity

	1997	1998	GWh 1999
Electricity production			
Total production *(5)*	**346,965r**	**360,390r**	**363,896**
Primary electricity			
Major power producers	**101,483r**	**104,497r**	**100,703**
Nuclear	98,146	100,140	96,281
Large scale hydro *(5)*	3,337r	4,357r	4,422
Small scale hydro	-	-	-
Wind	-	-	-
Autogenerators	**1,499r**	**1,757r**	**1,828**
Nuclear	-	-	-
Large scale hydro	668r	674r	698
Small scale hydro	164r	206r	232
Wind	667r	877r	898
Secondary electricity			
Major power producers	**221,004r**	**229,024r**	**233,537**
Coal	114,968r	118,595r	102,129
Oil	5,267r	3,445r	2,769
Gas	100,330r	106,338r	127,993
Renewables	439r	646r	646
Other	-	-	-
Autogenerators	**22,979r**	**25,112r**	**27,828**
Coal	4,659r	4,267r	3,943
Oil	2,587r	3,044r	2,782
Gas	9,185r	10,786r	13,372
Renewables	2,153r	2,592r	3,341
Other	4,395r	4,423r	4,390
Primary and secondary production *(6)*			
Nuclear	98,146	100,140	96,281
Hydro	4,169r	5,237r	5,352
Wind	667r	877r	898
Coal	119,627r	122,862r	106,072
Oil	7,854r	6,489r	5,551
Gas	109,515r	117,124r	141,365
Other renewables	2,592r	3,238r	3,987
Other	4,395r	4,423r	4,390
Total production	**346,965r**	**360,390r**	**363,896**

(1) Pumped storage production.
(2) Stock fall (+), stock rise (-).
(3) Total supply minus total demand.
(4) See paragraph 5.15.
(5) Excludes pumped storage production.
(6) These figures are the same as the electricity generated figures in Table 5.5 except that they exclude pumped storage production. Table 5.5 shows that electricity used on works is deducted to obtain electricity supplied. It is electricity supplied that is used to produce Chart 5.3 showing each fuel's share of electricity output (see paragraph 5.28).

5.2 Electricity supply and consumption

					GWh
	1995	1996	1997	1998	1999
Supply					
Production	335,872r	348,991r	346,965r	360,390r	363,896
Other sources (1)	1,552	1,556	1,486	1,624	2,902
Imports	16,336	16,792r	16,615	12,630	14,507
Exports	-23	-37	-41	-162	-263
Total supply	**353,737r**	**367,302r**	**365,025r**	**374,482r**	**381,042**
Statistical difference (2)	**+1,445r**	**+3,794r**	**-1,721r**	**+1,192r**	**+1,574**
Total demand	**352,292r**	**365,508r**	**366,746r**	**373,290r**	**379,468**
Transformation	-	-	-	-	-
Energy industry use	**28,991r**	**30,185r**	**28,519r**	**29,677r**	**30,817**
Electricity generation	17,515r	17,704r	16,503r	17,686r	17,456
Oil and gas extraction	745	772	674	537	408
Petroleum refineries	4,659r	5,237r	5,284r	5,192r	5,351
Coal and coke	2,000	1,793	1,528	1,342	1,358
Blast furnaces	780	864	904	921	948
Pumped storage	2,282	2,430	2,477	2,594	3,774
Other	1,010	1,385	1,149	1,405	1,522
Losses	**28,579**	**27,501**	**28,670**	**27,969**	**28,298**
Final consumption	**294,722r**	**305,822r**	**309,557r**	**315,644r**	**320,353**
Industry	**100,656r**	**103,115r**	**104,914r**	**105,547r**	**108,515**
Unclassified	-	-	-	-	-
Iron and steel	9,743r	10,214r	9,649r	9,699r	9,899
Non-ferrous metals	5,698r	5,581r	5,261r	5,698r	5,895
Mineral products	7,112r	7,292r	7,097r	7,051r	7,217
Chemicals	19,102r	19,329r	19,384r	19,493r	19,851
Mechanical engineering etc	8,195r	7,804r	8,333r	8,552r	8,845
Electrical engineering etc	5,204r	5,574r	6,113r	5,996r	6,006
Vehicles	6,238r	6,549r	5,642r	5,595r	5,616
Food, beverages etc	10,351r	11,276r	11,568r	11,765r	12,003
Textiles, leather, etc	3,325r	3,151r	3,736r	3,666r	3,751
Paper, printing etc	9,528r	9,516r	10,781r	10,651r	10,909
Other industries	14,360r	14,985r	15,803r	15,842r	17,095
Construction	1,800r	1,844r	1,547	1,539	1,428
Transport	**8,125r**	**8,118r**	**8,406r**	**8,372r**	**8,459**
Other	**185,941r**	**194,589r**	**196,237r**	**201,725r**	**203,379**
Domestic	102,210	107,513	104,455	109,610	110,408
Public administration	22,338r	22,939r	21,688r	21,891r	22,686
Commercial	57,601	60,315	66,286	66,352	66,448
Agriculture	3,792	3,822	3,808r	3,872r	3,837
Miscellaneous	-	-	-	-	-
Non energy use	-	-	-	-	-

(1) Pumped storage production.
(2) Total supply minus total demand.

5.3 Commodity balances 1997 to 1999
Public distribution system and other generators

	1997			1998			1999		
	Public distribution system	Other generators	Total	Public distribution system	Other generators	Total	Public distribution system	Other generators	Total
Supply									
Major power producers	322,487	-	322,487	333,521	-	333,521	334,240	-	334,240
Other generators	-	24,478	24,478	-	26,869	26,869	-	29,656	29,656
Other sources (1)	1,486	-	1,486	1,624	-	1,624	2,902	-	2,902
Imports	16,615	-	16,615	12,630	-	12,630	14,507	-	14,507
Exports	-41	-	-41	-162	-	-162	-263	-	-263
Transfers	+5,506	-5,506	-	+5,919	-5,919	-	+6,792	-6,792	-
Total supply	**346,053**	**18,972**	**365,025**	**353,532**	**20,950**	**374,482**	**358,178**	**22,864**	**381,042**
Statistical difference (2)	**-1,648**	**-73**	**-1,721**	**+1,329**	**-137**	**+1,192**	**+1,919**	**-345**	**+1,574**
Total demand	**347,701**	**19,045**	**366,746**	**352,203**	**21,087**	**373,290**	**356,259**	**23,209**	**379,468**
Transformation	**-**	**-**	**-**	**-**	**-**	**-**	**-**	**-**	
Energy industry use	**22,953**	**5,566**	**28,519**	**23,588**	**6,089**	**29,677**	**24,059**	**6,758**	**30,817**
Electricity generation	15,404	1,099	16,503	16,471	1,215	17,686	16,182	1,274	17,456
Oil and gas extraction	674	-	674	537	-	537	408	-	408
Petroleum refineries	2,057	3,227	5,284	1,646	3,546	5,192	1,214	4,137	5,351
Coke manufacture	1,336	192	1,528	1,150	192	1,342	1,174	184	1,358
Blast furnaces	-	904	904	-	921	921	-	948	948
Pumped storage	2,477	-	2,477	2,594	-	2,594	3,774	-	3,774
Other fuel industries	1,005	144	1,149	1,190	215	1,405	1,307	215	1,522
Losses	**28,558**	**112**	**28,670**	**27,854**	**115**	**27,969**	**28,195**	**103**	**28,298**
Final consumption	**296,190**	**13,367**	**309,557**	**300,761**	**14,883**	**315,644**	**304,005**	**16,348**	**320,353**
Industry	**93,494**	**11,420**	**104,914**	**92,941**	**12,606**	**105,547**	**94,471**	**14,044**	**108,515**
Iron and steel	8,391	1,258	9,649	8,282	1,417	9,699	8,486	1,413	9,899
Non-ferrous metals	3,461	1,800	5,261	3,718	1,980	5,698	3,767	2,128	5,895
Mineral products	7,037	60	7,097	6,996	55	7,051	7,159	58	7,217
Chemicals	14,771	4,613	19,384	14,514	4,979	19,493	14,874	4,977	19,851
Mechanical engineering etc	8,287	46	8,333	8,479	73	8,552	8,772	73	8,845
Electrical engineering etc	6,111	2	6,113	5,994	2	5,996	5,992	14	6,006
Vehicles	5,531	111	5,642	5,487	108	5,595	5,512	104	5,616
Food, beverages etc	10,730	838	11,568	10,652	1,113	11,765	10,019	1,984	12,003
Textiles, leather, etc	3,725	11	3,736	3,653	13	3,666	3,749	2	3,751
Paper, printing etc	8,374	2,407	10,781	8,230	2,421	10,651	8,165	2,744	10,909
Other industries	15,560	243	15,803	15,422	420	15,842	16,568	527	17,095
Construction	1,516	31	1,547	1,514	25	1,539	1,408	20	1,428
Transport	**6,881**	**1,525**	**8,406**	**6,790**	**1,582**	**8,372**	**6,862**	**1,597**	**8,459**
Of which National Rail (3)	2,700	-	2,700	2,700	-	2,700	2,700	-	2,700
Other	**195,815**	**422**	**196,237**	**201,030**	**695**	**201,725**	**202,672**	**707**	**203,379**
Domestic	104,455	-	104,455	109,610	-	109,610	110,408	-	110,408
Standard	58,123	-	58,123	60,741	-	60,741	60,908	-	60,908
Economy 7 and other off-peak	30,177	-	30,177	32,570	-	32,570	31,941	-	31,941
Prepayment (standard)	10,488	-	10,488	11,314	-	11,314	11,779	-	11,779
Prepayment (off-peak)	4,913	-	4,913	4,447	-	4,447	5,060	-	5,060
Sales under any other arrangement	754	-	754	538	-	538	720	-	720
Public administration	21,268	420	21,688	21,198	693	21,891	21,981	705	22,686
Public lighting (4)	2,679	-	2,679	2,182	-	2,182	1,962	-	1,962
Other public sector	18,589	420	19,009	19,016	693	19,709	20,019	705	20,724
Commercial	66,286	-	66,286	66,352	-	66,352	66,448	-	66,448
Shops	28,307	-	28,307	29,612	-	29,612	31,026	-	31,026
Offices	18,948	-	18,948	18,451	-	18,451	18,720	-	18,720
Hotels	7,370	-	7,370	7,761	-	7,761	7,623	-	7,623
Combined domestic/ commercial premises	2,023	-	2,023	1,669	-	1,669	1,671	-	1,671
Post and telecommunications	6,117	-	6,117	5,049	-	5,049	5,284	-	5,284
Unclassified	3,521	-	3,521	3,810	-	3,810	2,124	-	2,124
Agriculture	3,806	2	3,808	3,870	2	3,872	3,835	2	3,837

(1) Pumped storage production.
(2) Total supply minus total demand.
(3) See paragraph 5.15.

(4) Sales for public lighting purposes are increasingly covered by wider contracts that cannot distinguish the public lighting element.

5.4 Fuel used in generation[1]

	Unit	1995	1996	1997	1998	1999
					Original units of measurement	
Major power producers (2)						
Coal	M tonnes	57.94	53.42	45.34	46.63	38.99
Oil (3)	"	3.44	3.18	1.38r	0.77	0.79
Gas	GWh	133,065	176,702	223,691r	236,994r	281,306
Other generators (2)						
Transport undertakings:						
Gas	GWh	2,518	2,722	2,595r	2,555r	2,496
Undertakings in industrial and commercial sectors:						
Coal	M tonnes	2.10r	2.03r	1.93r	1.89r	1.49
Oil (4)	"	0.94r	0.90r	0.85r	0.91r	0.88
Gas (5)	GWh	18,810r	20,508r	23,904r	27,081r	29,610
					Million tonnes of oil equivalent	
Major power producers (2)						
Coal		35.021	32.400	27.713r	28.982r	24.146
Oil (3)		3.129	3.018	1.377r	0.783r	0.817
Gas		11.442	15.194	19.233r	20.378r	24.187
Nuclear		21.250	22.180	22.993	23.589r	22.680
Hydro (natural flow) (6)		0.340r	0.250	0.287r	0.375r	0.380
Other renewables (6)		0.133	0.131	0.139	0.147	0.193
Net imports		1.403	1.441	1.425r	1.072	1.225
Total major power producers (2)		**72.718r**	**74.614r**	**73.167r**	**75.326r**	**73.628**
Of which: conventional steam stations		40.273r	38.112r	31.874r	32.832r	27.983
combined cycle gas turbine stations		9.317	12.496	16.536r	17.311r	21.167
Other generators (2)						
Transport undertakings:						
Gas		0.217	0.234	0.223r	0.220r	0.215
Undertakings in industrial and commercial sectors:						
Coal		1.270r	1.230r	1.318r	1.264r	0.944
Oil (4)		1.020r	0.980r	0.923r	0.973r	0.943
Gas (5)		1.617r	1.763r	2.055r	2.329r	2.546
Hydro (natural flow) (6)		0.060r	0.050r	0.072r	0.076r	0.080
Other renewables (6)		0.668	0.735	0.901	1.140	1.318
Other fuels (7)		0.984r	1.032r	1.026	1.031r	1.040
Total other generators (2)		**5.836r**	**6.024r**	**6.518**	**7.033r**	**7.086**
All generating companies						
Coal		36.291r	33.630r	29.031r	30.246r	25.090
Oil (3)(4)		4.149r	3.998r	2.300r	1.756r	1.760
Gas (5)		13.276r	17.191r	21.511r	22.927r	26.948
Nuclear		21.250	22.180	22.993	23.589r	22.680
Hydro (natural flow) (6)		0.400r	0.300r	0.359r	0.451r	0.460
Other renewables (6)		0.801	0.866	1.040	1.287	1.511
Other fuels (7)		0.984r	1.032r	1.026r	1.031r	1.040
Net imports		1.403	1.441	1.425r	1.072	1.225
Total all generating companies		**78.554r**	**80.638r**	**79.685r**	**82.359r**	**80.714**

(1) A monthly update of fuel used in electricity generation by major power producers is given in Table 19 of Energy Trends, and a quarterly update of fuel used in electricity generation by all generating companies is given in Table 18 of Energy Trends.
(2) See paragraphs 5.57 and 5.58 for information on companies covered.
(3) Includes Orimulsion, oil used in gas turbine and diesel plant, and oil used for lighting up coal fired boilers.
(4) Includes refinery gas.
(5) Includes colliery methane.
(6) Renewable sources which are included under hydro and other renewables in this table are shown separately in Table 7.6 of Chapter 7.
(7) Main fuels included are coke oven gas, blast furnace gas, and waste products from chemical processes.

5.5 Electricity fuel use, generation and supply

GWh

| | Thermal sources | | | | | | | Non-thermal sources | | | Total |
	Coal	Oil	Gas	Nuclear	Renew-ables (1)	Other (3)	Total	Hydro-natural flow	Hydro-pumped storage	Other (4)	All sources
1997											
Major power producers (2)											
Fuel used	322,302	16,015	223,691	267,428	1,617	-	831,053	3,337	1,486	-	835,876
Generation	114,968	5,267	100,330	98,146	439r	-	319,150	3,337	1,486	-	323,973r
Used on works	4,909	378	1,177	8,805	50	-	15,319	38	47	-	15,404
Supplied (gross)	110,059r	4,889	99,153	89,341	389r	-	303,831	3,299	1,439	-	308,569r
Used in pumping											2,477
Supplied (net)											306,092r
Other generators (2))											
Fuel used	15,328	10,734	26,499	-	9.816	11,937	74,314	832	-	667	75,813
Generation	4,659r	2,587r	9,185r	-	2,153r	4,395r	22,979	832r	-	667	24,478r
Used on works	237	191	306	-	153	203	1,090	9	-	-	1,099r
Supplied	4,422r	2,396r	8,879r	-	2,000r	4,192r	21,889	823	-	667	23,379r
All generating companies											
Fuel used	337,630	26,749	250,190	267,428	11,433	11,937	905,367	4,169	1,486	667	911,689
Generation	119,627r	7,854r	109,515r	98,146r	2,592r	4,395r	342,129	4,169r	1,486	667	348,451r
Used on works	5,146	569	1,483	8,805	203	203	16,409	47	47	-	16,503r
Supplied (gross)	114,481r	7,285r	108,032r	89,341r	2,389r	4,192r	325,720	4,122r	1,439	667	331,948r
Used in pumping											2,477
Supplied (net)											329,471r
1998											
Major power producers (2)											
Fuel used	337,061	9,106	236,994	274,356	1,709	-	859,226	4,357	1,624	-	865,207
Generation	118,595	3,445	106,338	100,140	646r	-	329,164	4,357r	1,624	-	335,145r
Used on works	5,494	231	1,650	8,954	74	-	16,403	13	55	-	16,471
Supplied (gross)	113,101r	3,214	104,688	91,186	572r	-	312,761	4,344r	1,569	-	318,674r
Used in pumping											2,594
Supplied (net)											316,080r
Other generators (2)											
Fuel used	14,700	11,316	29,636	-	12,386	11,986	80,024	880	-	877	81,781
Generation	4,267r	3,044r	10,786r	-	2,592r	4,423r	25,112	880r	-	877	26,870r
Used on works	229	225	355	-	191	203	1,203	12	-	-	1,215r
Supplied	4,038r	2,819r	10,431r	-	2,401r	4,220r	23,909	868r	-	877	26,564r
All generating companies											
Fuel used	351,761	20,422	266,630	274,356	14,095	11,986	939,250	5,237	1,624	877	946,988
Generation	122,862r	6,489r	117,124r	100,140	3,238r	4,423r	354,276	5,237r	1,624	877	362,014r
Used on works	5,723	456	2,005	8,954	265	203	17,606	25	55	-	17,686r
Supplied (gross)	117,139r	6,033r	115,119r	91,186	2,973r	4,220r	336,670	5,212r	1,569	877	344,328r
Used in pumping											2,594
Supplied (net)											341,734r
1999											
Major power producers (2)											
Fuel used	280,818	9,505	281,306	263,784	2,244	-	837,657	4,422	2,902	-	844,981
Generation	102,129	2,769	127,993	96,281	646	-	329,818	4,422	2,902	-	337,142
Used on works	4,781	222	2,387	8,609	72	-	16,071	13	98	-	16,182
Supplied (gross)	97,348	2,547	125,606	87,672	574	-	313,747	4,409	2,804	-	320,960
Used in pumping											3,774
Supplied (net)											317,186
Other generators (2)											
Fuel used	10,978	10,968	32,106	-	14,433	12,094	80,579	930	-	898	82,407
Generation	3,943	2,782	13,372	-	3,341	4,390	27,828	930	-	898	29,656
Used on works	189	206	435	-	229	202	1,261	13	-	-	1,274
Supplied	3,754	2,576	12,937	-	3,112	4,188	26,567	917	-	898	28,382
All generating companies											
Fuel used	291,796	20,473	313,412	263,784	16,677	12,094	918,236	5,352	2,902	898	927,388
Generation	106,072	5,551	141,365	96,281	3,987	4,390	357,646	5,352	2,902	898	366,798
Used on works	4,970	428	2,822	8,609	301	202	17,332	26	98	-	17,456
Supplied (gross)	101,102	5,123	138,543	87,672	3,686	4,188	340,314	5,326	2,804	898	349,342
Used in pumping											3,774
Supplied (net)											345,568

(1) Thermal renewable sources are those included under biofuels in Chapter 7.

(2) See paragraphs 5.57 and 5.58 on companies covered.

(3) Other thermal sources include coke oven gas, blast furnace gas, and waste products from chemical processes.

(4) Other non-thermal sources include wind, and solar photovoltaics.

5.6 Electricity supply, electricity supplied (net), electricity available and electricity consumption

	GWh				
	1995	1996	1997	1998	1999
Total supply					
(as given in Tables 5.1 and 5.2)	353,737	367,302	365,025	374,482	381,042
less imports of electricity	-16,336	-16,792	-16,615	-12,630	-14,507
plus exports of electricity	+23	+37	+41	+162	+263
less electricity used in pumped storage	-2,282	-2,430	-2,477	-2,594	-3,774
less electricity used on works	-17,515	-17,704	-16,503	-17,686	-17,456
equals					
Electricity supplied (net)	317,627	330,413	329,471	341,734	345,568
(as given in Tables, 5.5, 5.11 and 5.12)					
Total supply					
(as given in Tables 5.1 and 5.2)	353,737	367,302	365,025	374,482	381,042
less electricity used in pumped storage	-2,282	-2,430	-2,477	-2,594	-3,774
less electricity used on works	-17,515	-17,704	-16,503	-17,686	-17,456
equals					
Electricity available	333,940	347,168	346,045	354,202	359,812
(as given in Tables 5.11)					
Final consumption					
(as given in Tables 5.2 and 5.3)	294,722	305,822	309,557	315,644	320,353
plus Iron and steel consumption counted as Energy industry use	+1,127	+1,202	+1,245	+1,266	+1,272
equals					
Final users	295,849	307,024	310,802	316,910	321,625
(as given in Tables 5.11)					

5.7 Plant capacity

					MW
	end-March 1996	end-Dec 1996	end-Dec 1997	end-Dec 1998	end-Dec 1999
Major power producers *(1)*					
Total declared net capability	**66,101**	**69,090**	**68,288**	**68,390**	**70,091**
Of which:					
Conventional steam stations:	38,242	38,230	37,395	35,081	35,427
Coal fired	27,494	25,796	25,796	25,324	25,581
Oil fired	5,214	3,989	4,069	2,829	2,829
Mixed or dual fired *(2)*	5,534	8,445	7,530	6,928r	7,017
Combined cycle gas turbine stations	9,034	12,052	12,252	14,638r	16,143
Nuclear stations	12,762	12,916	12,946	12,956	12,956
Gas turbines and oil engines	1,890	1,721	1,526	1,492	1,333
Hydro-electric stations:					
Natural flow	1,314	1,313	1,311	1,327	1,327
Pumped storage	2,788	2,788	2,788	2,788	2,788
Renewables other than hydro	71	70	70	108	117
Other generators *(1)*					
Total capacity of own generating plant *(3)*	**4,025r**	**4,181r**	**4,408r**	**4,763r**	**5,214**
Of which:					
Conventional steam stations *(4)*	3,234r	3,192r	3,223r	3,246r	3,334
Combined cycle gas turbine stations	343r	410r	551r	780r	1,052
Hydro-electric stations (natural flow)	118r	142r	145r	148r	150
Renewables other than hydro	330r	437r	489r	589r	678
All generating companies					
Total capacity *(3)*	**70,126r**	**73,271r**	**72,696r**	**73,153r**	**75,305**
Of which:					
Conventional steam stations *(4)*	41,476r	41,422r	40,618r	38,327r	38,761
Combined cycle gas turbine stations	9,377r	12,462r	12,803r	15,418r	17,195
Nuclear stations	12,762	12,916	12,946	12,956	12,956
Gas turbines and oil engines	1,890	1,721r	1,526r	1,492r	1,333
Hydro-electric stations:					
Natural flow	1,432r	1,455r	1,456r	1,475r	1,477
Pumped storage	2,788	2,788	2,788	2,788	2,788
Renewables other than hydro	401r	507r	559r	697r	795

(1) See paragraphs 5.57 and 5.58 for information on companies covered.
(2) Includes gas fired stations that are not Combined Cycle Gas Turbines.
(3) Capacity figures for other generators are all measured at the end of the calendar year.
(4) For other generators, conventional steam plant includes combined heat and power plants (electrical capacity only), but excludes combined cycle gas turbine plants, nuclear and hydro-electric stations and plants using renewable sources.

5.8 Capacity of other generators

MW

	end-December				
	1995	1996	1997	1998	1999
Capacity of own generating plant (1)					
Undertakings in industrial and commercial sector:					
Petroleum refineries	615r	648r	648r	671r	859
Iron and steel	359r	359r	359r	355r	355
Chemicals	921r	927r	1,025r	1,031r	1,082
Engineering and other metal trades	517r	506r	510r	509r	520
Food, drink and tobacco	207r	207r	217r	302r	392
Paper, printing and publishing	356r	369r	416r	468r	471
Other (2)	743r	859r	949r	1,144r	1,252
Total industrial and commercial sector	3,718r	3,875r	4,124r	4,480r	4,931
Undertakings in transport sector	306	306	283r	283r	283
Total other generators	**4,024r**	**4,181r**	**4,407r**	**4,763r**	**5,214**

(1) For combined heat and power plants the electrical capacity only is included. Further CHP capacity is included under major power producers in Table 5.7. A detailed analysis of CHP capacity is given in the tables of Chapter 6.

(2) Includes companies in the commercial sector.

5.9 Plant loads, demand and efficiency

Major power producers (1)

	Unit	1995/96	1996	1997	1998	1999
Simultaneous maximum load met (2)	MW	55,611	56,815	56,965	56,312	57,849
Plant load factor						
Conventional steam stations	Per cent	47.4	45.7r	38.4	40.2	36.5
Combined cycle gas turbine stations	"	71.2	71.0r	81.4	79,0r	83.6
Nuclear stations	"	73.7	76.3r	78.9	80.4	77.2
Gas turbines and oil engines	"	1.1	1.4	3.1	1.6	1.5
Hydro-electric stations:						
Natural flow	"	34.4	24.0r	28.7	36.6	37.9
Pumped storage	"	6.5	6.2	5.9	6.4	11.5
All plant	"	**52.2**	**52.3r**	**51.3**	**53.2**	**52.9**
System load factor	"	**65.4**	**65.5**	**65.4**	**67.2**	**66.1**
Thermal efficiency **(gross calorific value basis)**						
Combined cycle gas turbine stations	"	45.5	45.3	45.2	46.6	46.4
Conventional steam stations	"	36.2	36.6	36.5	36.2	36.5
Nuclear stations (3)	"	36.0	36.9	36.7	36.5	36.8

(1) See paragraphs 5.57 and 5.58 for information on companies covered.

(2) Data for 1996 to 1999 cover the years ending March 1997, March 1998, March 1999 and March 2000, respectively.

(3) All data are for calendar years.

5.10 Fuel input for electricity generation[1]

Million tonnes of oil equivalent

	Total all fuels	Coal	Oil (2)	Natural gas (3)	Electricity		Coke and breeze	Other fuels (4)
					Nuclear	Natural flow hydro		
1960	**37.30**	30.07	5.78	-	0.64	0.27	0.54	
1965	**51.72**	40.03	6.74	-	4.34	0.40	0.21	
1966	**54.41**	40.09	7.81	-	5.80	0.39	0.32	
1967	**55.05**	39.61	7.92	-	6.68	0.42	0.42	
1968	**58.22**	43.33	6.85	-	7.51	0.32	0.21	
1969	**62.04**	44.74	8.89	0.11	7.91	0.28	0.11	
1970	**63.84**	43.07	13.27	0.11	7.00	0.39	-	
1971	**66.46**	42.42	15.63	0.64	7.37	0.29	0.11	
1972	**68.37**	38.47	20.13	1.61	7.87	0.29	-	
1973	**70.93**	44.30	18.09	0.64	7.46	0.33	0.11	
1974	**69.01**	38.71	18.41	2.46	8.97	0.35	0.11	
1975	**66.25**	41.85	13.70	2.14	8.12	0.33	0.11	
1976	**66.97**	44.49	10.92	1.61	9.56	0.39	-	
1977	**69.32**	45.71	11.35	1.28	10.64	0.34	-	
1978	**69.64**	46.05	12.31	0.86	9.96	0.35	0.11	
1979	**72.80**	50.10	11.45	0.54	10.23	0.37	0.11	
1980	**69.46**	51.01	7.67	0.42	9.91	0.34	0.11	
1981	**65.98**	49.64	5.46	0.21	10.18	0.38	0.11	
1982	**65.98**	46.75	6.64	0.21	11.88	0.39	0.11	
1983	**66.37**	47.16	5.14	0.21	13.47	0.39	-	
1984	**69.18**	31.07	22.80	0.42	14.50	0.39	-	
1985	**71.54**	42.81	11.35	0.54	16.50	0.34	-	
1986	**70.46**	47.91	6.51	0.18	15.44	0.41	-	
1987 (5)	**70.50**	50.37	5.14	0.19	14.44	0.36	-	
1987 (5)	**74.31**	51.58	6.30	0.91	14.44	0.36	-	0.72
1988	**75.57**	49.83	7.01	0.97	16.57	0.42	-	0.77
1989	**75.27**	48.59	7.11	0.54	17.74	0.41	-	0.88
1990	**76.34**	49.84	8.40	0.56	16.26	0.44	-	0.84
1991	**76.87**	49.98	7.56	0.57	17.43	0.39	-	0.94
1992	**76.57**	46.94	8.07	1.54	18.45	0.46	-	1.09
1993	**75.40**	39.61	5.78	7.04	21.58	0.37	-	1.02
1994	**74.01r**	37.10r	4.11r	10.10r	21.20	0.44	-	1.06r
1995	**77.15r**	36.29r	4.15r	13.27r	21.25	0.40r	-	1.79r
1996	**79.20r**	33.63r	4.00r	17.19r	22.18	0.30r	-	1.90r
1997	**78.26r**	29.03r	2.30r	21.51r	22.99	0.36r	-	2.07r
1998	**81.29r**	30.25r	1.76r	22.93r	23.59r	0.45r	-	2.32r
1999	**79.49**	25.09	1.76	26.95	22.68	0.46	-	2.55

(1) Fuel inputs have been calculated on an energy supplied basis - see explanatory notes at paragraph 5.28.
(2) Includes oil used in gas turbine and diesel plant or for lighting up coal fired boilers, Orimulsion, and (from 1987) refinery gas.
(3) Includes colliery methane from 1987 onwards.
(4) Main fuels included are coke oven gas, blast furnace gas, waste products from chemical processes, refuse derived fuels and other renewable sources including wind.
(5) Data for all generating companies are only available from 1987 onwards, and the figures for 1987 to 1989 include a high degree of estimation. Before 1987 the data are for major power producers, transport undertakings and industrial hydro and nuclear stations only.

5.11 Electricity supply, availability and consumption

TWh

	Electricity supplied (net)	Purchases from other producers	Net imports (1)	Electricity available	Losses in transmission etc (2)	Electricity consumption					
						Total	Fuel industries	Final users			
								Industrial	Domestic	Other (3)	Total
1960	116.96	0.15	-	117.11	12.07	105.04	6.25	44.51	33.68	20.60	98.79
1965	169.57	0.21	0.10	169.88	15.17	154.71	7.55	58.54	57.23	31.39	147.16
1966	175.38	0.22	0.35	175.95	15.30	160.65	7.66	60.11	59.81	33.07	152.99
1967	181.14	0.21	0.16	181.51	15.76	165.75	7.75	60.91	62.35	34.74	158.00
1968	193.56	0.20	0.73	194.49	16.29	178.20	7.68	66.15	66.66	37.71	170.52
1969	206.37	0.20	0.58	207.15	17.29	189.86	6.88	70.34	72.19	40.45	182.98
1970	215.76	0.19	0.55	216.50	17.50	199.00	6.59	72.99	77.04	42.38	192.41
1971	222.92	0.53	0.12	223.57	19.01	204.56	6.60	73.43	80.67	43.86	197.96
1972	229.45	0.53	0.48	230.46	18.91	211.55	6.37	73.16	86.89	45.13	205.18
1973	245.42	0.59	0.06	246.07	19.59	226.48	6.67	80.07	91.30	48.44	219.81
1974	237.21	0.60	0.05	237.86	18.22	219.64	6.12	75.81	92.63	45.08	213.52
1975	237.76	0.70	0.08	238.54	19.47	219.07	6.29	75.36	89.21	48.21	212.78
1976	240.22	0.61	-0.10	240.73	18.73	222.00	6.39	80.84	85.12	49.65	215.61
1977	246.82	0.74	-	247.56	20.76	226.80	6.41	82.06	85.90	52.43	220.39
1978	252.65	0.66	-0.08	253.23	21.81	231.42	6.52	84.00	85.80	55.10	224.90
1979	264.34	0.63	-	264.97	22.97	242.00	6.78	87.55	89.67	58.00	235.22
1980	252.02	0.61	-	252.63	21.53	231.11	6.86	79.73	86.11	58.41	224.25
1981	246.60	0.74	-	247.34	20.13	227.21	6.86	77.03	84.44	58.88	220.35
1982	242.48	0.82	-	243.30	20.48	222.82	6.81	73.91	82.79	59.31	216.01
1983	246.15	1.15	-	247.30	21.21	226.09	6.69	74.17	82.95	62.28	219.40
1984	251.47	0.55	-	252.02	21.06	230.96	6.64	78.64	83.90	61.78	224.32
1985	263.56	0.92	-	264.48	22.63	241.85	7.76	79.53	88.23	66.33	234.09
1986(4)	266.81	1.10	4.26	272.17	22.83	249.34	7.68	79.93	91.83	69.68	241.44
1986(4)	278.48	-	4.26	282.73	22.91	259.82	9.51	88.80	91.83	69.68	250.31
1987	279.71	-	11.64	291.34	22.96	268.38	9.49	93.14	93.25	72.50	258.89
1988	285.71	-	12.83	297.85	23.35	274.50	9.16	97.14	92.36	75.84	265.34
1989	291.75	-	12.63	304.38	24.98	279.40	9.00	99.42	92.27	78.71	270.40
1990	297.50	-	11.94	309.41	24.99	284.42	9.99	100.64	93.79	80.00	274.43
1991	300.65	-	16.41	317.06	26.22	290.84	9.79	99.57	98.10	83.38	281.05
1992	298.55	-	16.69	315.24	23.79	291.45	9.98	95.28	99.48	86.71	281.47
1993	301.87	-	16.72	318.59r	22.84	295.75	9.62	96.84	100.46	88.83	286.13
1994	306.94r	-	16.89	323.83r	31.00r	292.83r	7.52	96.12r	101.41	87.78	285.31
1995	317.33r	-	16.61r	333.94r	30.02r	303.92r	8.07r	101.78r	102.21	91.86r	295.85
1996	330.41r	-	16.76r	347.17r	31.30r	315.87r	8.85r	104.32r	107.51	95.19r	307.02
1997	329.47r	-	16.57r	346.05r	26.95r	319.10r	8.29r	106.16r	104.46	100.19r	310.80
1998	341.73r	-	12.47	354.20r	29.16r	325.04r	8.13r	106.81r	109.61	100.49r	316.91
1999	345.57	-	14.24	359.81	29.87	329.94	8.32	109.79	110.41	101.43	321.63

(1) Net transfers between the Irish Republic and Northern Ireland (ceased in 1981 and recommenced in 1995) and between France and England (from 1986).

(2) Losses on the public distribution system (grid system and local networks) and other differences between data collected on sales and data collected on availability.

(3) Public administration, transport, agricultural and commercial sectors.

(4) Data for all generating companies are only available from 1986 onwards. Before 1986 the data are for major power producers, transport undertakings and industrial hydro and nuclear stations only.

5.12 Electricity generated and supplied 1960 to 1999

GWh

			Major power producers							
	Electricity generated	Electricity used on works	Electricity supplied (gross) (1)				Hydro (3)		Electricity used in pumping at pumped storage stations	Electricity supplied (net) (4)
			Total	Conventional thermal and other (2)	CCGT	Nuclear	Natural flow	Pumped storage		
1960	..	..	115,506	110,898	-	2,079	2,529	-	-	115,506
1961	..	..	124,182	118,597	-	2,399	3,186	-	-	124,182
1962	..	..	137,677	130,960	-	3,477	3,240		212	137,465
1963	..	..	149,340	140,332	-	5,949	3,059		488	148,852
1964	..	..	157,242	146,207	-	7,629	3,406		589	156,653
1965	..	..	168,774	150,711	-	14,145	3,918		616	168,158
1966	187,858	13,047	174,811	152,063	-	18,894	3,854		672	174,139
1967	194,596	13,610	180,986	154,197	-	21,754	4,180	855	1,117	179,869
1968	208,132	14,627	193,505	165,222	-	24,477	2,957	849	1,124	192,381
1969	222,256	15,462	206,794	177,109	-	25,771	2,794	1,120	1,513	205,281
1970	232,378	16,429	215,949	188,175	-	22,805	3,846	1,123	1,487	214,462
1971	240,080	17,143	222,937	195,181	-	24,013	2,835	908	1,209	221,728
1972	246,843	17,439	229,404	200,048	-	25,639	2,847	870	1,184	228,220
1973	263,140	18,157	244,983	216,796	-	24,310	3,214	663	882	244,101
1974	254,688	17,763	236,925	203,478	-	29,232	3,520	695	896	236,029
1975	255,084	17,136	237,948	207,159	-	26,463	3,186	1,140	1,430	236,518
1976	258,656	17,962	240,694	205,048	-	31,153	3,128	1,365	1,729	238,965
1977	265,649	18,468	247,181	207,904	-	34,660	3,320	1,297	1,608	245,573
1978	270,677	17,907	252,770	215,761	-	32,462	3,378	1,169	1,429	251,341
1979	283,186	18,744	264,442	226,329	-	33,335	3,617	1,161	1,424	263,018
1980	269,945	17,765	252,180	215,418	-	32,291	3,298	1,173	1,453	250,727
1981	263,658	16,983	246,675	208,589	-	33,191	3,906	989	1,196	245,479
1982	259,410	16,940	242,470	198,822	-	38,721	3,873	1,054	1,272	241,198
1983	264,589	17,380	247,209	197,600	-	43,911	3,882	1,816	2,337	244,872
1984	270,471	17,643	252,828	200,240	-	47,256	3,358	1,974	2,613	250,215
1985	284,712	18,903	265,809	205,906	-	53,767	3,435	2,701	3,494	262,315
1986	287,330	18,819	268,511	210,452	-	51,843	4,087	2,129	2,993	265,518
1987	287,701	18,740	268,961	215,290	-	48,205	3,460	2,006	2,804	266,157
1988	293,100	19,341	273,759	211,932	-	55,642	4,160	2,025	2,888	270,871
1989	297,890	19,315	278,575	209,169	-	63,602	3,992	1,812	2,572	276,003
1990	302,936	18,632	284,304	219,364	-	58,664	4,384	1,892	2,626	281,678
1991	305,704	19,142	286,562	218,260	309	62,761	3,767	1,465	2,109	284,453
1992	303,715	19,157	284,558	206,245	2,964	69,135	4,579	1,635	2,257	282,301
1993	305,433r	18,170	287,264	178,773r	22,611	80,979	3,513	1,388	1,948	285,316
1994	307,476r	16,696	290,780r	168,321r	36,815	79,962	4,265	1,417	2,051	288,729r
1995	315,510r	16,510	299,000r	164,324r	48,525	80,598	4,051	1,502	2,282	296,718r
1996	327,843r	16,674	311,169r	155,475r	65,604	85,820	2,763	1,507	2,430	308,739r
1997	323,973r	15,404	308,569r	127,808r	86,682	89,341	3,299	1,439	2,477	306,092r
1998	335,145r	16,471	318,674r	128,570r	93,005	91,186	4,344r	1,569	2,594	316,080r
1999	337,142	16,182	320,960	113,312	112,763	87,672	4,409	2,804	3,774	317,186

(1) Electricity generated less electricity used on works.
(2) Includes electricity supplied by gas turbines and oil engines. From 1988 also includes electricity produced by plants using renewable sources.

144

5.12 Electricity generated and supplied 1960 to 1999 (continued)

GWh

Other generators				All generating companies						
Electricity supplied (gross) (1)				Electricity supplied (gross)						
Total	Conventional thermal and other (2)	CCGT	Hydro natural flow	Total	Conventional thermal and other (2)	CCGT	Nuclear	Hydro	Electricity supplied (net) (4)	
13,565	12,974	-	591	129,071	123,872	-	2,079	3,120	129,071	1960
13,332	12,679	-	653	137,514	131,276	-	2,399	3,839	137,514	1961
13,331	12,661	-	670	151,008	143,621	-	3,477	3,910	150,796	1962
13,800	13,214	-	586	163,140	153,546	-	5,949	3,645	162,652	1963
14,298	13,699	-	599	171,540	159,906	-	7,629	4,005	170,951	1964
14,610	13,919	-	691	183,384	164,630	-	14,145	4,609	182,768	1965
14,070	13,426	-	644	188,881	165,489	-	18,894	4,498	188,209	1966
14,139	13,447	-	692	195,125	167,644	-	21,754	5,727	194,008	1967
14,540	13,945	-	595	208,045	179,167	-	24,477	4,401	206,921	1968
15,631	15,180	-	451	222,425	192,289	-	25,771	4,365	220,912	1969
15,674	14,996	-	678	231,623	203,171	-	22,805	5,647	230,136	1970
15,388	14,837	-	551	238,325	210,018	-	24,013	4,294	237,116	1971
15,746	15,175	-	571	245,150	215,223	-	25,639	4,288	243,966	1972
17,655	17,008	-	647	262,638	233,804	-	24,310	4,524	261,756	1973
17,222	16,660	-	562	254,147	220,138	-	29,232	4,777	253,251	1974
15,766	15,175	-	591	253,714	222,334	-	26,463	4,917	252,284	1975
17,013	16,414	-	599	257,707	221,462	-	31,153	5,902	255,978	1976
16,434	15,848	-	586	263,615	223,752	-	34,660	5,203	262,007	1977
16,034	15,387	-	647	268,804	231,148	-	32,462	5,194	267,375	1978
15,720	15,062	-	658	280,162	241,391	-	33,335	5,436	278,738	1979
14,132	13,509	-	623	266,312	228,927	-	32,291	5,094	264,859	1980
13,264	12,801	-	463	259,939	221,390	-	33,191	5,358	258,743	1981
12,613	11,943	-	670	255,083	210,765	-	38,721	5,597	253,811	1982
12,152	11,486	-	666	259,361	209,086	-	43,911	6,364	257,024	1983
11,319	10,685	-	634	264,148	210,925	-	47,256	5,966	261,535	1984
12,112	11,467	-	645	277,922	217,373	-	53,767	6,781	274,427	1985
12,957	12,278	-	679	281,469	222,730	-	51,843	6,895	278,476	1986
13,551	12,831	-	720	282,512	228,121	-	48,205	6,186	279,708	1987
14,840	14,085	-	755	288,599	226,018	-	55,642	6,939	285,711	1988
15,747	15,007	-	740	294,323	224,179	-	63,602	6,542	291,751	1989
15,824	14,738	280	806	300,128	234,101	280	58,664	7,082	297,502	1990
16,202	15,065	298	839	302,764	233,325	607	62,761	6,071	300,654	1991
16,246	15,020	394	832	300,804	221,265	3,358	69,135	7,046	298,547	1992
16,552	15,196	584	772	303,816r	193,969r	23,195	80,979	5,673	301,868	1993
18,207r	15,972r	1,486r	769	308,987r	184,293r	38,281r	79,962	6,451	306,936r	1994
20,909r	18,076r	2,100r	733r	319,909r	182,400r	50,625r	80,598	6,286r	317,627r	1995
21,674r	18,715r	2,374r	585r	332,843r	174,190r	67,978r	85,820	4,855r	330,413r	1996
23,379r	19,293r	3,263r	823r	331,948r	147,101r	89,945r	89,341	5,561r	329,471r	1997
25,654r	20,239r	4,547r	868r	344,328r	148,809r	97,552r	91,186	6,781r	341,734r	1998
28,382	21,258	6,207	917	349,342	134,570	118,970	87,672	8,130	345,568	1999

(3) Separate pumped storage figures are not available for the years 1962 to 1966.
(4) Electricity supplied (gross) less electricity used in pumping at pumped storage stations

Chapter 6
Combined heat and power

Introduction

6.1   This chapter sets out the contribution made by Combined Heat and Power (CHP) to the United Kingdom's energy requirements.   There have been some changes to the methodology this year and the data have been considerably improved, in preparation for the Government's new Quality Assurance programme for CHP (CHPQA).   Data for previous years have been revised in the light of the improved information on certain schemes obtained in 1999. An outline of the changes in methodology and data are included at the end of this chapter in paragraphs 6.12 and 6.28 to 6.33.

6.2   CHP is the simultaneous generation of usable heat and power (usually electricity) in a single process. Useful outputs can be more varied: increasingly, heat is being used to drive absorption chilling, and in some cases power can be mechanical power eg to drive a compressor.   The term CHP is synonymous with cogeneration and total energy, which are terms often used in other Member States of the European Community and in the United States.   CHP uses a variety of fuels and technologies across a wide range of sites, and scheme sizes.   The basic elements of a CHP plant comprise one or more prime movers (a reciprocating engine, gas turbine, or steam turbine) driving electrical generators, or other machinery, where the steam or hot water generated in the process is utilised via suitable heat recovery equipment for use either in industrial processes, or in community heating and space heating.

6.3   Whereas an electricity-only plant is typically large, and connected at very high voltage to the grid transmission system, a CHP plant is typically much smaller, attached to a site which consumes the heat and power produced (or a large proportion of it), is sized to make use of the available heat, and connected to the lower voltage distribution system (i.e. embedded).   Not only is CHP more efficient through utilisation of heat, it also avoids significant transmission and distribution losses, and can provide important network services such as black start, improvements to power quality, and the ability to continue to supply the site if the grid goes down.

6.4   CHP usually displaces boiler plant and electricity-only plant using a range of fuels and technologies. CHP typically achieves a 25 to 35 per cent reduction in primary energy usage compared with electricity-only generation and heat-only boilers. This can allow the host organisation to make substantial savings in costs and emissions where there is a suitable heat load.

6.5   There are four principal types of CHP system, steam turbine, gas turbine, combined cycle systems and reciprocating engines.   Each of these is defined in paragraph 6.34 below.

6.6   57 per cent of CHP electrical capacity is now either simple cycle or combined cycle gas turbine.   In terms of heat capacity, steam turbines and gas turbines (in simple or combined cycle) each account for just under half of capacity, with reciprocating engines accounting for only a few per cent.   In terms of numbers of installations, the dominant technology is reciprocating engines, though the average size of these installations is less than 0.5 MWe.   The majority of these schemes have been installed in the last ten years.

6.7   A small number of large schemes exist primarily for electricity generation, and are not sized to make use of all the useful heat. For such schemes the electricity generation and capacity qualifying as 'Good Quality' CHP is scaled back to recognise the reduced environmental benefits based on the actual heat supplied, and in line with the proposals for CHP Equivalent Generation Limit and CHP Equivalent Capability in the Government's forthcoming CHPQA programme[1].   The 'Good Quality' portion of the capacity, fuel use and output is counted here.   The electricity-only generation capacity and output of these schemes is included in the total electricity generating capacity in Chapter 5 of this Digest.   In total, there is 1,088 MWe of CHP which has been scaled back, whilst the total electrical capacity of these plants is 7,350 MWe.

Government policy towards CHP

6.8   The Government has confirmed its new target of at least 10,000 MWe of CHP by 2010 as part of its Climate Change Programme.   Capacity increased by 354 MWe during 1999 and the previous target of 5,000 MWe is now expected to be met during 2001. Chart 6.1 shows the recent growth in capacity.

6.9   There are a number of important issues, which affect the planning framework, or the cost effectiveness of CHP:

---

[1]   See the document "Quality Assurance for Combined Heat and Power, the Government's decisions following consultation" on www.chpqa.com or telephone CHPQA on 0800 585794

I. The Chancellor announced in November 1999 that 'Good Quality' CHP would be exempt from the Climate Change Levy (CCL). This applies to fuels used; heat supplied; and electricity generated where electricity is used on-site or exported to direct customers.

II. The treatment of electricity exports under the Climate Change Levy is in part determined by supply licensing and license exemptions. The DTI consulted on possible changes to facilitate CHP and other embedded generation[2] at the end of 1999 and will make a further announcement shortly.

III. The major industrial sectors may be in a position to negotiate an 80 per cent reduction in the levy if they reach an agreement with the Government on, and achieve, energy efficiency targets. CHP accounts for up to half the cost-effective savings in these sectors, and will benefit from such agreements.

IV. At present, businesses are able to claim capital allowances on investment in plant and machinery. Small and medium size enterprises can claim 40 per cent, and larger businesses can claim 25 per cent in the accounting period in which the expenditure is incurred. In both cases the balance of unrelieved expenditure is carried forward and written off at 25 per cent per annum on a reducing balance. As part of the forthcoming Climate Change Levy package, Enhanced Capital Allowances (ECAs) will give 100 per cent allowance in the first year on plant and machinery that meets energy efficiency criteria (including 'Good Quality' CHP). Even though the new scheme is tax neutral, there is a valuable cash flow benefit for the purchaser over and above any other savings that can be made.

V. Following a review of business rating, the Government has announced that plant and machinery which is 'Good Quality' CHP will be exempt from the assessment of the rateable value for a site or business.

VI. Assessment and Certification under CHPQA will determine 'Good Quality'. The Government has consulted on how CHPQA will operate, and has announced its decisions[3]. Schemes must apply before 2nd October 2000 if they are to achieve CCL exemption from April 2001, - the date of introduction of the levy. The programme is voluntary. The Standard Guidance Notes and Self-Assessment Forms are available from CHPQA (0800 585794 and www.chpqa.com).

VII. Support for those wanting to invest in CHP or improve their CHP scheme will be available from the new 'CHPclub', recently launched by the Environment Minister. Details are available from www.chpclub.com.

VIII. The Government has operated a 'stricter consents policy' since December 1997. Under this policy, new gas-fired generation would not normally be consented, although an exception was made for 'Good Quality' CHP projects. This policy was temporary whilst the distortions in the electricity market were removed. The Government has announced that the policy will be lifted as soon as the new electricity trading arrangements are in place, which it expects to be in the autumn of 2000. In making his announcement, the Secretary of State for Trade and Industry said "The Government strongly supports CHP and we will expect developers to be able to show that they have explored opportunities to use it. We shall discuss with developers information that needs to be submitted as part of notifications under section 14 of the Energy Act 1976 and applications under section 36 of the Electricity Act 1989."

IX. The New Electricity Trading Arrangements, (see VIII, above), which are designed to remove the distortions which in the Electricity Pool have acted to the disadvantage of flexible plant.

X. Under the Utilities Bill, the Government will set future new Energy Efficiency Standards of Performance (EESoPs) to be achieved by energy suppliers (both gas and electricity). The Government is consulting on the size of these obligations, which are expected to be substantially greater than current EESoPs set by the Regulator. CHP will be one option to help meet these obligations.

6.10 The Department of the Environment, Transport and the Regions (DETR) will be consulting on the Government's strategy to achieve its target of at least 10,000 MWe capacity by 2010.

6.11 A detailed study of the CHP potential in industry, commerce and the public sector was carried out for DETR by ETSU (a division of AEA Technology) in July 1997[4]. The cost effective potential for CHP in industry, commerce and the public sector was put at between 10,000 and 17,000 MWe, depending on assumptions made about future energy prices, users' required rate of return on investment, and other factors. There is an additional

2 Electricity (Class Exemption from the Requirements for a Licence) Order 1997 Proposed Amendments

3 Hansard, 17 May 2000 : Column: 137W and www.chpqa.com

4 Assessment of CHP potential ETSU - reference RYCA 18501113 available from ETSU, telephone 0800 585794.

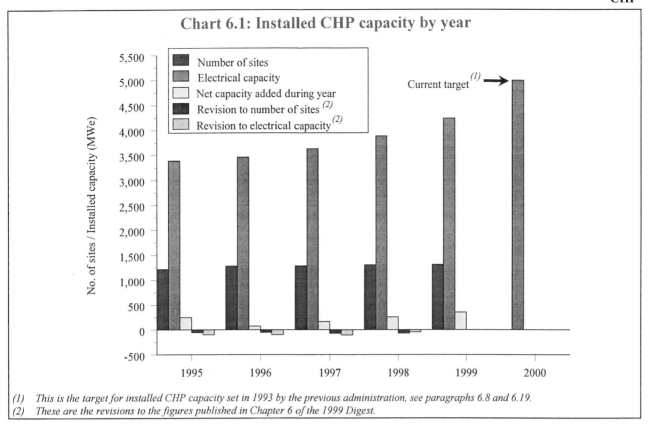

Chart 6.1: Installed CHP capacity by year

Legend:
- Number of sites
- Electrical capacity
- Net capacity added during year
- Revision to number of sites [2]
- Revision to electrical capacity [2]

Current target [1] →

(Y-axis: No. of sites / Installed capacity (MWe), ranging from -500 to 5,500)
(X-axis years: 1995, 1996, 1997, 1998, 1999, 2000)

(1) This is the target for installed CHP capacity set in 1993 by the previous administration, see paragraphs 6.8 and 6.19.
(2) These are the revisions to the figures published in Chapter 6 of the 1999 Digest.

2,000 to 3,000 MWe of cost effective potential for CHP in community heating, making the total economic potential between 12,000 and 20,000 MWe.

Improved survey quality

6.12   The improved survey outlined in paragraphs 6.28 to 6.33 below has resulted in the following changes.

- The methodologies developed for the CHPQA programme have been used here, and all 5 years recalculated on that basis. Specifically: larger schemes that are not 'Good Quality' for their entire capacity and output have been scaled back using CHP Equivalent Generation Limit and CHP Equivalent Capability; and mechanical drives powered by steam from CHP have been treated as equivalent electrical capacity, not as heat load. This reduces heat capacity and heat supplied and increased equivalent electrical capacity and electricity supplied.

- Additional capacity embedded in refinery sites has been included.

- The number of schemes below 250 kWe has been revised down since investigations revealed a number of scheme closures and duplicates in the database. These sites are not normally included in the annual survey. However, other small schemes may emerge through applications to CHPQA.

- The improved survey, and subsequent verification procedures, resulted in better data on sites that were audited. The survey achieved returns from 61 per cent of sites over 250 kWe, accounting for

66 per cent of all electricity generated by CHP. Data for other sites were interpolated from historical data for these sites and audited data for similar sites.

- The Government now has a much better understanding of what is installed and how it operates. Average operational data for sites audited are shown in Table 6D for each technology.

Progress towards the Government's targets

6.13   Based on these revised data, Chart 6.1 shows the change in installed CHP capacity over the last five years. Installed capacity at the end of 1999 stands at 4,239 MWe. Over the last decade, capacity has more than doubled, representing an average growth rate over the period of 7 per cent per annum. Growth over the last year has been just over 9 per cent, or 354 MWe.

New and retired capacity in 1999

6.14   Growth in any one year depends on the rate of retirement of old plant as well as the rate at which new plant is built. A particular theme has been the restructuring of some industrial sectors, particularly paper, chemicals, and oil refineries. Several sites have been targeted for closure, with production concentrated on other sites. Additional capacity in 1999 came from 26 new and 9 upgraded sites, less 20 retirements. In capacity terms, there was 305 MWe of new schemes, and a further 109 MWe of upgrades to existing schemes, making a total of 414 MWe of new

Table 6A: A summary of the recent development of CHP

	Unit	1995	1996	1997	1998	1999
Number of sites		1,220	1,282	1,287	1,307	1,313
Net number of sites added during year		53	62	5	20	6
Electrical capacity	MWe	3,390	3,463	3,628	3,885	4,239
Net capacity added during year		249	73	165	257	354
Heat capacity	MWth	15,511	15,263	15,112	15,340	15,093
Heat to power ratio (1)		4.58	4.41	4.17	3.95	3.56
Fuel input	GWh	107,012	106,843	106,778	108,613	109,894
Electricity generation	GWh	14,826	15,524	16,499	18,329	20,213
Heat generation	GWh	57,901	57,402	55,152	55,682	54,062
Overall efficiency (2)	Per cent	68.0	68.3	67.1	68.1	67.6
Load factor	Per cent	49.9	51.2	51.9	53.9	54.4

(1) Heat to power ratios are calculated from the installed capacity of the systems and not from the heat and electricity actually produced. Heat to power ratios for the individual types of installation can be derived from the information given in Tables 6.5 and 6.7.

(2) These are measured in gross terms; overall net efficiencies are some 5 percentage points higher.

capacity. 60 MWe of capacity was decommissioned either because a site closed, or the site no longer generates electricity. There are a number of important and dynamic trends in new installations:

- The growth in CHP installed under energy services arrangements continues. Around two-thirds of the CHP capacity installed over the last decade (worth around £800 million) has been installed under energy services arrangements. However this proportion is rising, and around 90 per cent of capacity coming on stream in the next few years may be operated under an energy services arrangement.

- Increasingly, with market liberalisation, there is a stronger incentive to design schemes to maximise electricity generation on a given heat load, and to export electricity to adjacent customers or to the local public electricity supplier (PES).

- Whilst much of the new capacity was in traditional applications such as refineries and fuel processing (an additional 188 MWe) chemicals (an additional 70 MWe), sugar (and additional 56 MWe) and paper (7.8 MWe), there were new schemes in sectors with less history of CHP including the emergence of a substantial new market for CHP in horticulture (an additional 41.5 MWe this year).

- Around 57 per cent of capacity is now gas turbine based, with around two-thirds of this in combined cycle mode. Continued growth in combined cycle gas turbine installations came predominantly from the upgrading of existing steam turbine based CHP schemes. In addition, there were a number of new smaller schemes employing simple cycle gas turbines and reciprocating engines.

6.15 As a consequence of the increase in capacity in 1999, and in the nature of that capacity - the gradual switch from steam turbine to gas turbine - the heat to power ratio continues to decline (Table 6A). Over the last five years, heat output has remained broadly stable, whilst electricity generated has increased by 66 per cent. The load factor increased to 54.4 per cent in 1999, equivalent to an average of around 4,765 hours of full load operation per annum. This figure is deceptive – it hides a range of schemes which were only commissioned towards the end of 1999, and effectively hardly ran at all, to schemes which ran all but a few hours of the year when they were down for maintenance. In addition, several large, partial schemes count towards the capacity figure. These schemes have the capability to operate in CHP mode, but several supply much lower amounts of heat than they are capable of. In these circumstances, much of the fuel used, and electricity generated figures are not counted as CHP, but counted as electricity-only plant, (see Chapter 5 of this Digest). The average load factor has increased by 10 per cent over the last 5 years, reflecting the fact that many new large installations on industrial sites - which dominate capacity - are designed to operate with 80 to 100 per cent load factors.

Schemes under development in 1999

6.16 Industrial restructuring continues with some sites in the paper and chemicals industries and at oil refineries likely to close, while schemes on other sites in these sectors are to be upgraded.

6.17 A large amount of capacity has recently received consent under Section 36 of the Electricity Act 1989, and clearance under Section 14 of the Energy Act 1976, and is currently under construction.

- In total 1,250 MWe has been consented during the period of the stricter consents policy. Of these:

- Nine schemes totalling 500 MWe are expected to be commissioned in 2000.

- A further six, adding up to another 550 MWe are expected in 2001. However, not all of these schemes will be additional, and may replace or upgrade some existing (steam turbine) capacity.

- At 31 May 2000, applications for another 1,000 MWe of further CHP capacity were with the Department of Trade and Industry for consideration. Even if all these schemes were built, they are unlikely to be commissioned before 2003. Again, not all of these schemes will be additional, and may replace or upgrade some existing (steam turbine) capacity.

6.18   The capacities shown in Table 6B are the total consented capacities. However, under normal heat supply conditions, where a scheme includes a pass-out steam turbine, electrical output declines marginally as heat supplied to site increases. The capacity counted towards the CHP target will be the electrical capacity under normal supply conditions.

Table 6B:	CHP schemes under development	Total consented capacity MWe
Likely to be commissioned in 2000		
Brunner Mond, Cheshire		140
BP Chemicals Saltend [1]		100
Rolls-Royce, Derby		62
Bridgewater Paper, Ellesmere Port		58
Hickson and Welch, Castleford		56
BPB Paper board Mugiemoss, Aberdeen		38
BP Grangemouth, Stirlingshire,		20
Slough Heat and Power, [2]		12
Aylesford Newsprint [3]		10
Total		**496**
Likely to be commissioned in 2001		
Shotton Paper, Flintshire,		214
BP Grangemouth, Stirlingshire,		130
Michelin, Stoke,		58
Sappi Mill, Blackburn,		57
EniChem, Southampton,		46
Enron/Phillips Petroleum, Seal Sands		44
Total new in 2001 before retirements		**549**

(1)   1,200 MWe overall capacity, of which ETSU estimate around 100 MWe is 'Good Quality' CHP.
(2)   12 MWe in addition to an existing 90 MWe.
(3)   Final phase of a larger scheme.

6.19   Capacity over the next few years is projected to reach around 4,600 MWe in 2000, 5,100 MWe in 2001 (fulfilling the Government's 5,000 MWe target) and around 5,800 MWe in 2002, net of new, upgraded and retired schemes. However, final decisions to proceed with new capacity are dependent on market conditions, especially the price that developers can achieve from exported power relative to the price of gas. Currently a number of schemes have been delayed, given the uncertain state of the market.

Installed capacity in 1999

6.20   The current installed capacity displays the following characteristics:

- CHP installations are dominated by schemes with an installed electrical capacity of less than 100 kWe (49 per cent of sites), and between 100 kWe and 999 kWe (34 per cent of sites), as shown in Table 6C. However, schemes larger than 10 MWe represent 81 per cent of the total electrical capacity.

- In terms of heat capacity the steam turbines continue to predominate with 55 per cent of total heat capacity. In terms of numbers, the largest segment is for reciprocating engines, though the average size of these installations is less than 1 MWe. Table 6.5 provides data on electrical capacity and Table 6.7 provides data on heat capacity for each type of CHP installation.

- The electricity generated and heat generated from each type of plant reflects the heat to power ratio. The average heat to power ratios in 1999 ranged from 6.41:1 for back pressure steam turbines to 1.58:1 for reciprocating engines (with an all installations average of 3.56:1), so back pressure steam turbines, generated 28½ per cent of heat but only 13 per cent of the electricity, while reciprocating engines generated 5½ per cent of the heat but 10½ per cent of the electricity. Table 6.4 provides data on electricity generated and Table 6.6 provides data on heat generated for each type of CHP installation.

6.21   Table 6A gives a summary of the overall CHP market. The electricity generated by CHP schemes was 20,214 GWh. This represents about 5½ per cent of the total electricity generated in the UK in 1999. Across all industry (including the fuel industries other than electricity generation) CHP's electrical output accounted for 15 per cent of electricity consumption, 3 percentage points more than in 1995. For the commercial and public sectors the figures are 2 per cent and 1½ per cent respectively. CHP plants supplied 54,062 GWh of heat, which is 2.7 times the amount of electricity generated, although the heat to power ratio itself (which is measured in terms of capacities) is higher than this figure at 3.56:1 and on a falling trend.

Table 6C: CHP installations by capacity size ranges in 1999

Electrical capacity size range	Number of installations	Share of total (per cent)	Total electricity capacity (MWe)	Share of total (per cent)
Less than 100 kWe	638	48.6	35.6	0.8
100 kWe - 999 kWe	447	34.0	113.4	2.7
1 MWe - 9.9 MWe	147	11.2	635.8	15.0
Greater than 10 MWe	81	6.2	3,454.0	81.5
Total	**1,313**	**100.0**	**4,238.9**	**100.0**

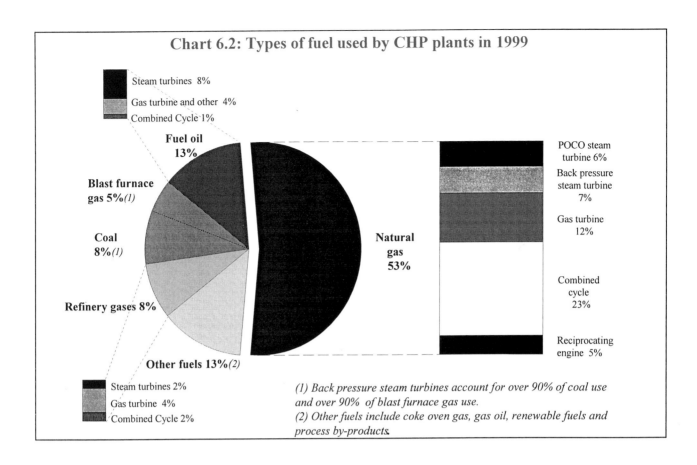

Chart 6.2: Types of fuel used by CHP plants in 1999

Steam turbines 8%
Gas turbine and other 4%
Combined Cycle 1%

Fuel oil 13%

Blast furnace gas 5%(1)

Coal 8%(1)

Refinery gases 8%

Natural gas 53%

Other fuels 13%(2)

Steam turbines 2%
Gas turbine 4%
Combined Cycle 2%

POCO steam turbine 6%

Back pressure steam turbine 7%

Gas turbine 12%

Combined cycle 23%

Reciprocating engine 5%

(1) Back pressure steam turbines account for over 90% of coal use and over 90% of blast furnace gas use.
(2) Other fuels include coke oven gas, gas oil, renewable fuels and process by-products.

Fuel used by types of CHP installation

6.22   Table 6.2 shows the fuel used to generate electricity and heat in CHP plants, (see paragraph 6.36, below for an explanation of the convention for dividing fuel between electricity and heat production). Table 6.3 gives the overall fuel used by types of CHP installation (which are explained in paragraph 6.34). Total fuel use is summarised in Chart 6.2. In 1999 natural gas dominates with 53 per cent of total fuel use, and this is set to increase as most new CHP schemes are fired by natural gas. CHP schemes accounted for 6 per cent of UK gas consumption in 1999 (see Table 4.3). Fuel oil use fluctuates annually, and depends on the extent of interruptions to gas supplies (which in turn depends on the weather, overall gas demand, and the extent to which individual gas contracts allow for interruptions in order to cope with peak demand).

6.23   Non-conventional fuels (gases, liquids or solids that are by-products or waste products from industrial processes, or are renewable fuels) account for nearly a quarter of fuel used in CHP. These are fuels which are not part of the mainstream electricity generating industry, and which would otherwise be flared or disposed of by some means. These fuels (with the exception of some waste gases) will always be burnt in boilers feeding steam turbines. In almost all cases, the technical nature of the combustion process (lower calorific value of the fuel, high moisture content of the fuel, the need to maintain certain combustion conditions to ensure complete disposal, etc) will always imply a lower efficiency. However, given that the use of such fuels avoids the use of fossil fuels, and since they would have to have been disposed of in some way in any case, the use of these fuels for the generation of power and heat is, in environmental terms, at very low cost.

CHP capacity, output and total fuel use by sector

6.24   Table 6.8 gives data on all operational schemes by economic sector. A definition of the sectors used in this table can be found in Chapter 1, paragraph 1.75 and Table 1E.

Table 6D: A summary of performance of CHP schemes surveyed in 1999 [1]

	Typical operating hours per annum (Full load equivalent)	Average electrical efficiency (% GCV)	Average heat efficiency (% GCV)	Average overall efficiency (% GCV)
Main prime mover in CHP plant				
Back pressure steam turbine	5,007	12.1	54.4	66.5
Pass out condensing steam turbine	4,139	14.2	54.3	68.5
Combined Cycle	3,823	20.9	46.9	67.8
Gas Turbine	5,676	23.4	41.0	64.4
Reciprocating engine	4,617	29.9	31.1	61.1

(1) Includes CHP element of partial CHP schemes.

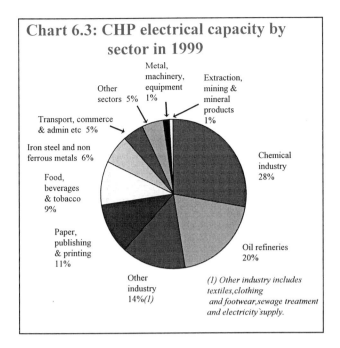

Chart 6.3: CHP electrical capacity by sector in 1999

Metal, machinery, equipment 1%
Other sectors 5%
Extraction, mining & mineral products 1%
Transport, commerce & admin etc 5%
Iron steel and non ferrous metals 6%
Food, beverages & tobacco 9%
Paper, publishing & printing 11%
Chemical industry 28%
Oil refineries 20%
Other industry 14%(1)

(1) Other industry includes textiles,clothing and footwear,sewage treatment and electricity`supply.

- Four industrial sectors account for two thirds of the CHP electrical capacity - chemicals (28 per cent of capacity), oil refineries (20 per cent), paper, publishing and printing (11 per cent) and food beverages and tobacco (9 per cent) as Chart 6.3 shows.

- In terms of numbers of sites the largest growth in CHP schemes has been in the "other" category which encompasses community heating, horticulture and leisure. This increase is primarily in the small reciprocating engine driven plant.

- Table 6.9 gives details of the quantities of fuels used in each sector.

CHP performance
6.25   Detailed performance statistics are available for 67 per cent of the surveyed CHP capacity in 1999. These are summarised in Table 6D. Electrical efficiency of reciprocating engines is typically 28-33 per cent (GCV). If only high-grade heat is recovered overall efficiency is around 50 per cent. If low-grade heat is recovered as well, efficiency exceeds 70 per cent. New combined cycle gas turbine CHP schemes have a large gas turbine and small steam turbine and may achieve electrical efficiencies of 45 per cent (GCV). Many existing CCGT CHP schemes have a small gas turbine, and large steam turbine. Here steam generated in the waste heat recovery boiler of the gas turbine is frequently supplemented with steam from additional boilers. The overall electrical efficiency of such schemes can be as low as 12 per cent. In some cases, these additional boilers use non-conventional fuels. In other cases, these schemes represent prime candidates for upgrade. In aggregate, however, the electrical efficiency of CCGT CHP is currently lower than gas turbine based CHP which achieve electrical efficiencies in the range 22 to 36 per cent. Steam turbines achieve electrical efficiencies in the range 5 to 20 per cent.

- 952 sites (7 per cent of capacity) are in the commercial, public and residential sectors, and 361 sites (93 per cent of capacity) are in the industrial and other sectors.

- Table 6E gives a summary of the 952 schemes installed in the commercial, public sector and residential buildings. The installed electrical capacity was 309 MWe in 1999 with an average heat to power ratio of 1.77:1. The vast majority of these schemes (98 per cent) are based on reciprocating engines fuelled with natural gas, though the larger schemes, where higher grade heat or steam is needed rather than just hot water for space heating, use gas turbines. In terms of electrical capacity the largest sectors are health (32 per cent) and residential heating (22 per cent). In terms of number of sites the market is dominated by three sectors: health, hotels and leisure. The schemes in Table 6E form a major part of the "Transport, commerce and administration" and "Other" sectors in Tables 6.8 and 6.9.

Table 6E: Number and capacity of CHP schemes installed in buildings by sector in 1999

Sector	Number of sites	Electrical capacity (MWe)	Heat capacity (MWth)
Leisure	327	26.41	45.07
Hotel	260	28.01	46.30
Health	204	99.47	175.17
Residential group heating	47	67.91	146.19
Offices	40	16.73	24.06
Education	28	1.85	2.98
Universities	17	21.67	49.02
Government estate	15	10.54	17.91
Retail	6	5.57	4.23
Other (1)	8	31.07	37.93
Total	**952**	**309.23**	**548.87**

(1) Other includes agriculture; airports; domestic buildings.

Emissions savings

6.26　The calculation of emissions savings from CHP is important, given the substantial contribution of CHP to the Climate Change Programme, but complex, given that CHP displaces a variety of fuels, technologies and sizes of plant. An annual article in Energy Trends has for several years included a methodology for deriving the CO_2 savings attributable to CHP. The next such article will outline a methodology for calculating CO_2 savings both from existing capacity (which displaces existing fossil plant) and future CHP capacity (which displaces a mix of current and future fossil-fired plant). In summary, the current stock of CHP saves around 4 MtC against the current average fossil mix.

Technical notes and definitions

6.27　These notes and definitions are in addition to the technical notes and definitions covering all fuels and energy as a whole in Chapter 1, paragraphs 1.45 to 1.78.

Changes in methodology and data for 1999

6.28　The data are summarised from the results of a long term project being undertaken by ETSU (part of AEA Technology (AEAT) Environment) on behalf of the Department of Trade and Industry (DTI), the Department of the Environment, Transport and the Regions (DETR) and the Statistical Office of the European Communities (Eurostat). Data are included for CHP schemes installed in all sectors of the UK economy.

6.29　The project continues to be overseen by a Steering Group that comprises officials from the DTI, DETR, Office for National Statistics (ONS), the Office of Gas and Electricity Markets (OFGEM) and the Combined Heat and Power Association (CHPA), all of whom have an interest in the collection of information on CHP schemes and the promotion of the wider use of CHP throughout the UK economy.

6.30　Data for 1999 were collected by a survey carried out by the Office for National Statistics (ONS) between December 1999 and March 2000 of companies (other than major power producers) who generate their own electricity, either in CHP schemes or in electricity-only schemes. Information on the CHP plant included in the major power producers category comes from surveys conducted by DTI as part of the electricity statistics system. DTI also collects directly fuel use relating to the iron and steel industry.

6.31　The survey forms were followed up with deeper verification and analysis by a number of Consultants, financed under the Government's Energy Efficiency Best Practice Programme, in preparation for the CHPQA programme. This checking included obtaining a description of installed equipment, verifying electrical efficiency and heat recovery, and the extent and quality of metering on sites.

6.32　In all, 327 questionnaires were sent out in the survey to companies that were believed to have CHP or electricity-only capacity greater than 250 kWe. 201 forms were returned, of which 12 schemes were no longer generating, and 24 others were not yet generating or were electricity-only schemes. Comprehensive and verified data were obtained on 165 operational CHP schemes. The survey covers a large proportion of the CHP capacity but a small minority of sites.

6.33　To provide operational data for CHP schemes that are smaller than 250 kWe, a variety of monitoring studies was used to obtain an average for the heat to power ratio for schemes, the load factor, the annual electrical and heat outputs and the fuel consumption for schemes.

Definitions of schemes

6.34　There are four principal types of CHP system, These are:

Steam turbine, where steam at high pressure is generated in a boiler. In **back pressure steam turbine systems**, the steam passes through a turbine before being exhausted from the turbine at the required pressure for the site. In **pass-out condensing steam turbine systems**, a proportion of the steam used by the turbine is extracted at an intermediate pressure from the turbine with the exhaust being condensed. (Condensing steam turbines without passout and which do not utilise steam are not included in these statistics as they are not CHP.) The boilers used in steam turbine schemes can burn a wide variety of fuels including coal, gas, oil, and waste-derived fuels. With the exception of waste-fired schemes, steam turbine plant has often been in service for several decades. Steam turbine plants capable of supplying useful steam have electrical efficiencies of between 5 and 15 per cent, depending on size, and thus between 70 per cent and 30 per cent of the fuel input is available as useful heat. Steam turbines used in CHP applications typically range in size from a few MWe to over 100 MWe.

Gas turbine systems, often aero-engine derivatives, where fuel (gas, or gas-oil) is combusted in the gas turbine and the exhaust gases are normally used in a waste heat boiler to produce usable steam, though the exhaust gases may be used directly in some process applications. Gas turbines typically range from 1 MWe upwards (although micro turbines of down to 30 kWe are being installed), achieving electrical efficiency of 23 to 30 per cent (depending on size) and with the potential to recover up to 50 per cent of the fuel input as useful heat. They have been common in CHP since the mid 1980's. The waste heat boiler can include supplementary or auxiliary firing using a wide range of fuels, and thus the heat to power ratio of the scheme can vary.

Combined cycle systems, where the plant comprises more than one prime mover. These are usually gas turbines where the exhaust gases are utilised in a waste heat recovery boiler, the steam from which is passed wholly or in part into one or more steam turbines. Additional boilers often supplement this heat. In rare cases reciprocating engines may be linked with steam turbines. Combined cycle is suited to larger installations of 7 MWe and over. They achieve higher electrical efficiency and a lower heat to power ratio than steam turbines or gas turbines used alone. Recently installed combined cycle gas turbine (CCGT) schemes have achieved an electrical efficiency approaching 45 per cent (GCV), with 20 per cent heat recovery, and a heat to power ratio of less than 1:1.

Reciprocating engine systems range from less than 100 kWe up to around 5 MWe, and are found on smaller industrial sites as well as in buildings. They are based on auto engine or marine engine derivatives converted to run on gas. Both compression ignition and spark ignition firing is used. Reciprocating engines operate at around 28 to 33 per cent electrical efficiency with around 50 per cent to 33 per cent of the fuel input available as useful heat. Reciprocating engines produce two grades of waste heat: high-grade heat from the engine exhaust and low-grade heat from the engine cooling circuits. Steam generation from reciprocating engines is comparatively rare. Commonly, schemes of less than 800 kWe have no heat dump facility and if heat demand falls both electrical and thermal outputs are reduced.

6.35   Scaling back CHP capacity and outputs
For statistical purposes, there are a number of circumstances in which a scheme can be considered in two portions, one element a 'Good Quality' CHP, and a second element an electricity-only scheme:

- a plant whose main purpose is electricity generation for export to others, but which offers limited steam sales,

- cases where steam can be supplied from more than one power station to a site (in which case, only one scheme can supply heat at a time and *actual* steam supplied is smaller than the design capability),

- where a scheme has several prime movers, some of which supply steam to site, whilst others, the steam output is condensed,

- the scheme may have been 'Good Quality' CHP, but the heat use has declined on site.

The methodology used to calculate the CHP outputs and fuel inputs as a proportion of the total, and to calculate CHP Equivalent Capacity is that described in the CHPQA response to consultation. In consequence of both better data and different methodology, previous data has been re-analysed to produce a consistent time-series.

Determining fuel consumption for heat and electricity

6.36   In order to provide a comprehensive picture of electricity generation in the United Kingdom and the fuels used to generate that electricity, the energy input to CHP plants must be allocated to either heat or electricity production. This allocation is notional and is not determinate. For sites surveyed, the allocation between heat and electricity generation is based on the disaggregation provided by the site in the survey return. Where sites do not provide this detail and for other sites for which ETSU has data, an algorithm related to the heat to power ratio is used. If insufficient information is available for either of these calculations to be made, the fuel consumption for electricity generation over and above that which is necessary to supply heat to the site is calculated by assuming that the CHP plant displaces heat-only-boiler plant with an overall efficiency of 75 per cent. The fuel that would be consumed in the heat-only-boiler plant to produce the same amount of heat that each CHP plant produces is subtracted from the CHP plant's fuel consumption. The balance is the amount of fuel assumed to be used for electricity generation.

6.37   This methodology is under review and other methods to split the fuel use between electricity and heat generation may be used in the future.

6.38   In CHP schemes that use **sewage waste**, the heat output from the generation of electricity is used in the digestion process to produce the sewage gas. Some definitions of CHP would exclude such schemes since the heat is not being put to a beneficial use outside the generating plant itself. However, since in many cases some of the heat is used in space heating buildings on the site, all sewage waste fired plant are included in the CHP statistics for the UK.

Contact: *Dr Mark Hinnells, ETSU*
mark.hinnells@aeat.co.uk
01235 433725

Mike Janes (Statistician), DTI
mike.janes@dti.gsi.gov.uk
020 7215 5186

6.1 CHP installations by capacity and size range

	1995	1996	1997	1998	1999
Number of sites	**1,220r**	**1,282r**	**1,287r**	**1,307r**	**1,313**
Less than 100 kWe	643	662	665	644	638
100 kWe to 999 kWe	379r	415r	411r	435r	447
1 MWe to 9.9 MWe	131r	138r	142r	151r	147
10.0 MWe and above	67r	67r	69r	77r	81
					MWe
Total capacity	**3,390r**	**3,463r**	**3,628r**	**3,885r**	**4,239**
Less than 100 kWe	35	37	37	36	36
100 kWe to 999 kWe	92	100	100	107r	113
1 MWe to 9.9 MWe	559r	583r	596r	637r	636
10.0 MWe and above	2,704r	2,743r	2,895r	3,105r	3,454

6.2 Fuel used to generate electricity and heat in CHP plants

					GWh
	1995	1996	1997	1998	1999
Fuel used to generate electricity *(1)*					
Coal	2,796r	2,170r	2,754r	2,313r	1,755
Fuel oil	2,564r	2,330r	2,809r	2,889r	3,009
Natural gas	10,414r	11,512r	13,456r	15,813r	18,040
Renewable fuels *(2)*	1,042r	1,120r	968r	1,135r	1,197
Other fuels *(3)*	6,737r	7,074r	6,196r	6,201r	6,444
Total all fuels	**23,553r**	**24,206r**	**26,183r**	**28,350r**	**30,445**
Fuel used to generate heat					
Coal	18,153r	15,779r	13,215r	10,741r	6,933
Fuel oil	14,378r	13,625r	11,665r	11,298r	11,136
Natural gas	27,128r	28,759r	33,082r	35,830r	39,670
Renewable fuels *(2)*	1,547r	1,672r	1,633r	1,754r	1,851
Other fuels *(3)*	22,253r	22,802r	21,000r	20,640r	19,859
Total all fuels	**83,459r**	**82,637r**	**80,595r**	**80,263r**	**79,449**
Overall fuel use					
Coal	20,949r	17,949r	15,969r	13,054r	8,688
Fuel oil	16,942r	15,955r	14,474	14,187r	14,145
Natural gas	37,542r	40,271r	46,538r	51,643r	57,710
Renewable fuels *(2)*	2,589r	2,792r	2,601r	2,889r	3,048
Other fuels *(3)*	28,990r	29,876r	27,196r	26,841r	26,303
Total all fuels	**107,012r**	**106,843r**	**106,778r**	**108,613r**	**109,894**

(1) The allocation of fuel use between heat generation and electricity generation is largely notional. See paragraph 6.36 for an explanation of the method used.
(2) Renewable fuels include: sewage gas; other biogases; clinical waste; municipal waste.
(3) Other fuels include: process by-products, coke oven gas, blast furnace gas, gas oil and uranium.

6.3 Fuel used by types of CHP installation

	1995	1996	1997	1998	1999
Coal					
Back pressure steam turbine	13,304r	10,494r	9,073r	8,037r	3,825
Gas turbine	132	132r	-r	-	-
Combined cycle	197r	264r	306r	501r	274
Reciprocating engine	30r	-r	-r	-	5
Pass out condensing steam turbine	7,286r	7,059r	6,590r	4,516r	4,584
Total coal	**20,949r**	**17,949r**	**15,969r**	**13,054r**	**8,688**
Fuel oil					
Back pressure steam turbine	6,450r	6,086r	3,719r	3,216r	2,797
Gas turbine	2,537r	2,487r	3,634r	3,523r	3,613
Combined cycle	1,402r	964r	1,237r	1,231r	1,230
Reciprocating engine	406r	296r	267r	373r	396
Pass out condensing steam turbine	6,147r	6,122r	5,617r	5,844r	6,109
Total fuel oil	**16,942r**	**15,955r**	**14,474r**	**14,187r**	**14,145**
Natural gas					
Back pressure steam turbine	9,956r	10,196r	10,704r	8,206r	7,129
Gas turbine	7,033r	8,562r	10,056r	11,562r	13,344
Combined cycle	12,207r	12,849r	16,774r	20,475r	25,269
Reciprocating engine	2,450r	2,939r	3,382r	4,547r	5,116
Pass out condensing steam turbine	5,895r	5,726r	5,622r	6,853r	6,851
Total natural gas	**37,541r**	**40,272r**	**46,538r**	**51,643r**	**57,709**
Renewable fuels *(1)*					
Back pressure steam turbine	719r	776r	776r	761r	746
Gas turbine	23r	-r	-	-	-
Combined cycle	-	-	-	-	-
Reciprocating engine	1,545r	1,714r	1,699r	1,655r	1,641
Pass out condensing steam turbine	302r	302r	126r	473r	661
Total renewable fuels	**2,589r**	**2,792r**	**2,601r**	**2,889r**	**3,048**
Other fuels *(2)*					
Back pressure steam turbine	11,585r	11,722r	10,861r	10,988r	10,810
Gas turbine	6,362r	6,460r	6,120r	5,624r	5,640
Combined cycle	2,585r	2,967r	4,014r	3,804r	3,672
Reciprocating engine	280r	342r	309r	293r	335
Pass out condensing steam turbine	8,179r	8,384r	5,892r	6,131r	5,847
Total other fuels	**28,991r**	**29,875r**	**27,196r**	**26,840r**	**26,304**
Total - all fuels					
Back pressure steam turbine	42,014r	39,274r	35,133r	31,208r	25,307
Gas turbine	16,087r	17,641r	19,810r	20,709r	22,597
Combined cycle	16,391r	17,044r	22,331r	26,011r	30,445
Reciprocating engine	4,711r	5,291r	5,657r	6,868r	7,493
Pass out condensing steam turbine	27,809r	27,593r	23,847r	23,817r	24,052
Total all fuels	**107,012r**	**106,843r**	**106,778r**	**108,613r**	**109,894**

(1) Renewable fuels include: sewage gas; other biogases; clinical waste; municipal waste.
(2) Other fuels include: process by-products, coke oven gas, blast furnace gas, gas oil and uranium.

6.4 CHP - electricity generated by fuel and type of installation

GWh

	1995	1996	1997	1998	1999
Coal					
Back pressure steam turbine	956r	775r	671r	603r	253
Gas turbine	36r	35r	-r	-	-
Combined cycle	32r	50r	58r	82r	43
Reciprocating engine	5r	-r	-r	-r	-
Pass out condensing steam turbine	1,104r	1,099r	975r	705r	686
Total coal	**2,113r**	**1,959r**	**1,704r**	**1,390r**	**982**
Fuel oil					
Back pressure steam turbine	542r	514r	335r	294r	262
Gas turbine	339r	329r	460r	465r	478
Combined cycle	146r	84r	116r	170r	144
Reciprocating engine	107r	59r	71r	114r	121
Pass out condensing steam turbine	749r	747r	777r	852r	894
Total fuel oil	**1,883r**	**1,733r**	**1,759r**	**1,895r**	**1,899**
Natural gas					
Back pressure steam turbine	862r	922r	936r	668r	605
Gas turbine	1,489r	1,942r	2,416r	2,842r	3,447
Combined cycle	2,717r	2,953r	3,771r	4,654r	6,576
Reciprocating engine	730r	849r	954r	1,347r	1,492
Pass out condensing steam turbine	613r	596r	712r	916r	908
Total natural gas	**6,411r**	**7,262r**	**8,789r**	**10,427r**	**13,028**
Renewable fuels (1)					
Back pressure steam turbine	111r	97r	111r	104r	96
Gas turbine	4r	-r	-	-	-
Combined cycle	-	-	-	-	-
Reciprocating engine	385r	416r	413r	411r	411
Pass out condensing steam turbine	44r	52r	34r	63r	59
Total renewable fuels	**544r**	**565r**	**558r**	**578r**	**566**
Other fuels (2)					
Back pressure steam turbine	1,471r	1,467r	1,444r	1,465r	1,398
Gas turbine	920r	941r	811r	781r	783
Combined cycle	252r	334r	366r	688r	491
Reciprocating engine	92r	121r	100r	94r	110
Pass out condensing steam turbine	1,120r	1,142r	968r	1,010r	957
Total other fuels	**3,855r**	**4,005r**	**3,689r**	**4,038r**	**3,739**
Total - all fuels					
Back pressure steam turbine	3,942r	3,775r	3,497r	3,134r	2,614
Gas turbine	2,788r	3,247r	3,687r	4,088r	4,708
Combined cycle	3,147r	3,421r	4,311r	5,594r	7,254
Reciprocating engine	1,319r	1,445r	1,538r	1,966r	2,134
Pass out condensing steam turbine	3,630r	3,636r	3,466r	3,546r	3,504
Total all fuels	**14,826r**	**15,524r**	**16,499r**	**18,328r**	**20,214**

(1) Renewable fuels include: sewage gas; other biogases; clinical waste; municipal waste.
(2) Other fuels include: process by-products, coke oven gas, blast furnace gas, gas oil and uranium.

6.5 CHP - electrical capacity by fuel and type of installation

MWe

	1995	1996	1997	1998	1999
Coal					
Back pressure steam turbine	246r	192r	161r	155r	74
Gas turbine	5	4r	-r	-	-
Combined cycle	5r	10r	11r	15r	9
Reciprocating engine	1r	-r	-r	-	-
Pass out condensing steam turbine	244r	237r	235r	157r	194
Total coal	**501r**	**443r**	**407r**	**327r**	**277**
Fuel oil					
Back pressure steam turbine	143r	126r	84r	78r	59
Gas turbine	48r	47r	67r	70r	74
Combined cycle	20r	12r	22r	25r	22
Reciprocating engine	31r	18r	20r	25r	25
Pass out condensing steam turbine	164r	163r	190r	202r	213
Total fuel oil	**406r**	**366r**	**383r**	**400r**	**393**
Natural gas					
Back pressure steam turbine	198r	202r	206r	147r	146
Gas turbine	245r	309r	388r	431r	540
Combined cycle	866r	923r	1,007r	1,226r	1,532
Reciprocating engine	183r	209r	225r	285r	324
Pass out condensing steam turbine	125r	121r	149r	201r	199
Total natural gas	**1,617r**	**1,764r**	**1,975r**	**2,290r**	**2,741**
Renewable fuels *(1)*					
Back pressure steam turbine	17r	15r	16r	16r	16
Gas turbine	-	-r	-	-	-
Combined cycle	-	-	-	-	-
Reciprocating engine	83r	89r	88r	92r	93
Pass out condensing steam turbine	12r	12r	7r	13r	16
Total renewable fuels	**112r**	**116r**	**111r**	**121r**	**125**
Other fuels *(2)*					
Back pressure steam turbine	231r	226r	221r	229r	219
Gas turbine	204r	208r	192r	160r	166
Combined cycle	38r	50r	97r	101r	75
Reciprocating engine	27r	35r	31r	26r	32
Pass out condensing steam turbine	254r	255r	211r	231r	211
Total other fuels	**754r**	**774r**	**752r**	**747r**	**703**
Total - all fuels					
Back pressure steam turbine	835r	761r	688r	625r	514
Gas turbine	502r	568r	647r	661r	780
Combined cycle	929r	995r	1,137r	1,367r	1,638
Reciprocating engine	325r	351r	364r	428r	474
Pass out condensing steam turbine	799r	788r	792r	804r	833
Total all fuels	**3,390r**	**3,463r**	**3,628r**	**3,885r**	**4,239**

(1) Renewable fuels include: sewage gas; other biogases; clinical waste; municipal waste.
(2) Other fuels include: process by-products and uranium.

6.6 CHP - heat generated by fuel and type of installation

GWh

	1995	1996	1997	1998	1999
Coal					
Back pressure steam turbine	7,138r	5,592r	4,628r	4,743r	2,630
Gas turbine	38r	38r	-r	-	-
Combined cycle	94r	162r	186r	328r	177
Reciprocating engine	8r	-r	-r	-	1
Pass out condensing steam turbine	3,925r	3,761r	3,638r	2,419r	2,425
Total coal	**11,203r**	**9,553r**	**8,452r**	**7,490r**	**5,233**
Fuel oil					
Back pressure steam turbine	4,182r	4,020r	2,333r	2,046r	1,775
Gas turbine	1,476r	1,455r	1,934r	1,812r	1,842
Combined cycle	751r	534r	680r	631r	661
Reciprocating engine	168	150r	102r	109r	110
Pass out condensing steam turbine	3,641r	3,636r	3,269r	3,387r	3,532
Total fuel oil	**10,218r**	**9,795r**	**8,318r**	**7,985r**	**7,920**
Natural gas					
Back pressure steam turbine	6,710r	7,044r	7,032r	5,540r	4,841
Gas turbine	3,047r	3,684r	4,223r	4,800r	5,333
Combined cycle	5,946r	6,148r	7,764r	9,565r	10,490
Reciprocating engine	1,051r	1,237r	1,396r	1,725r	2,047
Pass out condensing steam turbine	4,050r	3,952r	3,708r	4,467r	4,409
Total natural gas	**20,804r**	**22,065r**	**24,123r**	**26,097r**	**27,120**
Renewable fuels (1)					
Back pressure steam turbine	98r	80r	78r	66r	80
Gas turbine	6r	-r	-	-	-
Combined cycle	-	-	-	-	-
Reciprocating engine	607r	686r	682r	636r	639
Pass out condensing steam turbine	126r	126r	60r	249r	264
Total renewable fuels	**837r**	**892r**	**820r**	**951r**	**983**
Other fuels (2)					
Back pressure steam turbine	6,730r	6,680r	6,255r	6,313r	6,083
Gas turbine	3,168r	3,203r	2,876r	2,772r	2,742
Combined cycle	1,383r	1,490r	2,127r	1,770r	1,842
Reciprocating engine	105r	121r	112r	108r	106
Pass out condensing steam turbine	3,453r	3,603r	2,069r	2,196r	2,033
Total other fuels	**14,839r**	**15,097r**	**13,439r**	**13,159r**	**12,806**
Total - all fuels					
Back pressure steam turbine	24,858r	23,416r	20,326r	18,708r	15,409
Gas turbine	7,735r	8,380r	9,033r	9,384r	9,917
Combined cycle	8,174r	8,334r	10,757r	12,294r	13,170
Reciprocating engine	1,939r	2,194r	2,292r	2,578r	2,903
Pass out condensing steam turbine	15,195r	15,078r	12,744r	12,718r	12,663
Total all fuels	**57,901r**	**57,402r**	**55,152r**	**55,682r**	**54,062**

(1) Renewable fuels include: sewage gas; other biogases; clinical waste; municipal waste.
(2) Other fuels include: process by-products and uranium.

6.7 CHP - heat capacity by fuel and type of installation

MWth

	1995	1996	1997	1998	1999
Coal					
Back pressure steam turbine	2,092r	1,662r	1,379r	1,361r	712
Gas turbine	7r	7r	-	-	-
Combined cycle	36r	36r	40r	56r	34
Reciprocating engine	2r	-r	-r	-	-
Pass out condensing steam turbine	1,935r	1,853r	1,828r	1,171r	1,325
Total coal	**4,072r**	**3,558r**	**3,247r**	**2,588r**	**2,071**
Fuel oil					
Back pressure steam turbine	1,230r	1,156r	733r	646r	461
Gas turbine	210r	207r	294r	302r	319
Combined cycle	171r	109r	137r	144r	131
Reciprocating engine	48r	41r	31r	32r	31
Pass out condensing steam turbine	1,228r	1,222r	1,426r	1,527r	1,622
Total fuel oil	**2,887r**	**2,735r**	**2,621r**	**2,651r**	**2,564**
Natural gas					
Back pressure steam turbine	1,559r	1,629r	1,635r	1,303r	1,202
Gas turbine	721r	920r	1,044r	1,135r	1,234
Combined cycle	1,801r	1,939r	2,284r	2,756r	3,326
Reciprocating engine	314r	357r	407r	464r	519
Pass out condensing steam turbine	840r	820r	987r	1,431r	1,414
Total natural gas	**5,235r**	**5,665r**	**6,357r**	**7,089r**	**7,695**
Renewable fuels (1)					
Back pressure steam turbine	53r	50r	53r	52r	52
Gas turbine	2r	-r	-	-	-
Combined cycle	-	-	-	-	-
Reciprocating engine	138r	146r	150r	142r	152
Pass out condensing steam turbine	65r	65r	35r	73r	89
Total renewable fuels	**258r**	**261r**	**238r**	**267r**	**293**
Other fuels (2)					
Back pressure steam turbine	1,047r	998r	871r	913r	868
Gas turbine	821r	829r	762r	668r	658
Combined cycle	299r	315r	431r	438r	378
Reciprocating engine	34r	39r	38r	35r	35
Pass out condensing steam turbine	858r	863r	547r	691r	531
Total other fuels	**3,059r**	**3,044r**	**2,649r**	**2,745r**	**2,470**
Total - all fuels					
Back pressure steam turbine	5,981r	5,495r	4,671r	4,275r	3,296
Gas turbine	1,761r	1,963r	2,100r	2,105r	2,211
Combined cycle	2,307r	2,399r	2,892r	3,394r	3,868
Reciprocating engine	536r	583r	626r	673r	737
Pass out condensing steam turbine	4,926r	4,823r	4,823r	4,893r	4,981
Total all fuels	**15,511r**	**15,263r**	**15,112r**	**15,340r**	**15,093**

(1) Renewable fuels include: sewage gas; other biogases; clinical waste; municipal waste.
(2) Other fuels include: process by-products and uranium.

6.8 CHP capacity, output and total fuel use[1] by sector

	Unit	1995	1996	1997	1998	1999
Iron and steel and non ferrous metals						
Number of sites		7	7	7	6r	6
Electrical capacity	MWe	272r	272r	272r	268r	268
Heat capacity	MWth	552r	552r	552r	542r	541
Electrical output	GWh	1,463r	1,454r	1,501r	1,504r	1,502
Heat output	GWh	2,578r	2,670r	2,672r	2,675r	2,660
Fuel use	GWh	7,863r	8,129r	7,796r	7,806r	8,033
of which : for electricity	GWh	1,242r	1,268r	1,177r	1,177r	1,449
for heat	GWh	6,621r	6,861r	6,619r	6,629r	6,584
Chemicals						
Number of sites		53r	51r	52r	53r	52
Electrical capacity	MWe	1,019r	1,026r	1,123r	1,129r	1,180
Heat capacity	MWth	6,409r	6,081r	5,922r	6,063r	5,970
Electrical output	GWh	4,923r	5,026r	5,262r	5,756r	5,866
Heat output	GWh	20,464r	19,112r	19,100r	18,784r	18,214
Fuel use	GWh	36,848r	34,687r	36,223r	34,709r	33,522
of which : for electricity	GWh	6,854r	6,389r	8,467r	8,363r	9,020
for heat	GWh	29,994r	28,298r	27,756r	26,346r	24,502
Oil refineries						
Number of sites		12r	13r	13r	14r	16
Electrical capacity	MWe	594r	627r	627r	650r	838
Heat capacity	MWth	3,427r	3,546r	3,546r	3,630r	3,918
Electrical output	GWh	3,170r	3,465r	3,303r	3,613r	4,243
Heat output	GWh	17,629r	18,103r	15,888r	16,332r	16,754
Fuel use	GWh	29,997r	31,108r	28,119r	28,826r	30,640
of which : for electricity	GWh	6,406r	6,851r	6,086r	6,443r	7,105
for heat	GWh	23,591r	24,257r	22,033r	22,383r	23,535
Paper, publishing and printing						
Number of sites		34r	35r	35r	33	31
Electrical capacity	MWe	356r	369r	416r	468r	471
Heat capacity	MWth	1,542r	1,577r	1,572r	1,513r	1,379
Electrical output	GWh	1,812r	1,952r	2,580r	2,620r	3,068
Heat output	GWh	6,940r	7,266r	6,900r	6,240r	6,022
Fuel use	GWh	12,455r	12,697r	13,781r	13,011r	13,342
of which : for electricity	GWh	3,142r	3,446r	4,021r	3,993r	4,107
for heat	GWh	9,313r	9,251r	9,760r	9,018r	9,235
Food, beverages and tobacco						
Number of sites		37r	38	41r	45r	44
Electrical capacity	MWe	207r	207r	217r	302r	392
Heat capacity	MWth	1,446r	1,447r	1,431r	1,407r	1,331
Electrical output	GWh	886r	896r	991r	1,304r	2,194
Heat output	GWh	4,907r	4,914r	5,254r	5,655r	5,031
Fuel use	GWh	8,155r	8,112r	8,633r	9,709r	10,431
of which : for electricity	GWh	2,148r	2,098r	2,361r	2,753r	3,108
for heat	GWh	6,007r	6,014r	6,272r	6,956r	7,323
Metal products, machinery and equipment						
Number of sites		9r	9r	10r	9r	10
Electrical capacity	MWe	41r	30	34r	33r	34
Heat capacity	MWth	197	72r	74r	61r	69
Electrical output	GWh	176r	145r	166r	192r	200
Heat output	GWh	650r	339	357r	302r	323
Fuel use	GWh	1,188r	723r	768r	750r	823
of which : for electricity	GWh	284r	236r	235r	245r	279
for heat	GWh	904r	487r	533r	505r	544

6.8 CHP capacity, output and total fuel use[1] by sector (continued)

	Unit	1995	1996	1997	1998	1999
Mineral products, extraction, mining and agglomeration of solid fuels						
Number of sites		4r	6	6	7r	7
Electrical capacity	MWe	17r	29r	29r	35r	56
Heat capacity	MWth	122r	135r	135r	109r	150
Electrical output	GWh	96r	170r	169r	244r	309
Heat output	GWh	644r	688r	669r	678r	773
Fuel use	GWh	982r	1,276r	1,211r	1,367r	1,599
of which : for electricity	GWh	63r	309r	201r	249r	288
for heat	GWh	919r	967r	1,010r	1,118r	1,311
Textiles, clothing and footwear						
Number of sites		4	4	4	4	3
Electrical capacity	MWe	4	4	4	5	2
Heat capacity	MWth	55	55	31	31	13
Electrical output	GWh	11	11	11	14	2
Heat output	GWh	198	198	100	103	25
Fuel use	GWh	280	279	165	172	42
of which : for electricity	GWh	31	31	32	37	7
for heat	GWh	249	248	133	135	35
Sewage treatment						
Number of sites		126	126	125	125	120
Electrical capacity	MWe	88	94	94	107	108
Heat capacity	MWth	148	157	161	165	173
Electrical output	GWh	407	441	439	467	464
Heat output	GWh	654	735	732	715	713
Fuel use	GWh	1,652	1,828	1,818	1,870	1,842
of which : for electricity	GWh	717	776	691	804	801
for heat	GWh	935	1,052	1,127	1,066	1,041
Electricity supply						
Number of sites		4	4	4	5	5
Electrical capacity	MWe	452	452	452	464	464
Heat capacity	MWth	610	610	610	650	650
Electrical output	GWh	538	538	538	638	342
Heat output	GWh	684	684	684	877	508
Fuel use	GWh	1,455	1,455	1,455	1,860	1,111
of which : for electricity	GWh	679	679	679	885	510
for heat	GWh	776	776	776	975	601
Other industrial branches						
Number of sites		3r	4r	5r	10r	10
Electrical capacity	MWe	7r	12r	13r	26r	26
Heat capacity	MWth	12r	25r	37r	53r	54
Electrical output	GWh	28r	80r	59r	143r	158
Heat output	GWh	42r	137r	144r	251r	254
Fuel use	GWh	164r	354r	336r	671r	699
of which : for electricity	GWh	52r	107r	133r	262r	329
for heat	GWh	112r	247r	203r	409r	370
Transport, commerce and administration						
Number of sites		598r	633r	629r	619r	628
Electrical capacity	MWe	167r	173r	177r	184r	201
Heat capacity	MWth	289r	298r	331r	313r	330
Electrical output	GWh	732r	751r	858r	954r	982
Heat output	GWh	1,175r	1,206r	1,314r	1,240r	1,377
Fuel use	GWh	2,696r	2,762r	3,151r	3,328r	3,596
of which : for electricity	GWh	977r	1,003r	1,113r	1,491r	1,650
for heat	GWh	1,719r	1,759r	2,038r	1,837r	1,946

6.8 CHP capacity, output and total fuel use[1] by sector (continued)

	Unit	1995	1996	1997	1998	1999
Other (2)						
Number of sites		329r	352r	356r	377r	381
Electrical capacity	MWe	166r	168r	170r	214r	199
Heat capacity	MWth	702r	708r	710r	803r	515
Electrical output	GWh	584r	595r	622r	880r	883
Heat output	GWh	1,336r	1,350r	1,338r	1,830r	1,408
Fuel use	GWh	3,277r	3,433r	3,322r	4,534r	4,214
of which : for electricity	GWh	958r	1,014r	987r	1,607r	1,792
for heat	GWh	2,319r	2,419r	2,335r	2,927r	2,422
Total CHP usage by all sectors						
Number of sites		1,220r	1,282r	1,287r	1,307r	1,313
Electrical capacity	MWe	3,390r	3,463r	3,628r	3,885r	4,239
Heat capacity	MWth	15,511r	15,263r	15,112r	15,340r	15,093
Electrical output	GWh	14,826r	15,524r	16,499r	18,329r	20,213
Heat output	GWh	57,901r	57,402r	55,152r	55,682r	54,062
Fuel use	GWh	107,012r	106,843r	106,778r	108,613r	109,894
of which : for electricity	GWh	23,553r	24,207r	26,183r	28,350r	30,445
for heat	GWh	83,459r	82,636r	80,595r	80,263r	79,449

(1) The allocation of fuel use between electricity and heat is largely notional and the methodology is outlined in paragraph 6.36.
(2) Sectors included under Other are agriculture, community heating, leisure, landfill and incineration.

6.9 CHP - use of fuels by sector

GWh

	1995	1996	1997	1998	1999
Iron and steel and non ferrous metals					
Coal	180r	180	154	173r	89
Fuel oil	761r	739r	487r	478r	545
Gas oil	-r	-r	-	-r	-
Natural gas	107r	243r	415r	415r	533
Blast furnace gas	5,688r	5,764r	5,651r	5,651r	5,593
Coke oven gas	1,025r	1,111r	1,000r	1,000r	1,184
Renewable fuels (1)	13	-r	-	-	-
Other fuels (2)	88r	92r	89r	89r	89
Total iron and steel and non ferrous metals	**7,862r**	**8,129r**	**7,796r**	**7,806r**	**8,033**
Chemicals					
Coal	14,250r	11,472r	10,198r	7,876r	4,826
Fuel oil	5,043r	4,420r	3,656r	3,664r	3,886
Gas oil	-	2r	236r	80r	135
Natural gas	14,994r	15,911r	18,855r	20,085r	22,181
Renewable fuels (1)	23	-r	-	-	-
Other fuels (2)	2,538r	2,882r	3,277r	3,004r	2,494
Total chemical industry	**36,848r**	**34,687r**	**36,222r**	**34,709r**	**33,522**
Oil refineries					
Fuel oil	7,439r	7,297r	8,207r	8,128r	8,237
Gas oil	55r	55r	55r	72r	338
Natural gas	3,861r	4,985r	4,149r	5,198r	6,871
Refinery gas	13,306r	12,969r	9,906r	9,645r	9,386
Other fuels (2)	5,336r	5,802r	5,802r	5,783r	5,808
Total oil refineries	**29,997r**	**31,108r**	**28,119r**	**28,826r**	**30,640**
Paper, publishing and printing					
Coal	3,292r	3,246r	2,605r	2,189r	1,588
Fuel oil	777r	845r	497r	430r	482
Gas oil	-	10r	14r	17	61
Natural gas	8,386r	8,596r	10,662r	10,035r	11,209
Other fuels	-	-	3r	340r	2
Total paper, publishing and printing	**12,455r**	**12,697r**	**13,781r**	**13,011r**	**13,342**
Food, beverages and tobacco					
Coal	2,035	2,035	2,100	2,012r	1,844
Fuel oil	1,948r	1,902r	977r	691r	585
Gas oil	-	13r	17r	31r	26
Natural gas	4,135r	4,159r	5,537r	6,974r	7,973
Renewable fuels (1)	2	2	2	2	3
Other fuels (2)	36r	-r	-r	-r	-
Total food, beverages and tobacco	**8,156r**	**8,111r**	**8,633r**	**9,710r**	**10,431**
Metal products, machinery and equipment					
Coal	369	181r	136	33r	37
Fuel oil	110	-	-	90r	90
Natural gas	654r	542r	632r	627r	696
Other fuels (2)	56	-r	-r	-	-
Total metal products, machinery and equipment	**1,189r**	**723r**	**768r**	**750r**	**823**
Mineral products, extraction, mining and agglomeration of solid fuels					
Fuel oil	15	13r	15	-	-
Gas oil	-r	-r	-r	-r	31
Natural gas	817	857r	791r	961r	1,162
Coke oven gas	150r	406r	406r	406r	406
Other fuels (2)	-r	-r	-r	-r	-
Total mineral products, extraction, mining and agglomeration of solid fuels	**982r**	**1,276r**	**1,212r**	**1,367r**	**1,599**

6.9 CHP - use of fuels by sector (continued)

GWh

	1995	1996	1997	1998	1999
Textiles, clothing and footwear					
Fuel oil	-	-	2	8	7
Gas oil	-	-	-	-	2
Natural gas	280	280	163	164	33
Total textiles, clothing and footwear	**280**	**280**	**165**	**172**	**42**
Sewage treatment(3)					
Fuel oil	113	121	125	125	115
Natural gas	12	12	12	115	116
Renewable fuels (1)	1,513	1,695	1,681	1,630	1,612
Other fuels (2)	13	-	-	-	-
Total sewage treatment	**1,651**	**1,828**	**1,818**	**1,870**	**1,843**
Electricity supply					
Coal	177	177	177	177	177
Natural gas	1,020	1,020	1,020	1,425	676
Other fuels (2)	258	258	258	258	258
Total electricity supply	**1,455**	**1,455**	**1,455**	**1,860**	**1,111**
Other industrial branches					
Coal	30r	-r	-r	-r	-
Fuel oil	50r	-r	25r	81r	109
Gas oil	-r	-r	-r	-r	-
Natural gas	84r	354r	311r	585r	585
Renewable fuels (1)	-r	-r	-r	5r	5
Other fuels (2)	-r	-r	-r	-r	-
Total other industrial branches	**164r**	**354r**	**336r**	**671r**	**699**
Transport, commerce and administration					
Fuel oil	231r	164r	69r	79r	75
Gas oil	115r	186r	155r	137r	98
Natural gas	2,349r	2,411r	2,925r	3,104r	3,411
Renewable fuels (1)	-	-	-	5r	8
Other fuels (2)	1	1	2r	3r	4
Total transport, commerce and administration	**2,696r**	**2,762r**	**3,151r**	**3,328r**	**3,596**
Other (3)					
Coal	616r	658r	599r	594r	127
Fuel oil	455r	454	414	413r	14
Gas oil	282r	282r	282r	282r	345
Natural gas	843r	901r	1,066r	1,955r	2,264
Renewable fuels (1)	1,037r	1,095r	918r	1,247r	1,420
Other fuels (2)	44r	43r	43r	43r	43
Total other	**3,277r**	**3,433r**	**3,322r**	**4,534r**	**4,213**
Total - all sectors					
Coal	20,949r	17,949r	15,969r	13,054r	8,688
Fuel oil	16,942r	15,955r	14,474r	14,187r	14,145
Gas oil	452r	548r	759r	619r	1,036
Natural gas	37,542r	40,271r	46,538r	51,643r	57,710
Blast furnace gas	5,688r	5,764r	5,651r	5,651r	5,593
Coke oven gas	1,175r	1,517r	1,406r	1,406r	1,590
Refinery gas	13,306r	12,969r	9,906r	9,645r	9,386
Renewable fuels (1)	2,588r	2,792r	2,601r	2,889r	3,048
Other fuels (2)	8,370r	9,078r	9,474r	9,520r	8,698
Total CHP fuel use	**107,012r**	**106,843r**	**106,778r**	**108,614r**	**109,894**

(1) Renewable fuels include: sewage gas; other biogases; clinical waste; municipal waste.
(2) Other fuels include: process by-products and uranium.
(3) Sectors included under Other are agriculture, community heating, leisure, landfill and incineration.

Chapter 7
Renewable sources of energy

Introduction

7.1 This chapter provides information on the contribution of renewable energy sources to the United Kingdom's energy requirements for the last eleven years, to 1999.

7.2 The data summarise the results of an ongoing study undertaken by the ETSU (part of AEA Technology (AEAT) Environment), on behalf of the Department of Trade and Industry, to update a database containing information on all relevant renewable energy sources in the United Kingdom. This database is called RESTATS, the Renewable Energy STATisticS database. The study is partly financed by the Statistical Office of the European Communities (Eurostat).

7.3 The study started in 1989, when all relevant renewable energy sources were identified and, where possible, information was collected on the amounts of energy derived from each source. The renewable energy sources identified were the following: active solar heating; photovoltaics; onshore wind power; wave power; large and small scale hydro; biofuels; geothermal aquifers. The technical notes at the end of this chapter define each of these renewable energy sources.

7.4 The information contained in the database is collected by a number of methods. For larger projects, an annual survey is carried out in which questionnaires are sent to project managers. For technologies in which there are large numbers of small projects, the values given in this chapter are estimates based on information collected from a sub-sample of the projects. Further details about the data collection methodologies used in RESTATS, including the quality and completeness of the information, are given in the technical notes at the end of this chapter.

7.5 The main instruments for pursuing the development of renewables capacity have been the Non Fossil Fuel Obligation (NFFO) Orders for England and Wales and for Northern Ireland, and the Scottish Renewable Orders (SRO). In this chapter the term "NFFO Orders" is used is refer to these instruments collectively.

7.6 For projects contracted under NFFO Orders in England and Wales, details of capacity and generation were provided by the Non Fossil Purchasing Agency (NFPA). Information on the Scottish and Northern Ireland NFFO Orders were provided by the Scottish Executive and Northern Ireland Electricity, respectively.

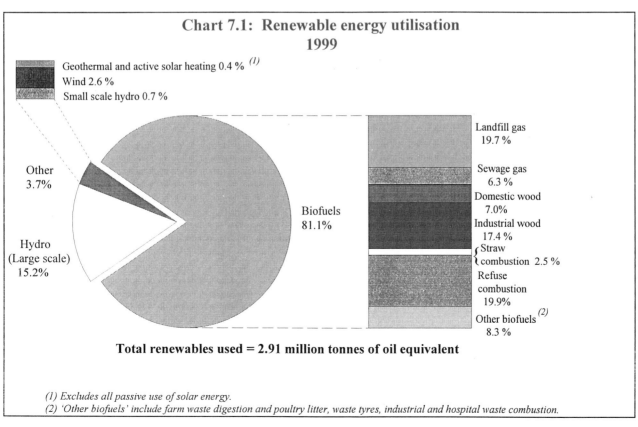

Chart 7.1: Renewable energy utilisation 1999

Geothermal and active solar heating 0.4 % [1]
Wind 2.6 %
Small scale hydro 0.7 %

Other 3.7%

Hydro (Large scale) 15.2%

Biofuels 81.1%

Landfill gas 19.7 %
Sewage gas 6.3 %
Domestic wood 7.0%
Industrial wood 17.4 %
Straw combustion 2.5 %
Refuse combustion 19.9%
Other biofuels [2] 8.3 %

Total renewables used = 2.91 million tonnes of oil equivalent

(1) Excludes all passive use of solar energy.
(2) 'Other biofuels' include farm waste digestion and poultry litter, waste tyres, industrial and hospital waste combustion.

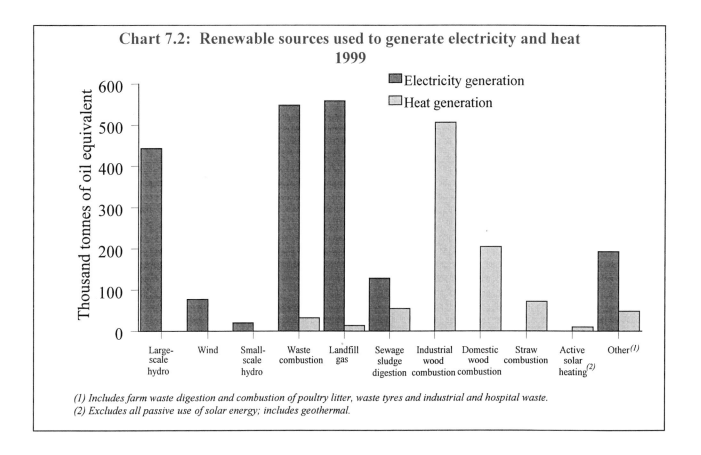

Chart 7.2: Renewable sources used to generate electricity and heat 1999

- Electricity generation
- Heat generation

Thousand tonnes of oil equivalent (y-axis: 0, 100, 200, 300, 400, 500, 600)

Categories: Large-scale hydro, Wind, Small-scale hydro, Waste combustion, Landfill gas, Sewage sludge digestion, Industrial wood combustion, Domestic wood combustion, Straw combustion, Active solar heating[2], Other[1]

(1) Includes farm waste digestion and combustion of poultry litter, waste tyres and industrial and hospital waste.
(2) Excludes all passive use of solar energy; includes geothermal.

7.7 Commodity balances for renewable energy sources covering each of the last three years form the first three tables (Tables 7.1 to 7.3). These are followed by the 5 year table showing capacity of and electricity generation from renewable sources (Table 7.4), a table summarising all the renewable orders (Table 7.5) and a long term trends table covering the use of renewables to generate electricity and heat (Table 7.6).

7.8 Unlike in the commodity balance tables in other chapters of the Digest, Tables 7.1 to 7.3 have zero statistical differences. This is because the data for each category of fuel are, in the main, taken from a single source where there is less likelihood of differences due to timing or measurement.

Commodity balances for renewables in 1999 (Table 7.1), 1998 (Table 7.2) and 1997 (Table 7.3).

7.9 Eleven different categories of renewable fuels are identified in the commodity balances. Two of these categories are themselves groups of renewables because a more detailed disaggregation could disclose data for individual companies. The largest contribution is from biofuels, with large scale hydro electricity production contributing the majority of the remainder as Chart 7.1 shows. Only just under 4 per cent of renewable energy comes from renewable sources other than biofuels and large scale hydro.

These include solar, wind, small scale hydro and geothermal aquifers.

7.10 Sixty eight per cent of the renewable energy produced in 1999 was transformed into electricity. This is an increase from 65 per cent in 1998, 60 per cent in 1997 and 55 per cent in 1996. Chart 7.2 shows renewable energy sources in 1999 split between those that produce electricity and those that produce heat. While municipal solid waste and industrial wood appear to dominate the picture when fuel inputs are being measured, hydro electricity dominates when the output of electricity is being measured as Table 7.4 shows. This is because on an energy supplied basis (see Chapter 5, paragraph 5.27) hydro (and wind) inputs are assumed to be equal to the electricity produced. For landfill gas, sewage sludge, municipal solid waste and other renewables a substantial proportion of the energy content of the input is lost in the process of conversion to electricity. Chart 7.2 also shows the contribution from renewables to heat generation. Here only a small share comes from geothermal and active solar heating and from various wastes, while the main contribution is from wood burning.

7.11 Overall, renewable sources, excluding passive uses of solar energy, provided 1.3 per cent of the United Kingdom's total primary energy requirements in 1999, up from 1.2 per cent in 1998 and 1.0 per cent in 1997 and more than double the proportion recorded 7 years earlier in 1992.

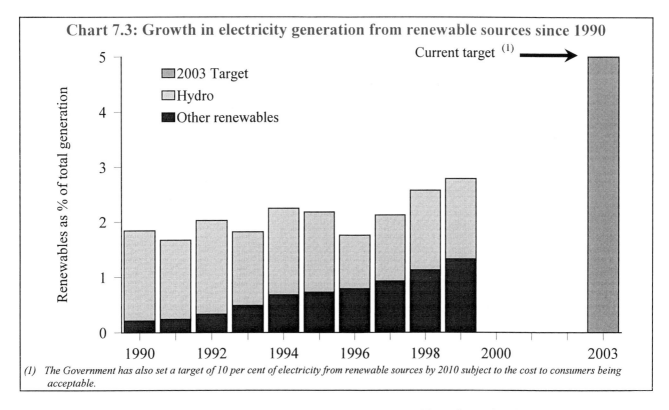

Chart 7.3: Growth in electricity generation from renewable sources since 1990

(1) The Government has also set a target of 10 per cent of electricity from renewable sources by 2010 subject to the cost to consumers being acceptable.

Capacity of, and electricity generated from renewable sources (Table 7.4)

7.12 Table 7.4 shows the capacity of, and the amounts of electricity generated from, each renewable source. Total electricity generation from renewables in 1999 amounted to 10,237 GWh, 50 per cent of which was from large scale hydro generation. Renewables provided 2.8 per cent of the electricity generated in the United Kingdom. Chart 7.3 shows the growth in the proportion of electricity produced from renewable sources and progress toward the target of 5 per cent set in 1999 for 2003.

7.13 In 1999 the electricity generated from renewable sources in the UK was 9½ per cent greater than in 1998. Generation from renewables, other than large scale hydro, in 1999 was 18½ per cent higher than in 1998 and almost double the level in 1995.

7.14 There was an increase of 44 per cent in electricity generation from landfill gas as a result of new projects being brought on-line under NFFO. There was an increase in the total generation from onshore wind of 2 per cent in 1999 which resulted from new NFFO schemes coming on stream during the year.

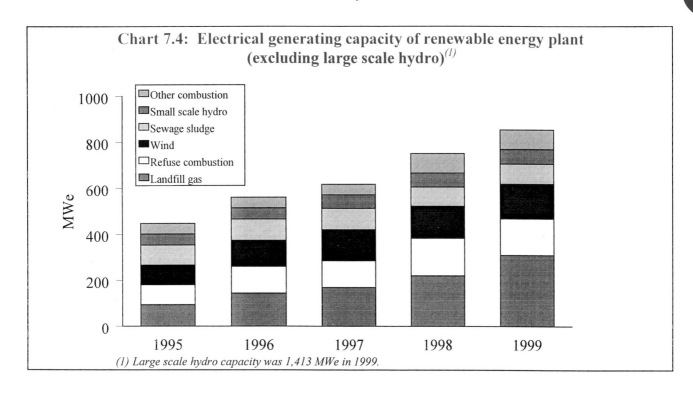

Chart 7.4: Electrical generating capacity of renewable energy plant (excluding large scale hydro)[1]

(1) Large scale hydro capacity was 1,413 MWe in 1999.

7.15 There was a rise of 6 per cent generation from sewage sludge digestion during 1999, but the increase in generation from municipal solid waste was less than 1 per cent following the 45 per cent increase in the previous year.

7.16 Chart 7.4 (which covers all renewables capacity except large scale hydro) shows how the electricity generation capacity from all significant renewable sources has risen steadily in the five years from 1995. This trend in the capacity of new and renewable sources will continue as further projects already contracted under NFFO Orders come on line.

7.17 In 1999, 47 per cent of electricity from renewables (excluding large-scale hydro) was generated under NFFO contracts. If ex-NFFO sites (NFFO 1 and 2 in England and Wales – see paragraphs 7.19 to 7.26, below) are included the proportion increases to 71 per cent. Table 7.4, however, includes both electricity generated outside of these contracts and electricity from large-scale hydro schemes and thus reports on total electricity generation from renewables. All electricity generated from renewables is also reported within the tables of Chapter 5 of this Digest (eg Table 5.5).

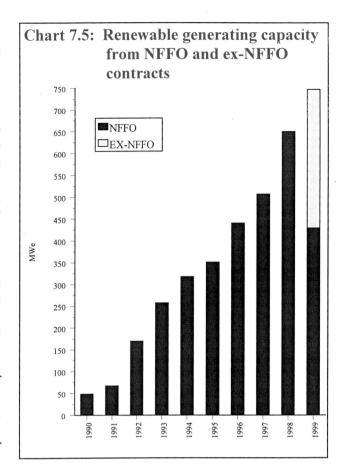

Chart 7.5: Renewable generating capacity from NFFO and ex-NFFO contracts

7.18 Plant load factors in Table 7.4 have been calculated in terms of installed capacity and express the average hourly quantity of electricity generated as a percentage of the average capacity at the beginning and end of the year. The overall figure is heavily influenced by the availability of hydro capacity during the year which is in turn influenced by the amount of rainfall during the preceding period. Plant load factors for all generating plant in the UK are shown in Table 5.9.

Renewable orders and operational capacity (Table 7.5)

7.19 In 1990, the first year of NFFO, projects contracted within NFFO accounted for about 34 per cent of the total capacity (excluding large-scale hydro); by 1998, this figure had risen to 86 per cent, but this has dropped to 47 per cent in 1999 due to the expiry of NFFO I and 2 contracts. Some 746.83 MW of new renewables generation capacity (measured in terms of declared net capacity, DNC - see paragraph 7.66) had begun to operate by the end of 1999 as shown in Chart 7.5.

(a)   Non Fossil Fuel Obligation (NFFO)

7.20 The 1989 Electricity Act empowered the Secretary of State to make orders requiring the Regional Electricity Companies in England and Wales (the RECs) to secure specified amounts of electricity from renewable energy sources.

7.21 Five NFFO Orders have been made, of which the first in 1990 was set for a total of 102 MW DNC. This first order resulted in contracts for 75 projects for 152 MW DNC and provided a premium price for the electricity produced which was funded from a levy on electricity sales in England and Wales. (The bulk of this levy was used to support electricity from nuclear stations.)

7.22 The second Order, made in late 1991, was set for 457 MW DNC. This resulted in 122 separate contracts (for a total of 472 MW DNC) between the generators and the Non-Fossil Purchasing Agency (NFPA) which acted on behalf of the RECs. For landfill gas, sewage gas and waste-derived generation contracts were awarded at around 6p/kWh, while for wind-based generation a price of 11p/kWh was established. These prices reflected the limited period for the recovery of capital costs. The levy (which funded both nuclear and renewables until 1998) was due to be removed in 1998 but is now retained for the renewables NFFO Orders only.

7.23 The third Order covers the period 1995 to 2014; this was for 627 MW DNC of contracted capacity at an average price of 4.35 p/kWh. The lower bid prices reflect the longer term contracts which are now available together with further developments which have led to improvements in the technologies. Taking into account factors such as the failure to gain planning permission it is estimated that about 300-400 MW DNC are likely to go forward for commissioning.

7.24   The fourth Order was announced in February 1997. Contracts have been let to 195 projects with a total DNC of 843 MW, at an average price of 3.46 p/kWh.

7.25   The fifth and largest Order was announced in September 1998.   Contracts have been let to 261 projects with a total DNC of 1,177.1 MW, at an average price of 2.71 p/kWh.

7.26   Since the expiry of the NFFO 1 and 2 contracts on 31 December 1998, these projects are longer included in the monitoring of NFFO Orders and DTI no longer receive any status/output data on them from the NFPA. For some of these projects operational data have been obtained from other sources, while for the others estimates have been made based on output in 1998. Thirty four existing operational projects were contracted under the first Order and 30 under the second Order; other contracts in both Orders were for new projects.

7.27   As at the end of December 1999, 73 projects in the third Order were operational, with total capacities of 251 MW DNC. There were 49 schemes with a capacity of 105 MW DNC commissioned from the fourth Order projects and 13 schemes totalling 18 MW DNC from the fifth Order. These 135 schemes now make up a good proportion of the electrical capacity from renewable sources (excluding large scale hydro capacity). Table 7.5 sets out the technologies and capacities of schemes in all five Orders.

(b)   Scottish Renewable Order (SRO)
7.28   In Scotland, the first Renewables Order was made in 1994 for approximately 76 MW DNC of new capacity and comprising 30 schemes. Four generation technology bands were covered; 12 wind, 15 hydro, 2 waste-to-energy and 1 biomass.   At the end of December 1999, 13 schemes were commissioned for 32 MW DNC.

7.29   A second SRO was launched in 1995 and was made in March 1997 for 112 MW DNC of new capacity comprising 26 schemes, nine of which were waste to energy projects, nine were hydro projects, seven were wind projects and one was a biomass

project. Under this Order, at the end of 1999 there were 3 commissioned schemes with 7 MW DNC.

7.30   A third SRO was laid before Parliament in February 1999 for 145.4 MW DNC of new capacity comprising 53 schemes. Sixteen of these are waste to energy projects, five are hydro projects, twenty-eight are wind projects, one is a biomass project and three are wave energy projects. All these projects are now contracted but have yet to be commissioned. Table 7.5 sets out the technologies and capacities of schemes in all three Scottish Orders.

(c)   Northern Ireland Non Fossil Fuel Obligation (NI NFFO)
7.31   In Northern Ireland a first Order was made in March 1994 for approximately 16 MW DNC comprising 20 schemes. The contracted schemes are spread throughout Northern Ireland and are divided into three technology bands.   There are 6 wind schemes of around 2 MW DNC each, totalling 12.7 MW DNC; 5 sewage gas projects totalling 0.56 MW DNC; and 9 small-scale hydro schemes totalling 2.4 MW DNC.   At the end of 1999, 13 schemes were commissioned for 14.6 MW DNC.

7.32   A second NI Order was made in 1996 for 10 schemes, totalling 16 MW DNC. These comprised 2 wind schemes, 2 hydro schemes, 2 biomass, 1 biogas, 2 landfill gas and 1 municipal and industrial waste scheme, as shown in Table 7.5. At the end of 1999 four schemes were commissioned for 1 MW DNC.

Renewable sources used to generate electricity and heat (Table 7.6).
7.33   Between 1998 and 1999 there was an increase of 9 per cent in the use of renewable sources of energy. Nearly all of the increase is accounted for by biofuels. The fastest rates of growth were recorded by landfill gas and poultry litter where growth was 42 and 106 per cent respectively between 1998 and 1999. Generation from landfill gas and poultry litter are respectively now more than 3 times and twice the levels five years earlier in 1994.

7.34   There was a further recovery in 1999 in the use of municipal solid waste (MSW) to generate heat following the low 1997 figure when one site was unavailable during a refit needed to bring the plant up to EU standards. In 1999 electricity generation levels were not significantly different from those in 1998.

Technical notes and definitions

7.35 Energy derived from renewable sources is included in the aggregate energy tables in Chapter 1 of this Digest. The main energy tables (Tables 7.1 to 7.3) present figures in the common unit of energy, the tonne of oil equivalent, which is defined in paragraph 1.45. The gross calorific values and conversion factors used to convert the data from original units are given on page 248 of Annex A and inside the back cover flap. The statistical methodologies and conversion factors are in line with those used by the International Energy Agency and the Statistical Office of the European Communities. Primary electricity contributions from hydro and wind are expressed in terms of an electricity supplied model (see Chapter 5, paragraph 5.28) and electrical capacities are quoted as Declared Net Capacity (DNC), taking into account the intermittent nature of the power output from some renewable sources (see paragraph 7.66, below).

7.36 The various renewable energy sources are described in the following paragraphs. This section also provides details of the quality of information provided within each renewables area, and the progress made to improve the quality of this information.

Use of existing solar energy

7.37 Nearly all buildings make use of some passive solar energy because they have windows or roof lights which allow in natural light and provide a view of the surroundings. This existing use of passive solar energy is making a substantial contribution to the energy demand in the UK building stock. Passive solar design, in which buildings are designed to enhance solar energy use, results in additional savings in energy. A study in 1990, on behalf of the Department of Trade and Industry, estimated that this existing use saves 12.6 million tonnes of oil equivalent per year in the United Kingdom. This figure reflects an estimate of the net useful energy flow (heat and lighting) across windows and other glazing in the United Kingdom building stock. The figure is very approximate and, as in previous years, has therefore not been included in the tables in this chapter.

Active solar heating

7.38 Active solar heating employs solar collectors to heat water mainly for domestic hot water systems but also for swimming pools and other applications. A recent study for ETSU, on behalf of the Department of Trade and Industry, has provided improved estimates for this issue of the Digest. For 1999 an estimated 45 GWh for domestic hot water generation replaces gas (80 per cent) or electric (20 per cent) heating; for swimming pools, an estimated 34 GWh generation for 1999 replaces gas (45 per cent), oil (45 per cent) or electricity (10 per cent).

Photovoltaics

7.39 Photovoltaics is the direct conversion of solar radiation into direct current electricity by the interaction of light with the electrons in a semiconductor device or cell. It is estimated that the electrical declared net capacity from photovoltaics is increasing at approximately 8.5 kWe per year. The figures for the use of photovoltaics are too small to be included in most tables of this chapter, although the capacity to generate electricity is now sufficiently large to feature separately in Table 7.4.

Onshore wind power

7.40 A wind turbine extracts energy from the wind by means of a rotor fitted with aerodynamic-section blades using the lifting forces on the blades to turn the rotor primary shaft. This mechanical power is used to drive an electrical generator via a step-up gearbox. The figures included for generation are based on the installed capacities of wind turbines, together with an average load factor for the United Kingdom or, where figures are available, on actual generation.

7.41 There have been a total of 302 wind projects awarded contracts under NFFO. Many of these are new projects, so this has resulted in a considerable increase in electricity generation from wind since 1990. At the end of 1999, there were 66 wind generation projects operational under NFFO. More are anticipated to be commissioned over the next 3-4 years. The figures for wind in this chapter cover all known schemes in the United Kingdom. Wind capacity in 1999 was nearly three times the capacity in 1994, while electricity generated from wind has increased by more than threefold over the same period. This is attributed to improvement in technologies together with the better siting of wind farms.

Wave power

7.42 Waves in the oceans are created by the interaction of winds with the surface of the sea. Because of the direction of the prevailing winds and the size of the Atlantic Ocean, the United Kingdom has wave power levels which are amongst the highest in the world. Since 1985, the Department of Trade and Industry's shoreline programme has concentrated on an oscillating water column device on the Hebridean island of Islay. This experimental prototype came on line in late 1991 and has now been decommissioned. There are currently 3 wave schemes contracted under the third SRO for a declared net capacity of 2 MW, with one already under construction.

Large scale hydro

7.43 In hydro schemes the turbines that drive the electricity generators are powered by the direct action of water either from a reservoir or from the run of the river. Large scale hydro covers plants belonging to companies with an aggregate hydro capacity of 5 MWe and over. Most of the plants are located in Scotland and Wales and mainly draw their water from high level reservoirs with their own natural catchment areas. Figures from the schemes are provided to the Department of Trade and Industry. The data excludes pumped storage stations (see paragraph 5.59). The coverage of these large scale hydro figures is the same as that used in the tables in the Chapter 5 of this Digest. This year the large scale hydro generation figure increased by 2 per cent on 1998, which in turn was 25½ per cent up on 1997 and that was 22½ per cent up on the 1996 figure due to increases in rainfall following the dry years of 1995 and 1996.

Small scale hydro

7.44 Electricity generation schemes belonging to companies with an aggregate hydro capacity below 5 MWe are classified as small scale. These are schemes being used for either domestic/farm purposes or for sale to the local regional electricity company. Data given for generation are actual figures where available, but otherwise are estimated using a typical load factor (55 per cent), or the design load factor, where known. The estimated figures for 1998 and 1997 have been calculated using a load factor of 36 per cent, while those for 1996 have been calculated using a load factor of 28 per cent to reflect the reduction of generation due to the reduced rainfall. A new survey of small scale hydro sites was carried out in 1999 giving a more detailed picture of the current situation. 146 small scale hydro schemes were contracted within NFFO. Twenty two of these were existing schemes. Four new schemes have come on line in 1999.

Geothermal aquifers

7.45 Aquifers containing water at elevated temperatures occur in some parts of the United Kingdom at between 1,500 and 3,000 metres below the surface. This water can be pumped to the surface and used, for example, in community heating schemes.

Biofuels

(a)   Landfill gas

7.46 Landfill gas is a methane-rich biogas formed from the decomposition of organic material in landfill. The gas can be used to fuel reciprocating engines or turbines to generate electricity or used directly in kilns and boilers. In other countries, the gas has been cleaned to pipeline quality or used as a vehicle fuel. Data on landfill gas exploitation are provided from LAMMCOS, the LAndfill gas Monitoring, Modelling and COmmunication System. This is a landfill gas database maintained by ETSU and containing information on all existing landfill gas exploitation schemes. Landfill gas exploitation has benefited considerably from the NFFO and this can be seen from the large rise in the amount of electricity generated

since 1992. Further commissioning of landfill gas projects under NFFO will continue to increase the amount of electricity generated from this technology. In 1999, 34 new schemes came on line under NFFO.

(b)   Sewage sludge digestion

7.47 In all sewage sludge digestion projects, some of the gas produced is used to maintain the optimum temperature for digestion. In addition, many use combined heat and power (CHP) systems. The electricity generated is either used on site or sold under the NFFO. Information from these projects was provided from the CHAPSTAT Database, which is compiled and maintained by ETSU on behalf of the Department of Trade and Industry. (See Chapter 6).

(c)   Domestic wood combustion

7.48 Domestic wood use includes the use of logs in open fires, "AGA"-type cooker boilers and other wood burning stoves. The figure given is an approximate estimate based on a survey carried out in 1989. A new survey to provide current information will be carried out in 2000.

(d)   Industrial wood combustion

7.49 In 1997, the industrial wood figure (which includes sawmill residues, furniture manufacturing waste etc.) was included as a separate category for the first time. This was due to the availability of better data as a result of a survey carried out in 1996 on wood fired combustion plants above 400 kW thermal input. From this survey, in which 194 sites were identified, an estimate was made for heat generation.

(e)   Coppice

7.50 Short rotation willow coppice development is now becoming well established with demonstration projects underway in Northern Ireland and England.

7.51 Under Northern Ireland's second Non-Fossil Fuel Renewable Energy order for electricity, four projects were live at the end of 1999. These include a 200 kW-electricity generator is installed at Blackwater Valley Museum in Co. Armagh. It is currently fuelled on wood chips produced from sawmill residues but, in the future, will make use of willow coppice. Another similar unit has been installed at Brook Hall Estate in Londonderry on a large arable farm where 100 kW of electricity can be generated through a gasifier, engine and generator fuelled off forest residues. Willow coppice is being grown on the farm and will be used as a fuel when it is ready for harvest.

7.52 In England, Project ARBRE in South Yorkshire is now under construction and will generate 8 MWe under NFFO 3. Some 500 hectares of short rotation coppiced willow have been established with a further 700 planted in spring 2000. Coppiced willow is expected to make up approximately 70 per cent of the plant's fuel requirement with the balance coming from forestry residues.

(f) Straw combustion

7.53 Straw can be burnt in high temperature boilers, designed for the efficient and controlled combustion of solid fuels and biomass to supply heat, hot water and hot air systems. There are large numbers of these small-scale batch feed whole bale boilers. The figures given are estimates based partly on 1990 information and partly on a survey of straw-fired boilers carried out in 1993-94.

(g) Waste combustion

7.54 Domestic, industrial and commercial wastes represent a significant resource for materials and energy recovery. Wastes may be combusted, as received, in purpose built incinerators or processed into a range of refuse derived fuels (RDFs) for both on-site and off-site utilisation. The paragraphs below describe various categories of waste combustion in greater detail.

7.55 **Municipal solid waste combustion:** Information was provided from the refuse incinerator operators in the United Kingdom that practice energy recovery. This included both direct combustion of unprocessed MSW and the combustion of refuse derived fuel (RDF). In the latter, process waste can be partially processed to produce coarse RDF which can then be burnt in a variety of ways. By further processing the refuse, including separating off the fuel fraction, compacting, drying and densifying, it is possible to produce an RDF pellet. This pellet has around 60 per cent of the gross calorific value of British coal.

7.56 Information on projects in this area was obtained from data collected, using the RESTATS questionnaire, for 1999.

7.57 90 MSW schemes, including CHP, have been contracted under NFFO, with 14 currently being operational.

7.58 **General/industrial waste combustion:** Certain wastes produced by industry and commerce can be used as a source of energy for industrial processes or space heating. These wastes include general waste from factories such as paper, cardboard, wood and plastics. Schemes burning general or industrial waste were identified through contact with equipment manufacturers. Data collected from the

1994 survey were used to derive estimates for 1998. This information was then used to derive an estimate of the overall energy contribution from the industry, which has not significantly altered for 1999.

7.59 **Specialised waste combustion:** Specialised wastes arise as a result of a particular activity or process. Materials in this category include scrap tyres, hospital wastes, poultry litter and farm waste digestion. All these data are included under the 'others' category.

7.60 One tyre incinerator is known to be operating with energy recovery. This is a large plant generating electricity.

7.61 Information on hospital waste incineration is based 1999 RESTATS survey carried by ETSU on behalf of the Department of Trade and Industry.

7.62 One poultry litter combustion project started generating electricity in 1992; a second began in 1993. Both of these are NFFO projects. In addition, a small-scale CHP scheme began generating towards the end of 1990 however this has now closed due to new emissions regulations. A further NFFO scheme started generating in 1998 at Thetford. This is now close to full capacity and has almost doubled the total figure for poultry litter generation between 1998 and 1999.

7.63 The Agricultural Development and Advisory Service (ADAS) carried out a review of farm waste digestion in the United Kingdom for the Department of Trade and Industry during 1991-92. Included in this review was a survey that identified all the farm waste digestors in the United Kingdom and provided an estimate of the thermal energy they produced. Information was collected from these projects during the 1993 survey and this information was also used to derive estimates for 1995. In 1996 farm waste digestion was again surveyed and a new estimate derived from the information gathered. There is a farm digestion project generating electricity under the NFFO; its output is included in the 'Other' category (Table 7.6) and under 'Poultry litter, farm waste digestion, and tyres' (Tables 7.1, 7.2 and 7.3) but it has now ceased to operate. Data collected from the 1996 survey were used to derive estimates for 1997, 1998 and 1999.

Combined Heat and Power

7.64 A Combined Heat and Power (CHP) plant is an installation where there is a simultaneous generation of usable heat and power (usually electricity) in a single process. Some CHP installations are fuelled either wholly or partially by renewable sources of energy.

The main renewable sources that are used for CHP are biofuels particularly sewage gas.

7.65 Chapter 6 of this Digest summarises information on the contribution made by CHP to the United Kingdom's energy requirements in 1999 using the results of a study undertaken to identify all CHP schemes. Included in Tables 6.1 to 6.9 of that chapter is information on the contribution of renewable sources to CHP generation in each year from 1995 to 1999. The information contained in those tables is therefore a subset of the data contained within the tables presented in this chapter.

Capacity and load factor

7.66 The electrical capacities are given in Table 7.5 as DNC (Declared Net Capacity), i.e. the maximum continuous rating of the generating sets in the stations, less the power consumed by the plant itself, and reduced by a specified factor to take into account the intermittent nature of the energy source e.g. 0.43 for wind. DNC represents the nominal maximum capability of a generating set to supply electricity to consumers.

7.67 Plant load factors have been calculated in terms of installed capacity (i.e. the maximum continuous rating of the generating sets in the stations) and express the average hourly quantity of electricity generated as a percentage of the average capacity at the beginning and end of the year.

Contact: *Steve Dagnall, ETSU*
steve.dagnall@aeat.co.uk
01235 433580

Mike Janes, DTI, Statistician
mike.janes@dti.gsi.gov.uk
020-7215 5186

7.1 Commodity balances 1999

Renewables and waste

	Wood waste	Wood	Straw	Poultry litter, farm waste digestion and tyres	Sewage gas	Landfill gas	Municipal solid waste
Supply							
Production	506	204	72	193	189	572	580
Other sources	-	-	-	-	-	-	-
Imports	-	-	-	-	-	-	-
Exports	-	-	-	-	-	-	-
Marine bunkers	-	-	-	-	-	-	-
Stock change *(1)*	-	-	-	-	-	-	-
Transfers	-	-	-	-	-	-	-
Total supply	**506**	**204**	**72**	**193**	**189**	**572**	**580**
Statistical difference *(2)*	-	-	-	-	-	-	-
Total demand	**506**	**204**	**72**	**193**	**189**	**572**	**580**
Transformation	-	-	-	193	135	558	548
Electricity generation	-	-	-	193	135	558	548
Major power producers	-	-	-	193	-	-	-
Autogenerators	-	-	-	-	135	558	548
Petroleum refineries	-	-	-	-	-	-	-
Coke manufacture	-	-	-	-	-	-	-
Blast furnaces	-	-	-	-	-	-	-
Patent fuel manufacture	-	-	-	-	-	-	-
Other	-	-	-	-	-	-	-
Energy industry use	-	-	-	-	-	-	-
Electricity generation	-	-	-	-	-	-	-
Oil and gas extraction	-	-	-	-	-	-	-
Petroleum refineries	-	-	-	-	-	-	-
Coal extraction	-	-	-	-	-	-	-
Coke manufacture	-	-	-	-	-	-	-
Blast furnaces	-	-	-	-	-	-	-
Patent fuel manufacture	-	-	-	-	-	-	-
Pumped storage	-	-	-	-	-	-	-
Other	-	-	-	-	-	-	-
Losses	-	-	-	-	-	-	-
Final consumption	**506**	**204**	**72**	-	**54**	**14**	**32**
Industry	**506**	-	-	-	-	**14**	**5**
Unclassified	506	-	-	-	-	14	5
Iron and steel	-	-	-	-	-	-	-
Non-ferrous metals	-	-	-	-	-	-	-
Mineral products	-	-	-	-	-	-	-
Chemicals	-	-	-	-	-	-	-
Mechanical engineering etc	-	-	-	-	-	-	-
Electrical engineering etc	-	-	-	-	-	-	-
Vehicles	-	-	-	-	-	-	-
Food, beverages etc	-	-	-	-	-	-	-
Textiles, leather, etc	-	-	-	-	-	-	-
Paper, printing etc	-	-	-	-	-	-	-
Other industries	-	-	-	-	-	-	-
Construction	-	-	-	-	-	-	-
Transport	-	-	-	-	-	-	-
Air	-	-	-	-	-	-	-
Rail	-	-	-	-	-	-	-
Road	-	-	-	-	-	-	-
National navigation	-	-	-	-	-	-	-
Pipelines	-	-	-	-	-	-	-
Other	-	**204**	**72**	-	**54**	-	**27**
Domestic	-	204	-	-	-	-	..
Public administration	-	-	-	-	54	-	..
Commercial	-	-	-	-	-	-	..
Agriculture	-	-	72	-	-	-	-
Miscellaneous	-	-	-	-	-	-	..
Non energy use	-	-	-	-	-	-	-

(1) Stock fall (+), stock rise (-).
(2) Total supply minus total demand.

7.1 Commodity balances 1999 (continued)

Renewables and waste

Thousand tonnes of oil equivalent

General industrial and hospital waste	Geothermal & active solar heat	Hydro	Wind	Total renewables	
					Supply
48	11	460	77	2,912	Production
-	-	-	-	-	Other sources
-	-	-	-	-	Imports
-	-	-	-	-	Exports
-	-	-	-	-	Marine bunkers
-	-	-	-	-	Stock change (1)
-	-	-	-	-	Transfers
48	11	460	77	2,912	**Total supply**
-	-	-	-	-	**Statistical difference** (2)
48	11	460	77	2,912	**Total demand**
-	-	460	77	1,971	**Transformation**
-	-	460	77	1,971	Electricity generation
-	-	380	-	573	Major power producers
-	-	80	77	1,398	Autogenerators
-	-	-	-	-	Petroleum refineries
-	-	-	-	-	Coke manufacture
-	-	-	-	-	Blast furnaces
-	-	-	-	-	Patent fuel manufacture
-	-	-	-	-	Other
-	-	-	-	-	**Energy industry use**
-	-	-	-	-	Electricity generation
-	-	-	-	-	Oil and gas extraction
-	-	-	-	-	Petroleum refineries
-	-	-	-	-	Coal extraction
-	-	-	-	-	Coke manufacture
-	-	-	-	-	Blast furnaces
-	-	-	-	-	Patent fuel manufacture
-	-	-	-	-	Pumped storage
-	-	-	-	-	Other
-	-	-	-	-	**Losses**
48	11	-	-	941	**Final consumption**
10	-	-	-	535	**Industry**
10	-	-	-	535	Unclassified
-	-	-	-	-	Iron and steel
-	-	-	-	-	Non-ferrous metals
-	-	-	-	-	Mineral products
-	-	-	-	-	Chemicals
-	-	-	-	-	Mechanical engineering etc
-	-	-	-	-	Electrical engineering etc
-	-	-	-	-	Vehicles
-	-	-	-	-	Food, beverages etc
-	-	-	-	-	Textiles, leather, etc
-	-	-	-	-	Paper, printing etc
-	-	-	-	-	Other industries
-	-	-	-	-	Construction
-	-	-	-	-	**Transport**
-	-	-	-	-	Air
-	-	-	-	-	Rail
-	-	-	-	-	Road
-	-	-	-	-	National navigation
-	-	-	-	-	Pipelines
38	11	-	-	406	**Other**
-	..	-	-	..	Domestic
38	..	-	-	..	Public administration
-	..	-	-	..	Commercial
-	-	-	-	72	Agriculture
-	..	-	-	..	Miscellaneous
-	-	-	-	-	**Non energy use**

7.2 Commodity balances 1998

Renewables and waste

Thousand tonnes of oil equivalent

	Wood waste	Wood	Straw	Poultry litter, farm waste digestion and tyres	Sewage gas	Landfill gas	Municipal solid waste
Supply							
Production	506	204	72	147r	180r	403r	574r
Other sources	-	-	-	-	-	-	-
Imports	-	-	-	-	-	-	-
Exports	-	-	-	-	-	-	-
Marine bunkers	-	-	-	-	-	-	-
Stock change (1)	-	-	-	-	-	-	-
Transfers	-	-	-	-	-	-	-
Total supply	506	204	72	147r	180r	403r	574r
Statistical difference (2)	-	-	-	-	-	-	-
Total demand	506	204	72	147r	180r	403r	574r
Transformation	-	-	-	147r	126	389r	550r
Electricity generation	-	-	-	147r	126	389r	550r
Major power producers	-	-	-	147r	-	-	-
Autogenerators	-	-	-	-	126	389r	550r
Petroleum refineries	-	-	-	-	-	-	-
Coke manufacture	-	-	-	-	-	-	-
Blast furnaces	-	-	-	-	-	-	-
Patent fuel manufacture	-	-	-	-	-	-	-
Other	-	-	-	-	-	-	-
Energy industry use	-	-	-	-	-	-	-
Electricity generation	-	-	-	-	-	-	-
Oil and gas extraction	-	-	-	-	-	-	-
Petroleum refineries	-	-	-	-	-	-	-
Coal extraction	-	-	-	-	-	-	-
Coke manufacture	-	-	-	-	-	-	-
Blast furnaces	-	-	-	-	-	-	-
Patent fuel manufacture	-	-	-	-	-	-	-
Pumped storage	-	-	-	-	-	-	-
Other	-	-	-	-	-	-	-
Losses	-	-	-	-	-	-	-
Final consumption	506	204	72	-	54r	14	24r
Industry	506	-	-	-	-	14	-
Unclassified	506	-	-	-	-	14	-
Iron and steel	-	-	-	-	-	-	-
Non-ferrous metals	-	-	-	-	-	-	-
Mineral products	-	-	-	-	-	-	-
Chemicals	-	-	-	-	-	-	-
Mechanical engineering etc	-	-	-	-	-	-	-
Electrical engineering etc	-	-	-	-	-	-	-
Vehicles	-	-	-	-	-	-	-
Food, beverages etc	-	-	-	-	-	-	-
Textiles, leather, etc	-	-	-	-	-	-	-
Paper, printing etc	-	-	-	-	-	-	-
Other industries	-	-	-	-	-	-	-
Construction	-	-	-	-	-	-	-
Transport	-	-	-	-	-	-	-
Air	-	-	-	-	-	-	-
Rail	-	-	-	-	-	-	. -
Road	-	-	-	-	-	-	-
National navigation	-	-	-	-	-	-	-
Pipelines	-	-	-	-	-	-	-
Other	-	204	72	-	54r	-	24r
Domestic	-	204	-	-	-	-	..
Public administration	-	-	-	-	54r	-	..
Commercial	-	-	-	-	-	-	..
Agriculture	-	-	72	-	-	-	-
Miscellaneous	-	-	-	-	-	-	..
Non energy use	-	-	-	-	-	-	-

(1) Stock fall (+), stock rise (-).
(2) Total supply minus total demand.

7.2 Commodity balances 1998 (continued)

Renewables and waste

Thousand tonnes of oil equivalent

General industrial and hospital waste	Geothermal & active solar heat	Hydro	Wind	Total renewables	
					Supply
51r	10r	450r	75r	2,672r	Production
-	-	-	-	-	Other sources
-	-	-	-	-	Imports
-	-	-	-	-	Exports
-	-	-	-	-	Marine bunkers
-	-	-	-	-	Stock change *(1)*
-	-	-	-	-	Transfers
51r	10r	450r	75r	2,672r	**Total supply**
-	-	-	-	-	**Statistical difference** *(2)*
51r	10r	450r	75r	2,672r	**Total demand**
-	-	450r	75r	1,737r	**Transformation**
-	-	450r	75r	1,737r	Electricity generation
-	-	374r	-	521r	Major power producers
-	-	76r	75r	1,216r	Autogenerators
-	-	-	-	-	Petroleum refineries
-	-	-	-	-	Coke manufacture
-	-	-	-	-	Blast furnaces
-	-	-	-	-	Patent fuel manufacture
-	-	-	-	-	Other
-	-	-	-	-	**Energy industry use**
-	-	-	-	-	Electricity generation
-	-	-	-	-	Oil and gas extraction
-	-	-	-	-	Petroleum refineries
-	-	-	-	-	Coal extraction
-	-	-	-	-	Coke manufacture
-	-	-	-	-	Blast furnaces
-	-	-	-	-	Patent fuel manufacture
-	-	-	-	-	Pumped storage
-	-	-	-	-	Other
-	-	-	-	-	**Losses**
51r	10r	-	-	935r	**Final consumption**
10	-	-	-	530	**Industry**
10	-	-	-	530	Unclassified
-	-	-	-	-	Iron and steel
-	-	-	-	-	Non-ferrous metals
-	-	-	-	-	Mineral products
-	-	-	-	-	Chemicals
-	-	-	-	-	Mechanical engineering etc
-	-	-	-	-	Electrical engineering etc
-	-	-	-	-	Vehicles
-	-	-	-	-	Food, beverages etc
-	-	-	-	-	Textiles, leather, etc
-	-	-	-	-	Paper, printing etc
-	-	-	-	-	Other industries
-	-	-	-	-	Construction
-	-	-	-	-	**Transport**
-	-	-	-	-	Air
-	-	-	-	-	Rail
-	-	-	-	-	Road
-	-	-	-	-	National navigation
-	-	-	-	-	Pipelines
41r	10r	-	-	405r	**Other**
-	..	-	-	..	Domestic
41r	..	-	-	..	Public administration
-	..	-	-	..	Commercial
-	-	-	-	72	Agriculture
-	..	-	-	..	Miscellaneous
-	-	-	-	-	**Non energy use**

7.3 Commodity balances 1997

Renewables and waste

Thousand tonnes of oil equivalent

	Wood waste	Wood	Straw	Poultry litter, farm waste digestion and tyres	Sewage gas	Landfill gas	Municipal solid waste
Supply							
Production	506	204	72	139r	192r	316r	424r
Other sources	-	-	-	-	-	-	-
Imports	-	-	-	-	-	-	-
Exports	-	-	-	-	-	-	-
Marine bunkers	-	-	-	-	-	-	-
Stock change (1)	-	-	-	-	-	-	-
Transfers	-	-	-	-	-	-	-
Total supply	**506**	**204**	**72**	**139r**	**192r**	**316r**	**424r**
Statistical difference (2)	-	-	-	-	-	-	-
Total demand	**506**	**204**	**72**	**139r**	**192r**	**316r**	**424r**
Transformation	-	-	-	**139r**	**134r**	**301**	**410r**
Electricity generation	-	-	-	139r	134r	301	410r
Major power producers	-	-	-	139r	-	-	-
Autogenerators	-	-	-	-	134r	301	410r
Petroleum refineries	-	-	-	-	-	-	-
Coke manufacture	-	-	-	-	-	-	-
Blast furnaces	-	-	-	-	-	-	-
Patent fuel manufacture	-	-	-	-	-	-	-
Other	-	-	-	-	-	-	-
Energy industry use	-	-	-	-	-	-	-
Electricity generation	-	-	-	-	-	-	-
Oil and gas extraction	-	-	-	-	-	-	-
Petroleum refineries	-	-	-	-	-	-	-
Coal extraction	-	-	-	-	-	-	-
Coke manufacture	-	-	-	-	-	-	-
Blast furnaces	-	-	-	-	-	-	-
Patent fuel manufacture	-	-	-	-	-	-	-
Pumped storage	-	-	-	-	-	-	-
Other	-	-	-	-	-	-	-
Losses	-	-	-	-	-	-	-
Final consumption	**506**	**204**	**72**	**-**	**58r**	**15r**	**14**
Industry	**506**	-	-	-	-	**15r**	-
Unclassified	506	-	-	-	-	15r	-
Iron and steel	-	-	-	-	-	-	-
Non-ferrous metals	-	-	-	-	-	-	-
Mineral products	-	-	-	-	-	-	-
Chemicals	-	-	-	-	-	-	-
Mechanical engineering etc	-	-	-	-	-	-	-
Electrical engineering etc	-	-	-	-	-	-	-
Vehicles	-	-	-	-	-	-	-
Food, beverages etc	-	-	-	-	-	-	-
Textiles, leather, etc	-	-	-	-	-	-	-
Paper, printing etc	-	-	-	-	-	-	-
Other industries	-	-	-	-	-	-	-
Construction	-	-	-	-	-	-	-
Transport	-	-	-	-	-	-	-
Air	-	-	-	-	-	-	-
Rail	-	-	-	-	-	-	-
Road	-	-	-	-	-	-	-
National navigation	-	-	-	-	-	-	-
Pipelines	-	-	-	-	-	-	-
Other	-	**204**	**72**	**-**	**58r**	-	**14**
Domestic	-	204	-	-	-	-	..
Public administration	-	-	-	-	58r	-	..
Commercial	-	-	-	-	-	-	..
Agriculture	-	-	72	-	-	-	-
Miscellaneous	-	-	-	-	-	-	..
Non energy use	-	-	-	-	-	-	-

(1) Stock fall (+), stock rise (-).
(2) Total supply minus total demand.

7.3 Commodity balances 1997 (continued)

Renewables and waste

Thousand tonnes of oil equivalent

General industrial and hospital waste	Geothermal & active solar heat	Hydro	Wind	Total renewables	
					Supply
51r	10	358r	57	2,329r	Production
-	-	-	-	-	Other sources
-	-	-	-	-	Imports
-	-	-	-	-	Exports
-	-	-	-	-	Marine bunkers
-	-	-	-	-	Stock change *(1)*
-	-	-	-	-	Transfers
51r	10	358r	57	2,329r	**Total supply**
-	-	-	-	-	**Statistical difference** *(2)*
51r	10	358r	57	2,329r	**Total demand**
-	-	358r	57	1,399r	**Transformation**
-	-	358r	57	1,399r	Electricity generation
-	-	287r	-	426r	Major power producers
-	-	71r	57	973r	Autogenerators
-	-	-	-	-	Petroleum refineries
-	-	-	-	-	Coke manufacture
-	-	-	-	-	Blast furnaces
-	-	-	-	-	Patent fuel manufacture
-	-	-	-	-	Other
-	-	-	-	-	**Energy industry use**
-	-	-	-	-	Electricity generation
-	-	-	-	-	Oil and gas extraction
-	-	-	-	-	Petroleum refineries
-	-	-	-	-	Coal extraction
-	-	-	-	-	Coke manufacture
-	-	-	-	-	Blast furnaces
-	-	-	-	-	Patent fuel manufacture
-	-	-	-	-	Pumped storage
-	-	-	-	-	Other
-	-	-	-	-	**Losses**
51r	10	-	-	930r	**Final consumption**
10	-	-	-	531r	**Industry**
10	-	-	-	531r	Unclassified
-	-	-	-	-	Iron and steel
-	-	-	-	-	Non-ferrous metals
-	-	-	-	-	Mineral products
-	-	-	-	-	Chemicals
-	-	-	-	-	Mechanical engineering etc
-	-	-	-	-	Electrical engineering etc
-	-	-	-	-	Vehicles
-	-	-	-	-	Food, beverages etc
-	-	-	-	-	Textiles, leather, etc
-	-	-	-	-	Paper, printing etc
-	-	-	-	-	Other industries
-	-	-	-	-	Construction
-	-	-	-	-	**Transport**
-	-	-	-	-	Air
-	-	-	-	-	Rail
-	-	-	-	-	Road
-	-	-	-	-	National navigation
-	-	-	-	-	Pipelines
41r	10	-	-	399r	**Other**
-	..	-	-	..	Domestic
41r	..	-	-	..	Public administration
-	..	-	-	..	Commercial
-	-	-	-	72	Agriculture
-	..	-	-	..	Miscellaneous
-	-	-	-	-	**Non energy use**

7.4 Capacity of, and electricity generated from, renewable sources[1]

	1995	1996	1997	1998	1999
Declared Net Capacity (MWe)					
Onshore wind	85.1	113.0	135.4	139.4r	150.5
Solar photovoltaics	0.2	0.3	0.5	0.6	1.2
Hydro:					
Small scale	48.6r	49.1r	58.5r	61.6r	63.6
Large scale (2)	1,383.0	1,406.2	1,397.0r	1,413.0	1,413.0
Biofuels:					
Landfill gas	94.7r	145.7r	169.4r	220.6r	309.0
Sewage sludge digestion	87.2r	87.2r	86.8r	89.8r	91.3
Municipal solid waste combustion (3)	86.8	115.0	115.0	162.1r	158.6
Other (4)	45.5	45.5	45.6	84.2	84.2
Total biofuels	314.2r	393.4r	416.8r	556.7	643.0
Total	**1,831.1r**	**1,962.0r**	**2,008.1r**	**2,171.3r**	**2,271.4**
Generation (GWh)					
Onshore wind (5)	392r	488r	667r	877r	897
Solar Photovoltaics	-	-	-	-	1
Hydro:					
Small scale (5)	166	118r	164r	206r	232
Large scale (2)	4,672	3,275	4,005r	5,031r	5,120
Biofuels:					
Landfill gas	562r	708r	918r	1,185r	1,703
Sewage sludge digestion	410r	410r	408r	386	410
Municipal solid waste combustion (3)	747	777r	929r	1,348r	1,359
Other (4)	334	326	338r	318	515
Total biofuels	2,053r	2,221r	2,593r	3,237r	3,987
Total	**7,283r**	**6,101r**	**7,428r**	**9,352r**	**10,237**
Load factors (per cent) (6)					
Onshore wind (5)	25.3r	25.4r	27.2r	30.7r	29.8
Hydro	37.7r	25.3r	30.3r	38.0r	38.5
Biofuels	65.2r	60.8r	61.7	65.7r	65.8
Total	**41.5r**	**32.1r**	**36.4r**	**43.4r**	**44.5**

(1) Includes some waste of fossil fuel origin.
(2) Excluding pumped storage stations. Capacities are as at the end of December except for the capacities of installations of major power producing companies which for 1995 are recorded as at the end-March of the following year.
(3) Includes combustion of refuse derived fuel pellets.
(4) Includes the use of farm waste digestion, waste tyre combustion and poultry litter combustion. ·
(5) Actual generation figures are given where available, but otherwise are estimated using a typical load factor or the design load factor, where known.
(6) Load factors are calculated based on installed capacity rather than DNC - see paragraph 7.67.

7.5 Renewable orders and operational capacity

Technology band	Contracted projects		Live projects operational at 31 December 1999	
	Number	Capacity MW	Number	Capacity MW
England and Wales				
NFFO - 1 (1990)				
Hydro	26	11.85	21	10.00
Landfill gas	25	35.50	19	30.78
Municipal and industrial waste	4	40.63	4	40.63
Other	4	45.48	4	45.48
Sewage gas	7	6.45	6	5.98
Wind	9	12.21	7	11.66
Total	**75**	**152.12**	**61**	**144.53**
NFFO - 2 (late 1991)				
Hydro	12	10.86	10	10.46
Landfill gas	28	48.45	26	46.39
Municipal and industrial waste	10	271.48	2	31.50
Other	4	30.15	1	12.50
Sewage gas	19	26.86	18	19.06
Wind	49	84.43	25	53.83
Total	**122**	**472.23**	**82**	**173.74**
NFFO - 3 (1995)				
Energy crops and agricultural and forestry waste - gasification	3	19.06		
Energy crops and agricultural and forestry waste - other	6	103.81	1	38.50
Hydro	15	14.48	7	10.08
Landfill gas	42	82.07	42	82.07
Municipal and industrial waste	20	241.87	6	77.42
Wind - large	31	145.92	8	34.76
Wind - small	24	19.71	9	7.93
Total	**141**	**626.92**	**73**	**250.76**
NFFO - 4 (1997)				
Hydro	31	13.22	5	1.42
Landfill gas	70	173.68	43	103.30
Municipal and industrial waste - CHP	10	115.29		
Municipal and industrial waste - fluidised bed combustion	6	125.93		
Wind - large	48	330.36		
Wind - small	17	10.33	1	0.63
Anaerobic digestion of agricultural waste	6	6.58		
Energy crops and forestry waste gasification	7	67.34		
Total	**195**	**842.73**	**49**	**105.35**
NFFO - 5 (1998)				
Hydro	22	8.87		
Landfill gas	141	313.73	11	16.58
Municipal and industrial waste	22	415.75		
Municipal and industrial waste - CHP	7	69.97		
Wind - large	33	340.16		
Wind - small	36	28.67	2	1.69
Total	**261**	**1,177.15**	**13**	**18.27**
NFFO Total	**794**	**3,271.15**	**278**	**692.64**
Scotland				
SRO - 1 (1994)				
Biomass	1	9.80		
Hydro	15	17.25	4	3.22
Waste to Energy	2	3.78	2	3.78
Wind	12	45.60	7	25.13
Total	**30**	**76.43**	**13**	**32.13**
SRO - 2 (1997)				
Biomass	1	2.00		
Hydro	9	12.36		
Waste to Energy	9	56.05	3	6.7
Wind	7	43.63		
Total	**26**	**114.04**	**3**	**6.7**

7.5 Renewable orders and operational capacity (continued)

	Technology band	Contracted projects		Live projects operational at 31 December 1999	
		Number	Capacity MW	Number	Capacity MW
Scotland (continued)					
SRO - 3 (1999)	Biomass	1	12.90		
	Hydro	5	3.90		
	Waste to Energy	16	49.11		
	Wave	3	2.00		
	Wind - large	11	63.43		
	Wind - small	17	14.06		
	Total	**53**	**145.40**		
SRO Total		**109**	**335.87**	**16**	**38.83**
Northern Ireland					
NI NFFO - 1 (1994)	Hydro	9	2.37	7	1.89
	Sewage gas	5	0.56		
	Wind	6	12.66	6	12.66
	Total	**20**	**15.59**	**13**	**14.55**
NI NFFO - 2 (1996)	Biogas	1	0.25		
	Biomass	2	0.30	2	0.30
	Hydro	2	0.25	1	0.08
	Landfill gas	2	6.25		
	Municipal and industrial waste	1	6.65		
	Wind	2	2.57	1	0.43
	Total	**10**	**16.27**	**4**	**0.81**
NI NFFO Total		**30**	**31.86**	**17**	**15.36**
All Renewables Obligations		**933**	**3,638.88**	**311**	**746.83**

7.6 Renewable sources used to generate electricity and heat[1]

Thousand tonnes of oil equivalent

	Onshore wind[2]	Hydro Small scale	Hydro Large scale[3]	Landfill gas	Sewage sludge digestion[4]	Municipal solid waste combustion[5]	Other[6]	Total biofuels	Total
Used to generate electricity									
1989	0.7	10.9	400.7	45.6	90.5	143.4	-	279.5	**691.8**
1990	0.8	10.9	436.8	45.6	103.6	110.7	-	260.0	**708.5**
1991	0.7r	12.2	385.4	68.2	107.6	111.9		288.3	**686.6r**
1992	2.8r	12.8	454.1	123.6	107.6	136.3	0.6	385.1	**854.8r**
1993	18.7r	13.6	356.2	146.6	123.8	189.0	17.5	518.4	**907.0r**
1994	29.5	13.6	424.3	169.5	118.3	304.8	114.5	707.1	**1,174.6**
1995	33.7r	14.2r	401.7	184.3r	134.6r	315.3	133.2	767.4r	**1,217.0r**
1996	41.9r	10.1r	281.6	232.1r	134.6r	325.9r	131.3	823.9r	**1,157.5r**
1997	57.3r	14.1r	344.4r	301.1r	133.7r	409.9r	138.2r	982.8r	**1,398.7r**
1998	75.4r	17.7r	432.6r	388.8r	126.5	549.9r	146.6r	1,211.8r	**1,737.6r**
1999	77.2	19.9	440.2	558.4	134.6	548.2	192.9	1,434.2	**1,971.5**

	Active solar heating[7]	Landfill gas	Sewage sludge digestion[4]	Wood combustion - domestic[8]	Wood combustion - industrial	Straw combustion[9]	Municipal solid waste combustion[5]	Other[10]	Total biofuels	Geothermal aquifers[11]	Total
Used to generate heat											
1989	7.7	30.0	33.9	174.1	-	71.7	49.9	22.6r	382.1r	0.8	**390.6r**
1990	6.4	34.2	34.6	174.1	-	71.7	49.3	23.1r	387.0r	0.8	**394.2r**
1991	6.8	36.3	43.5	174.1	-	71.7	53.2	23.5	402.2	0.8	**409.8**
1992	7.1	31.5	43.5	204.2	-	71.7	49.0	31.3	431.1	0.8	**439.0r**
1993	7.4r	15.0	34.0	204.2	236.8	71.7	44.8	37.3	643.7	0.8	**651.9r**
1994	7.7r	18.9	52.1	204.2	455.1	71.7	46.8	43.6r	892.4r	0.8	**901.0r**
1995	8.1r	15.1r	58.5r	204.2	498.1	71.7	48.5r	51.4r	947.4r	0.8	**956.4r**
1996	8.5r	16.6r	58.5r	204.2	505.5	71.7	50.6r	46.2r	953.2r	0.8	**962.6r**
1997	9.0	15.5r	58.2r	204.2	506.1	71.7	14.3r	51.0r	920.9r	0.8	**930.7r**
1998	9.4r	13.6	54.1r	204.2	506.1	71.7	24.1r	50.7r	924.4r	0.8	**934.7r**
1999	10.0	13.6	54.2	204.2	506.1	71.7	32.0	48.0	929.8	0.8	**940.5**

	Active solar heating[7]	Onshore wind	Hydro	Biofuels	Geothermal aquifers[11]	Total
Total use of renewable sources						
1989	7.7	0.7	411.6	661.6r	0.8	**1,082.4r**
1990	6.4	0.8	447.7	647.0r	0.8	**1,102.7r**
1991	6.8	0.7r	397.6	690.5	0.8	**1,096.4r**
1992	7.1	2.8r	466.9	816.1	0.8	**1,293.9r**
1993	7.4r	18.7	369.8	1,162.1	0.8	**1,558.9r**
1994	7.7r	29.5r	437.9	1,599.5r	0.8	**2,075.6r**
1995	8.1r	33.7r	415.9r	1,714.8r	0.8	**2,173.4r**
1996	8.5r	41.9r	291.7r	1,777.1r	0.8	**2,120.0r**
1997	9.0	57.3r	358.5r	1,903.7r	0.8	**2,329.4r**
1998	9.4r	75.4r	450.3r	2,136.2r	0.8	**2,672.3r**
1999	10.0	77.2	460.1	2,363.9	0.8	**2,912.0**

(1) Includes some waste of fossil fuel origin.
(2) For wind and hydro, the figures represent the energy content of the electricity supplied, but for biofuels the figures represent the energy content of the fuel used.
(3) Excluding pumped storage stations.
(4) No estimate is made for digestors where gas is used to heat the sludge.
(5) Includes combustion of refuse derived fuel pellets.
(6) Includes electricity from farm waste digestion , poultry litter combustion and waste tyre combustion.

(7) Based on a survey carried out in 1995 and updated using data from the Solar Trade Association.
(8) An approximate estimate of domestic combustion based on a survey carried out in 1989; a moisture content of 50% is assumed.
(9) An approximate estimate based on a limited survey carried out in 1994 and on information collected in 1990.
(10) Includes heat from waste tyre combustion, hospital waste combustion, general industrial waste combustion and farm waste digestion.
(11) Based on information collected by the 1994 RESTATS questionnaire.

Chapter 8
Foreign trade

Introduction

8.1   This section brings together detailed figures on imports and exports of fuels and related materials, generally in both quantity and value terms. Table 8.1 gives an overall view for all fuels, Tables 8.3 to 8.5 present more detailed figures for crude oil, petroleum products and coal and other solid fuels. Table 8.2 presents a long term view of the value of imports and exports of fuels from 1965 to 1999.

8.2   The information in this section is largely derived from returns made to HM Customs and Excise, and corresponds to that published in the *Overseas Trade Statistics of the United Kingdom* (O.T.S.). The figures for 1999 are provisional.

Imports and exports of fuel and related materials   (Table 8.1)

8.3   This table presents import, export and net export figures in quantity and value terms broken down by the main fuel groups for the years 1995 to 1999.

8.4   To allow the values of imports and exports to be compared, additional series are included presenting import values on a "free on board" (f.o.b.) basis. Import values are normally recorded in "cost, insurance and freight" (c.i.f.) prices whereas f.o.b. prices are always used for export values. This approach is similar to that used by the Office for National Statistics in the overall trade figures when they compile the Balance of Payments. Fuller descriptions of the c.i.f. and f.o.b. methods of valuing imports and exports are given in paragraph 8.25 of the Technical Notes.

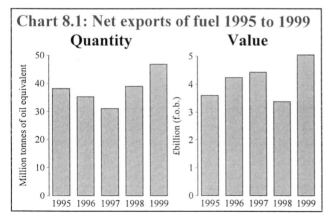

Chart 8.1: Net exports of fuel 1995 to 1999

8.5   Chart 8.1 illustrates the recent trends in the trade balance in fuels, both in terms of value and quantity. Trends in the value of the trade balance since 1970 can be seen in Chart 8.3, whilst figures for 1965 onwards are given in Table 8.2.

8.6   In 1999 the United Kingdom was a net exporter of fuels, in financial terms, with a surplus, on a balance of payments (f.o.b.) basis, of £5.0 billion, £1.6 billion higher than the surplus in 1998. The surplus of crude oil and petroleum products in 1999 was just over £4½ million compared to just under £3¼ million in 1998.   These increases reflect an increase in the volume of net exports and a rise in crude oil and petroleum product prices during 1999. At the end of December 1999, prices were nearly treble the low level seen in February 1999.

8.7   In volume terms the United Kingdom was also a net exporter of fuels in 1999, with net exports amounting to 46.8 million tonnes of oil equivalent (mtoe). This compares with surpluses of 39.0 mtoe in 1998 and 38.2 mtoe in 1995. There was an increase in imports of oil products in 1999, due to the need to make up for refinery closures in the United Kingdom.

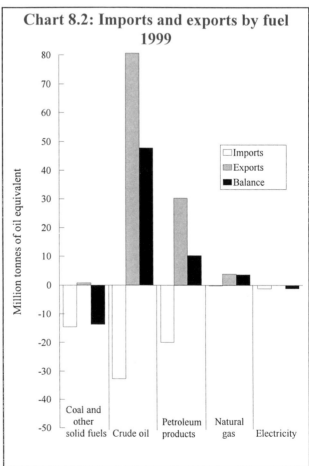

Chart 8.2: Imports and exports by fuel 1999

8.8   The figures for trade in individual fuels in 1999 are illustrated in Chart 8.2. This shows the extent to which the United Kingdom's trading position for all fuels is dominated by petroleum. The United Kingdom continues to be a net exporter of fuels largely as a result of high exports and low imports of crude oil and petroleum products.

Long term trends:
Value of imports and exports of fuels 1965 to 1999 (Table 8.2)

8.9   Values of imports (c.i.f.) and exports (f.o.b.) broken down by the main fuel groups are given in Table 8.2 which is based on Table 8.1 with the series extended back to 1965. Import values on a f.o.b. basis are also included back to 1970, enabling net exports to be presented on a comparable f.o.b. basis over the same period.

8.10   Although between 1989 and 1992 the United Kingdom was a net importer of fuels in volume terms, there has been a financial surplus in fuels, on a balance of payments (f.o.b.) basis, in every year since 1981, except for 1991. This is because the unit values of our exports have tended to be higher than that of our imports.

8.11   As can be seen in Chart 8.3 the United Kingdom's trade in fuels was dominated by imports until exports started to grow substantially in the mid-1970s, when production from the North Sea started coming on line, achieving a trade surplus in 1981. This surplus has been sustained, in value terms, since 1981, except for a small deficit in 1991, and amounted to just under £58¼ billion over the period 1981 to 1999. However, these surpluses were reduced by the fall in oil prices in 1986, and then by the fall in North Sea production following the Piper Alpha accident in 1988 and the resulting safety work. Although the trade surplus had increased steadily from 1992 to 1997, there was a slight fall in 1998 due to the fall in the price of crude oil. Prices of crude oil and petroleum products increased in 1999 giving it, in current price terms, the highest net surplus in recent years. However, this was still lower than in the early 1980s when oil prices were higher.

UK imports and exports of crude oil and petroleum products (Table 8.3)

8.12   The data in this table outline the pattern of trade in oil in the United Kingdom.   Table 8.3 shows quantities in thousands of tonnes, of crude oil and refined petroleum products, and unit values per tonne, with import values on a c.i.f. basis and export values on a f.o.b. basis. The total values of crude oil imports, on a f.o.b. basis, are shown in Table 8.1.

8.13   The United Kingdom has been a net exporter of oil since 1981. Broadly the level of crude oil exports reflects North Sea production. Exports were reduced because the Piper Alpha accident reduced production from 1988 and then production levels remained lower until 1992 as higher levels of maintenance and safety work prolonged platform shutdowns. Chart 8.4 shows the level of imports, exports and net exports in f.o.b. value terms from 1995 to 1999.

8.14   The main product imported into the United Kingdom is fuel oil.   As well as being used by electricity generators and by industry, the figures include imports of partly refined oil for further processing in UK refineries.   The main product exported is motor spirit.   UK refinery capacity is geared towards production of motor spirit, due to the increased investment that took place in the 1970s to increase such capacity to take advantage of the then increasing demand for motor spirit in both the UK and the rest of the world.   Since that time demand for motor spirit has stabilised and in fact, in recent years, has reduced slightly.   This is a consequence of a switch to diesel-engined vehicles and increases in engine efficiency, offsetting increasing numbers of vehicles and mileage. As such, the refinery capacity in the UK exceeds the level of demand in the UK, with the excess going for export.

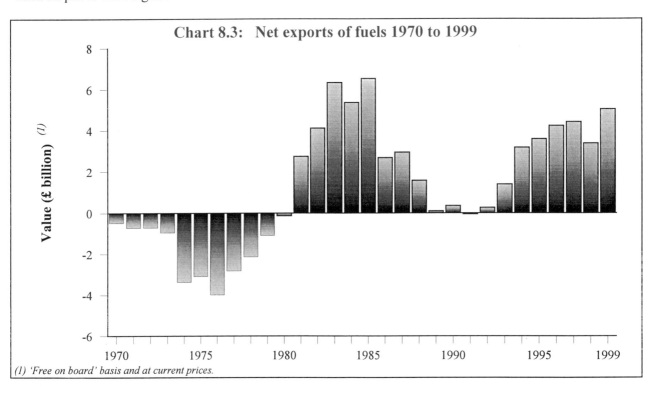

Chart 8.3:   Net exports of fuels 1970 to 1999

(1) 'Free on board' basis and at current prices.

8.15   In recent years there have been several closures of oil refineries in the UK (Gulf Oil's Milford Haven refinery in November 1997 and Shell's Shellhaven refinery in December 1999).   The effect of these closures in reducing the UK's excess refinery capacity is visible in the trade figures.   Imports of petroleum products have grown from 17.1 million tonnes in 1998 to 19.0 in 1999, due to increased imports of aviation turbine fuel and DERV fuel needed to meet UK demand for these fuels that cannot be met by UK refineries.   Similarly, there has been a slight downward trend in exports of petroleum products in recent years, with total exports of petroleum products in 1999 being only slightly higher than in 1998, but 12 per cent down on 1997.   Exports of motor spirit have decreased by almost 15 per cent between 1998 and 1999 as the surplus UK production previously exported now goes to meet UK demand. Other main exported products are gas oil/diesel oil, and fuel oil, which again includes partly refined oil for further processing.   Just under 75 per cent of UK petroleum product exports in 1999 went to other EU countries.   The largest customers are France, Germany, the Netherlands, and Spain.   The USA, who receive over 15 per cent of petroleum products exports, are the largest non-EU customer.

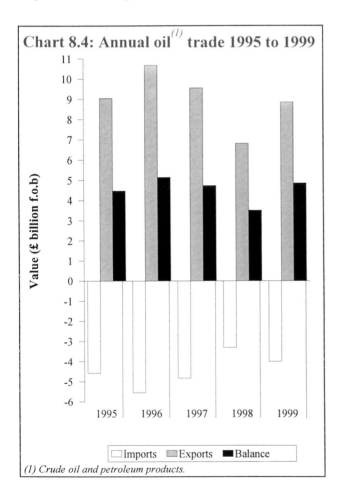

Chart 8.4: Annual oil$^{(1)}$ trade 1995 to 1999

Value (£ billion f.o.b)

☐ Imports ▨ Exports ■ Balance

(1) Crude oil and petroleum products.

UK imports and exports of crude oil by country (Table 8.4)

8.16   The data in Table 8.4 show details of trade in crude oil by country.   The import data are on a 'country of origin' (or production) basis as far as possible.   Since the introduction of 'Intrastat' at the start of 1993, recording of country of origin for Intra-EU trade has been optional, so a small amount may be recorded as country of consignment i.e. the country from which the goods were consigned to the United Kingdom as opposed to the true country of origin.   This change has had little impact, as virtually all of the UK's imported crude oil is supplied direct from countries outside the EU, in particular Norway.

8.17   Norway supplied just under 75 per cent of the United Kingdom's imports of crude oil in 1999.   The Middle East accounted for just under 8 per cent of imports, coming mainly from Saudi Arabia.   Over half of the remaining imports came from Mexico, Russia and Venezuela.   In 1999, two-thirds of United Kingdom exports of crude oil went to EU countries. Most of the non-EU export trade was with the United States of America.   Whilst the bulk of the exports to Germany are for refining and consumption there, the exports to the Netherlands include oil destined for onward trade to other countries.

8.18   In most years, the average value per tonne of crude oil exported from the UK is higher than that for imported crude oil.   However, in recent years (for example 1993 and 1995) this difference has been reducing, and, in 1996 the import value per tonne was greater than the export figure.   In 1997 and 1998, the impact of the higher than usual oil prices for imports from Norway was reduced, and so the difference resumed its normal pattern.   1999 saw a significant increase in prices of crude oil on international markets due to the actions of OPEC member countries to reduce world supply, leading to the differential increasing to £5.36 per tonne in 1999. This action by OPEC countries is also the reason why the volume of imports of crude oil into the UK in 1999 was 17 per cent lower than in 1998, while the value of imports increased by 5 per cent.

Imports and exports of solid fuels (Table 8.5)

8.19   Table 8.5 gives a breakdown of imports and exports of steam coal, coking coal, anthracite and other solid fuels by country of origin or destination. The imports and exports data are provided by HM Customs and Excise, but where there have been apparent misclassifications by the importers of the types of coal (e.g. because the country of origin does not produce that type of coal) the DTI has made adjustments.

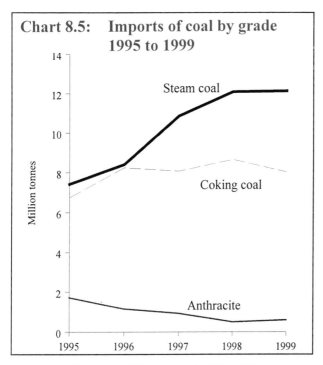

Chart 8.5:   Imports of coal by grade 1995 to 1999

Steam coal

Coking coal

Anthracite

Million tonnes

1995   1996   1997   1998   1999

8.20   In 1999, the UK imported 20.7 million tonnes of coal, 2½ per cent less than in 1998. 38½ per cent of coal imports were of coking coal, of which only limited amounts are produced in the United Kingdom. The figures for imports of coal by grade are illustrated in Chart 8.5.

8.21   In 1999, 65 per cent of the United Kingdom's imports of coal came from just three countries: Australia, USA and Colombia. A further 27 per cent of coal imports came from three additional countries, Canada (coking coal), South Africa (mainly steam coal) and Poland. Steam coal imports came mainly from Colombia (38 per cent), South Africa (22 per cent) and Australia (14 per cent); imports of steam coal from the USA were substantially lower in 1999 accounting for 6 per cent of the total compared with 26 per cent in 1998. All but 1 per cent of UK coking coal imports came from Australia (53 per cent), the USA (28 per cent) and Canada (18 per cent). Imports of coal by country of origin are illustrated in Chart 8.6.

8.22   Exports of coal amounted to 0.8 million tonnes in 1999, a decrease of 19 per cent on 1998. The UK's largest export markets in 1999 were the Irish Republic (36 per cent), Norway (20 per cent), France (13 per cent) and Belgium and Spain (both 9 per cent).

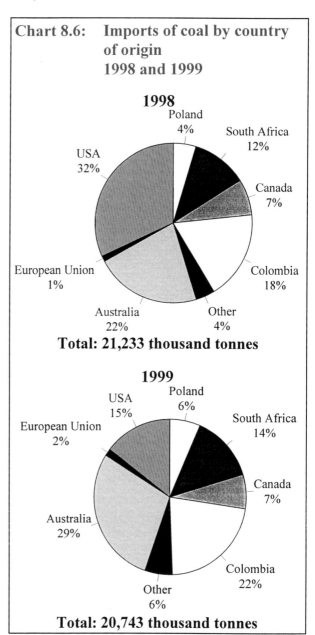

Chart 8.6:   Imports of coal by country of origin 1998 and 1999

1998

Poland 4%
South Africa 12%
Canada 7%
Colombia 18%
Other 4%
Australia 22%
European Union 1%
USA 32%

Total: 21,233 thousand tonnes

1999

USA 15%
Poland 6%
South Africa 14%
Canada 7%
Colombia 22%
Other 6%
Australia 29%
European Union 2%

Total: 20,743 thousand tonnes

Technical notes and definitions

8.23 The figures of imports and exports quoted are largely derived from notifications to HM Customs and Excise, and may differ from those for actual arrivals and shipments, derived from alternative and/or additional sources, in the sections of the Digest dealing with individual fuels. Data in Table 8.1 also include unpublished revisions to Customs data which cannot be introduced into Tables 8.3 to 8.5.

8.24 All quantity figures in Table 8.1 have been converted to million tonnes of oil equivalent to allow data to be compared and combined. This unit is a measure of the energy content of the individual fuels; it is also used in the Energy section of this Digest where further explanation can be found. (See Chapter 1, paragraphs 1.45 to 1.46). The quantities of imports and exports recorded in the Overseas Trade Statistics in their original units of measurement, are converted to tonnes of oil equivalent using weighted gross calorific values and standard conversion factors appropriate to each division of the Standard International Trade Classification (SITC). The electricity figures are expressed in terms of the energy content of the electricity traded.

8.25 Except as noted in Table 8.1, values of imports are quoted "c.i.f." (cost, insurance and freight); briefly this value is the price which the goods would fetch at that time, on sale in the open market between buyer and seller independent of each other, with delivery to the buyer at the port of importation, the seller bearing freight, insurance, commission and all other costs, etc., incidental to the sale and delivery of the goods with the exception of any duty or tax chargeable in the United Kingdom. Values of exports are "f.o.b." (free on board), which is the cost of the goods to the purchaser abroad, including packing, inland and coastal transport in the United Kingdom, dock dues, loading charges and all other costs, charges and expenses accruing up to the point where the goods are deposited on board the exporting vessel or at the land boundary of Northern Ireland.

8.26 Figures of the value of net exports in Tables 8.1 and 8.2 are derived from exports and imports measured on a Balance of Payments (B.O.P) basis. From 1997, sales of services in ports were reclassified as a trade on goods and allocated by the Office for National Statistics to the value of exports of petroleum products as most of these sales were for fuel. This means exports as recorded by HM Customs and Excise will differ from those recorded by the Office for National Statistics on a B.O.P basis. Table 8.1 shows figures on both basis.

8.27 Figures correspond to the following items of S.I.T.C (Rev 3).

Coal	321.1 and 321.2
Other solid fuels	322.1 and 325 (part)
Crude oil	333
Petroleum products	334, 335, 342 and 344 (plus Orimulsion reclassified to division 278 during 1994)
Natural gas	343
Electricity	351

8.28 Figures for trade within the European Union given in Tables 8.4 and 8.5 cover trade between the other 14 Member States belonging to the Union in 1999.

8.29 In 1993 the Single European Market was created. At that time a new system for recording the trade in goods between member states called INTRASTAT was introduced. As part of this system allows small traders to only have an obligation to report their annual trade and some trading supply returns late, it is necessary to include adjustments for unrecorded trade. This is particularly true of 1993, the first year of the system, and particularly true for coal imports in that year.

8.30 Quarterly data on imports and exports of fuels and related materials are provided in Table 25 of DTI's monthly statistical bulletin *Energy Trends*. See Annex F for more information about *Energy Trends*.

Contact: *Roger Barty*
          Caroline Barrington
          020 7215 5187

8.1 Imports and exports of fuels [1]

Quantity

	1995	1996	1997	1998	1999(2)
Imports					
Coal and other solid fuel	11.5	12.7	14.2	15.1	14.6
Crude oil	44.1	44.8	45.3	39.5	32.7
Petroleum products	17.4	17.8	15.3	17.9	20.0
Natural gas	1.3	1.4	1.3	0.4	0.3
Electricity	1.4	1.4	1.4	1.1	1.2
Total imports	75.7	78.2	77.5	74.0	68.8
Exports					
Coal and other solid fuel	0.9	1.0	1.1	0.9	0.8
Crude oil	86.4	83.4	76.6	80.4	80.4
Petroleum products	25.7	27.8	29.4	30.1	30.4
Natural gas	0.9	1.4	1.7	1.5	4.1
Electricity	-	-	-	-	-
Total exports	113.9	113.5	108.7	113.0	115.7
Net exports					
Coal and other solid fuel	-10.6	-11.8	-13.2	-14.2	-13.8
Crude oil	42.4	38.6	31.3	40.8	47.6
Petroleum products	8.2	10.0	14.1	12.3	10.4
Natural gas	-0.4	0.0	0.3	1.1	3.8
Electricity	-1.4	-1.4	-1.4	-1.0	-1.2
Total net exports	38.2	35.3	31.1	39.0	46.8

Value

	1995	1996	1997	1998	1999(2)
Imports - O.T.S basis (c.i.f.)					
Coal and other solid fuel	601	694	714	687	599
Crude oil	3,236	4,035	3,647	2,170	2,273
Petroleum products	1,542	1,821	1,441	1,416	1,961
Natural gas	105	117	103	43	27
Electricity	408	391	406	374	396
Total imports	5,892	7,058	6,311	4,691	5,246
Exports (f.o.b.)					
Coal and other solid fuel	70	82	82	69	61
Crude oil	6,428	7,426	6,322	4,485	5,993
Petroleum products	2,621	3,268	3,239	2,328	2,854
Natural gas	54	65	80	80	227
Electricity	0	2	1	3	9
Total exports	9,174	10,843	9,724	6,966	9,145
Net exports - O.T.S basis					
Coal and other solid fuel	-531	-612	-632	-618	-537
Crude oil	3,192	3,391	2,676	2,315	3,720
Petroleum products	1,080	1,446	1,798	912	892
Natural gas	-51	-52	-23	37	200
Electricity	-408	-389	-405	-371	-387
Total net exports	3,281	3,784	3,413	2,275	3,889
Imports - B.O.P. basis (f.o.b.)					
Oil (3)	4,568	5,544	4,825	3,305	4,005
Other fuels	1,003	1,061	1,084	842	699
Total imports	5,571	6,604	5,909	4,148	4,704
Net exports - B.O.P. basis					
Oil (3)	4,481	5,150	5,352	4,070	5,426
Other fuels	-879	-912	-921	-690	-402
Total net exports	3,602	4,238	4,431	3,380	5,025

Source: H.M. Customs and Excise

(1) See Energy Trends Table 25 for the latest quarterly figures.
(2) Provisional.
(3) Crude oil and petroleum products.

8.2 Value of imports and exports of fuels, 1965 to 1999[1][2]

£ million

		1965	1970	1971	1972	1973	1974
Imports (c.i.f.)	Coal and other solid fuels	1	2	46	57	27	66
	Crude oil	425	687	930	914	1,296	3,726
	Petroleum products *(3)*	176	242	259	257	389	823
	Natural gas	7	11	10	9	9	8
	Electricity	1	2	-	2	-	-
Total imports		609	943	1,245	1,240	1,721	4,623
Exports (f.o.b.)	Coal and other solid fuels	28	29	22	17	27	65
	Crude oil	-	8	10	21	23	29
	Petroleum products *(4)*	106	170	204	201	320	681
Total exports		134	208	236	238	370	775
Imports (f.o.b.)	Oil *(5)*	-	673	902	886	1,283	4,066
	Other fuels *(6)*	-	17	49	62	34	75
Total imports		-	690	951	948	1,317	4,141
Net exports[8]	Oil *(5)*	-	-495	-688	-665	-940	-3,356
(B.O.P basis)	Other fuels	-	12	-27	-46	-7	-10
Total net exports		-	**-482**	**-715**	**-710**	**-947**	**-3,366**

		1975	1976	1977	1978	1979	1980
Imports (c.i.f.)	Coal and other solid fuels	110	86	84	82	148	228
	Crude oil	3,371	4,445	3,971	3,506	3,678	4,292
	Petroleum products *(3)*	810	1,089	1,128	1,023	1,591	1,856
	Natural gas	14	21	44	188	356	521
	Electricity	1	-	-	-	-	-
Total imports		4,306	5,641	5,226	4,799	5,773	6,896
Exports (f.o.b.)	Coal and other solid fuels	84	72	80	90	100	180
	Crude oil	30	178	918	1,236	2,710	4,220
	Petroleum products *(4)*	705	1,004	1,086	1,038	1,500	2,017
Total exports		819	1,254	2,083	2,364	4,309	6,417
Imports (f.o.b.)	Oil *(5)*	3,796	5,117	4,759	4,243	4,928	5,869
	Other fuels *(6)*	114	118	139	260	470	683
Total imports		3,910	5,235	4,898	4,503	5,398	6,552
Net exports[8]	Oil *(5)*	-3,061	-3,935	-2,755	-1,969	-718	368
(B.O.P basis)	Other fuels	-30	-46	-60	-170	-370	-503
Total net exports		**-3,091**	**-3,981**	**-2,815**	**-2,139**	**-1,089**	**-135**

		1981	1982	1983	1984	1985	1986
Imports (c.i.f.)	Coal and other solid fuels	171	218	264	651	716	456
	Crude oil	4,112	3,951	3,308	3,993	4,341	2,440
	Petroleum products *(3)*	2,173	2,413	2,506	4,360	4,071	2,079
	Natural gas	699	815	977	1,307	1,511	1,320
	Electricity	-	-	-	-	-	80
Total imports		7,156	7,398	7,055	10,311	10,639	6,375
Exports (f.o.b.)	Coal and other solid fuels	372	330	239	88	178	190
	Crude oil	7,096	8,542	10,111	12,173	13,006	6,281
	Petroleum products *(4)*	2,148	2,365	2,776	3,047	3,611	2,200
Total exports		9,616	11,236	13,126	15,308	16,794	8,671
Imports (f.o.b.)	Oil *(5)*	6,043	6,126	5,605	8,046	8,120	4,186
	Other fuels *(6)*	826	988	1,183	1,884	2,142	1,797
Total imports		6,869	7,114	6,788	9,930	10,262	5,983
Net exports[8]	Oil *(5)*	3,201	4,781	7,282	7,174	8,497	4,295
(B.O.P basis)	Other fuels	-454	-658	-944	-1,796	-1,964	-1,607
Total net exports		**2,747**	**4,122**	**6,338**	**5,378**	**6,532**	**2,686**

(1) See Energy Trends Table 25 for the latest quarterly figures.
(2) See the notes to the Foreign Trade section of this and earlier editions of the Digest.
(3) Includes petroleum products not used as fuel, eg lubricants, and liquefied petroleum gases other than natural gas.
(4) Includes petroleum products not used as fuel, eg lubricants, and liquefied petroleum gases, and small quantities of natural gas.
(5) Crude oil and petroleum products.
(6) Data prior to 1985 include small quantities of non-fuel products (eg peat). These items are excluded from the c.i.f. import data and the export data.
(7) Provisional.
(8) Net exports are the difference between exports and imports on a Balance of Payments (B.O.P) basis – see Table 8.1 for figures in the period 1995 to 1999.

8.2 Value of imports and exports of fuels, 1965 to 1999[1][2] (continued)

£ million

		1987	1988	1989	1990	1991	1992
Imports (c.i.f.)	Coal and other solid fuels	390	472	513	630	734	744
	Crude oil	2,703	2,044	3,079	4,033	3,887	3,745
	Petroleum products (3)	1,880	1,546	1,889	2,427	2,063	1,711
	Natural gas	878	692	615	519	472	397
	Electricity	242	268	305	225	343	369
Total imports		6,094	5,022	6,400	7,834	7,500	6,965
Exports (f.o.b.)	Coal and other solid fuels	109	96	109	119	97	63
	Crude oil	6,765	4,515	4,024	5,172	4,370	4,413
	Petroleum products (4)	1,893	1,646	2,039	2,455	2,640	2,401
	Natural gas	-	-	-	-	-	2
	Electricity	-	-	-	25	-	-
Total exports		8,767	6,257	6,172	7,771	7,107	6,879
Imports (f.o.b.)	Oil (5)	4,363	3,310	4,707	6,119	5,697	5,227
	Other fuels	1,455	1,365	1,364	1,299	1,468	1,393
Total imports		5,818	4,675	6,071	7,418	7,165	6,620
Net exports[8]	Oil (5)	4,295	2,851	1,356	1,508	1,313	1,589
(B.O.P basis)	Other fuels	-1,346	-1,269	-1,255	-1,155	-1,371	-1,330
Total net exports		2,949	1,582	101	353	-58	259

		1993	1994	1995	1996	1997
Imports (c.i.f.)	Coal and other solid fuels	731	598	601	694	714
	Crude oil	4,078	3,241	3,236	4,035	3,647
	Petroleum products (3)	1,766	1,689	1,542	1,821	1,441
	Natural gas	327	231	105	117	103
	Electricity	426	388	408	391	406
Total imports		7,328	6,148	5,892	7,058	6,311
Exports (f.o.b.)	Coal and other solid fuels	73	75	70	82	82
	Crude oil	5,147	6,095	6,428	7,426	6,322
	Petroleum products (4)	3,149	2,776	2,621	3,268	3,239
	Natural gas	28	45	54	65	80
	Electricity	-	-	-	2	1
Total		8,397	8,991	9,174	10,843	9,724
Imports (f.o.b.)	Oil (5)	5,610	4,689	4,568	5,544	4,825
	Other fuels	1,387	1,122	1,003	1,061	1,084
Total imports		6,997	5,810	5,571	6,604	5,909
Net exports[8]	Oil (5)	2,686	4,182	4,481	5,150	5,352
(B.O.P basis)	Other fuels	-1,286	-1,001	-879	-912	-921
Total net exports		1,400	3,181	3,602	4,238	4,431

		1998	1999(7)
Imports (c.i.f.)	Coal and other solid fuels	687	599
	Crude oil	2,170	2,273
	Petroleum products (3)	1,416	1,961
	Natural gas	43	27
	Electricity	374	396
Total imports		4,691	5,256
Exports (f.o.b.)	Coal and other solid fuels	69	61
	Crude oil	4,485r	5,993
	Petroleum products (4)	2,328	2,854
	Natural gas	80	227
	Electricity	3	9
Total exports		6,966r	9,145
Imports (f.o.b.)	Oil (5)	3,305	4,005
	Other fuels	842	699
Total imports		4,148	4,704
Net exports[8]	Oil (5)	4,070r	5,426
(B.O.P basis)	Other fuels	-690	-402
Total net exports		3,380r	5,025

8.3 Imports and exports of crude oil and petroleum products

	1995		1996		1997	
	Quantity (Thousand tonnes)	Value per tonne (£)	Quantity (Thousand tonnes)	Value per tonne (£)	Quantity (Thousand tonnes)	Value per tonne (£)
Imports (c.i.f.)						
Crude oil	**40,349**	**80.21**	**40,496**	**98.79**	**41,383**	**88.01**
Refined petroleum products *(1)*						
Petroleum gases *(2)*	549	112.99	529	129.02	510	139.20
Motor spirit and aviation spirit	1,378	120.76	1,985	139.08	1,377	136.60
Other light oils and spirit *(3)*	895	117.96	825	134.98	877	130.68
Aviation turbine fuel (kerosene)	1,481	108.73	1,428	143.20	1,612	132.57
Other kerosene	32	145.19	30	152.73	162	142.79
Gas oil/diesel oil	1,620	107.10	1,645	127.31	1,837	130.43
Fuel oil *(4)*	8,657	70.31	8,175	81.27	7,948	71.99
Lubricating oils	412	259.86	519	226.24	409	237.48
Petroleum coke	889	36.96	1,071	42.91	1,055	40.21
Other *(5)*	1,581	63.71	113	118.87	295	193.26
Total refined petroleum products	**17,494**	**90.96**	**16,320**	**105.47**	**16,082**	**100.66**
Exports (f.o.b)						
Crude oil	**79,065**	**81.29**	**75,642**	**97.83**	**71,796**	**90.20**
Refined petroleum products *(1)*						
Petroleum gases *(2)*	3,342	120.31	3,393	141.61	4,172	142.40
Motor spirit and aviation spirit	5,865	113.43	6,861	130.21	9,557	123.46
Other light oils and spirit *(3)*	1,872	118.52	1,432	132.40	2,036	139.41
Aviation turbine fuel (kerosene)	909	109.99	695	138.52	834	117.44
Other kerosene	49	276.18	52	145.15	138	132.14
Gas oil/Diesel oil	5,553	97.96	4,976	120.37	3,799	107.94
Fuel oil *(4)*	4,001	69.28	4,965	78.77	5,982	67.34
Lubricating oils	872	251.84	1,325	233.20	3,637	157.70
Petroleum coke	612	157.94	652	184.22	584	188.21
Other	469	182.21	133	157.84	384	167.64
Total refined petroleum products	**23,544**	**111.52**	**24,489**	**127.11**	**31,123**	**120.00**

(1) Excludes pitch, mineral tars and natural gas.
(2) Includes small quantities of unidentified non-petroleum gases.
(3) Includes wide-cut gasoline, white spirit and petroleum naphthas.
(4) Includes partly refined oil for further processing.
(5) Includes Orimulsion.

8.3 Imports and exports of crude oil and petroleum products (continued)

	1998		1999	
	Quantity (Thousand tonnes)	Value per tonne (£)	Quantity (Thousand tonnes)	Value per Tonne (£)
Imports (c.i.f.)				
Crude oil	**36,136**	**60.05**	**29,924**	**75.95**
Refined petroleum products *(1)*				
Petroleum gases *(2)*	479	108.16	523	113.56
Motor spirit and aviation spirit	1,460	114.88	1,859	120.38
Other light oils and spirit *(3)*	1,104	96.42	936	109.41
Aviation turbine fuel (kerosene)	2,703	94.21	4,198	108.63
Other kerosene	119	110.83	146	69.95
Gas oil/diesel oil	2,824	93.41	4,408	120.14
Fuel oil *(4)*	6,945	54.01	5,476	62.61
Lubricating oils	445	191.56	510	235.41
Petroleum coke	786	49.20	526	43.86
Other *(5)*	295	167.09	424	148.41
Total refined petroleum products	**17,160**	**81.93**	**19,006**	**101.57**
Exports (f.o.b)				
Crude oil	**73,429**	**61.04**	**73,702**	**81.31**
Refined petroleum products *(1)*				
Petroleum gases *(2)*	3,604	88.63	4,465	112.32
Motor spirit and aviation spirit	7.015	94.45	5,975	107.64
Other light oils and spirit *(3)*	1,612	78.76	2,139	118.77
Aviation turbine fuel (kerosene)	795	88.78	717	103.29
Other kerosene	160	110.89	145	166.16
Gas oil/Diesel oil	5,665	77.30	5,394	98.92
Fuel oil *(4)*	5,720	47.02	5,266	61.22
Lubricating oils	1,149	177.35	1,765	144.92
Petroleum coke	788	139.11	643	153.14
Other	572	110.38	783	115.72
Total refined petroleum products	**27,080**	**84.22**	**27,291**	**102.51**

Source: H.M. Customs and Excise

8.4 Imports and exports of crude oil by country

	1995 Quantity Thousand tonnes	1995 Value (£million)	1995 Value per tonne (£)	1996 Quantity Thousand tonnes	1996 Value (£million)	1996 Value per tonne (£)	1997 Quantity Thousand tonnes	1997 Value (£million)	1997 Value per tonne (£)
Imports (c.i.f.)									
Middle East									
Abu Dhabi	-	-	-	-	-	-	39	4.2	108.58
Dubai	-	-	-	-	-	-	-	-	-
Iran	1,064	83.8	78.71	937	86.3	92.03	-	-	-
Kuwait	1,366	104.1	76.17	1,494	132.8	88.90	1,550	124.1	80.09
Oman	-	-	-	-	-	-	-	-	-
Saudi Arabia	1,390	107.7	77.48	847	80.5	95.09	1,842	138.0	74.89
Other countries	1,418	109.4	77.12	1,893	174.2	99.02	1,282	105.1	82.03
Total Middle East	**5,239**	**404.9**	**77.29**	**5,171**	**473.9**	**91.64**	**4,713**	**371.5**	**78.82**
Algeria	2,572	203.7	79.18	1,746	175.5	100.55	341	35.7	104.50
Angola	261	18.5	70.71	-	-	-	80	6.8	84.68
Latvia	538	40.2	74.63	595	57.2	96.16	1,084	91.5	84.38
Libya	577	41.9	72.66	238	21.0	88.29	-	-	-
Lithuania	33	2.0	61.43	-	-	-	-	-	-
Mexico	800	55.8	69.81	723	60.3	83.44	796	59.2	74.39
Netherlands	522	38.3	73.31	691	62.2	90.02	-	-	-
Nigeria	744	61.2	82.28	1,568	174.5	111.29	214	20.6	96.24
Norway	26,793	2,225.9	83.08	26,493	2,698.0	101.84	30,863	2,807.4	90.97
Russia	585	45.5	77.76	1,053	95.2	90.42	567	49.2	86.87
Venezuela	1,519	86.0	56.61	1,340	90.6	67.61	1,445	197.9	136.98
Other countries	165	12.6	76.22	879	92.3	105.07	1,280	2.4	18.75
Total Non Middle East	**35,110**	**2,831.6**	**80.65**	**35,325**	**3,526.9**	**99.84**	**36,671**	**3,270.8**	**89.19**
Total imports	**40,349**	**3,236.5**	**80.21**	**40,496**	**4,000.7**	**98.79**	**41,383**	**3,642.2**	**88.01**
Exports (f.o.b.)(1)									
European Union									
Belgium and Luxembourg	226	18.7	82.60	221	21.7	98.43	970	85.5	88.24
Denmark	60	4.7	78.26	70	7.1	101.00	-	-	-
Finland	3,492	287.6	82.38	3,688	352.3	95.51	1,173	109.8	93.60
France	9,638	784.4	81.39	13,744	1374.4	100.00	14,660	1329.3	90.67
Germany	15,006	1,227.8	81.82	14,976	1458.9	97.42	15,658	1,433.9	91.57
Greece	-	-	-	-	-	-	-	-	-
Irish Republic	274	2.6	82.50	303	28.8	94.97	86	7.1	82.35
Italy	3,339	274.0	82.04	1,465	137.8	94.06	657	61.0	92.91
Netherlands	9,264	725.9	78.36	13,198	1263.9	95.76	16,539	1,464.2	88.52
Portugal	215	18.9	87.66	383	42.6	111.07	781	69.0	88.36
Spain	3,860	308.9	80.02	1,680	166.8	99.27	3,740	340.0	90.92
Sweden	823	68.0	82.53	1,282	114.8	89.55	990	84.2	85.01
Total EU	**46,199**	**3,741.4**	**80.99**	**51,010**	**4,969**	**97.41**	**55,254**	**4,984.0**	**90.33**
Canada	2,074	168.2	81.05	1,953	184.3	94.35	1,049	95.4	90.99
Norway	1,223	83.2	68.02	151	16.1	106.14	1,176	109.5	93.14
U.S.A.	21,897	1,794.4	81.94	18,369	1814.4	98.77	12,295	1,105.1	89.88
Other countries	7,671	640.3	83.47	4,158	416.7	100.22	2,022	182.1	90.05
Total exports	**79,065**	**6,427.4**	**81.29**	**75,642**	**7400.4**	**97.83**	**71,796**	**6,476.2**	**90.20**

(1) Includes re-exports.

8.4 Imports and exports of crude oil by country (continued)

	1998			1999		
	Quantity Thousand tonnes	Value (£million)	Value per tonne (£)	Quantity Thousand tonnes	Value (£million)	Value per tonne (£)
Imports (c.i.f.)						
Middle East						
Abu Dhabi	-	-	-	-	-	-
Dubai	65	4.3	65.95	-	-	-
Iran	38	3.0	78.05	-	-	-
Kuwait	1,042	51.5	49.47	-	-	-
Oman	-	-	-	-	-	-
Saudi Arabia	2,288	113.8	49.74	1,050	67.6	64.61
Other countries	1,853	112.8	60.85	1,257	83.8	66.66
Total Middle East	**5,285**	**285.3**	**53.98**	**2,307**	**151.4**	**65.63**
Algeria	631	41.8	66.31	1,061	96.3	90.73
Angola	-	-	-	-	-	-
Latvia	-	-	-	29	3.2	111.44
Libya	-	-	-	-	-	-
Lithuania	-	-	-	-	-	-
Mexico	890	43.2	48.50	875	57.7	66.02
Netherlands	-	-	-	-	-	-
Nigeria	560	34.8	62.14	460	34.7	75.50
Norway	24,898	1,565.4	62.87	22,156	1,730.4	78.10
Russia	2,100	124.0	59.07	1,101	82.3	74.73
Venezuela	1,340	51.2	38.23	1,071	50.2	46.86
Other countries	433	23.8	54.97	865	66.6	76.99
Total Non Middle East	**30,851**	**1,884.2**	**61.08**	**27,617**	**2,121.4**	**76.82**
Total imports	**36,136**	**2,169.5**	**60.05**	**29,924**	**2,272.8**	**75.95**
Exports (f.o.b.)(1)						
European Union						
Belgium and Luxembourg	1,037	58.7	56.61	1,189	97.2	81.78
Denmark	-	-	-	-	-	-
Finland	656	39.0	59.42	701	52.7	75.19
France	14,320	851.4	59.46	13,103	1,044.9	79.75
Germany	17,237	1,077.8	62.53	11,661	977.4	83.82
Greece	-	-	-	74	6.2	83.63
Irish Republic	4	0.2	56.93	70	3.6	51.00
Italy	820	42.4	51.71	1,234	92.1	74.62
Netherlands	15,362	966.1	62.89	14,531	1,179.0	81.14
Portugal	1,237	73.8	59.66	1,403	106.5	75.90
Spain	3,511	211.5	60.24	3,655	286.4	78.35
Sweden	790	61.8	78.15	635	51.6	81.25
Total EU	**54,974**	**3,382.7**	**61.53**	**48,256**	**3,897.5**	**80.77**
Canada	804	48.2	59.95	623	42.1	67.58
Norway	799	47.5	59.42	99	5.6	56.39
U.S.A.	13,600	807.8	59.40	20,893	1,735.9	83.09
Other countries	3,251	196.0	60.30	3,832	311.7	81.36
Total exports	**73,429**	**4,482.2**	**61.04**	**73,702**	**5,992.9**	**81.31**

Source: HM Customs and Excise

8.5 Imports and exports of solid fuel

Thousand tonnes

1995	Imports (1)				Exports			
	Steam coal	Coking coal	Anthracite	Other solid fuel	Steam coal	Coking coal	Anthracite	Other solid fuel
European Union								
Austria	-	-	-	-	-	-	-	-
Belgium/Luxembourg	-	-	13	-	16	-	70	18
Denmark	-	-	-	-	30	-	1	3
Finland	-	-	-	-	-	-	1	6
France	-	-	-	18	110	-	32	23
Germany	6	-	126	24	5	-	5	2
Irish Republic	14	1	10	7	233	3	115	4
Italy	-	-	-	-	-	-	-	-
Netherlands (2)	11	-	93	22	14	-	13	-
Spain	-	-	-	9	24	-	36	-
Sweden	3	-	7	-	11	-	9	57
Total European Union	**34**	**1**	**249**	**80**	**444**	**3**	**283**	**113**
Australia	142	3,194	305	-	-	-	-	-
Canada	-	1,306	58	-	-	-	-	-
Colombia	2,440	38	208	232	-	-	-	-
Indonesia	57	-	7	-	-	-	-	-
Norway	2	-	-	-	87	-	32	269
People's Republic of China	-	-	77	125	-	-	-	-
Poland	716	396	162	60	-	-	-	-
Republic of South Africa	1,005	-	309	-	-	-	-	-
Russia	42	27	91	60	-	-	-	-
United States of America	2,937	1,792	130	-	-	-	-	-
Venezuela	-	-	101	-	-	-	-	-
Vietnam	-	-	-	-	-	-	-	-
Other countries	52	-	18	125	2	-	7	39
Total all countries	**7,426**	**6,754**	**1,715**	**682**	**533**	**3**	**322**	**422**
Value of imports (cif)/export (fob) (£m) (3)	213.8	246.8	92.2	36.6	19.7	0.1	21.5	28.8
Value per tonne (£)	28.79	36.53	51.80	63.8	36.96	30.16	66.74	68.28

1996

1996	Imports (1)				Exports			
	Steam coal	Coking coal	Anthracite	Other solid fuel	Steam coal	Coking coal	Anthracite	Other solid fuel
European Union								
Austria	-	-	-	-	5	-	-	-
Belgium/Luxembourg	5	-	14	4	14	-	90	2
Denmark	-	-	-	-	13	-	6	4
Finland	-	-	-	-	-	-	-	-
France	-	-	23	22	99	-	60	28
Germany	3	-	148	30	16	-	3	1
Irish Republic	16	-	24	6	324	3	153	7
Italy	-	-	-	-	-	-	-	-
Netherlands (2)	71	-	182	5	37	-	4	-
Spain	-	-	-	11	12	-	42	-
Sweden	20	-	13	7	-	-	8	47
Total European Union	**114**	**-**	**405**	**85**	**520**	**3**	**366**	**80**
Australia	-	3,748	-	2	-	-	-	-
Canada	-	1,410	-	28	-	-	-	-
Colombia	2,726	-	98	-	-	-	-	-
Indonesia	19	3	5	-	-	-	-	-
Norway	6	-	-	4	66	-	20	234
People's Republic of China	-	-	30	482	-	-	-	-
Poland	563	149	138	49	-	-	-	-
Republic of South Africa	1,323	-	236	-	-	-	-	-
Russia	-	-	75	29	-	-	-	-
United States of America	3,644	2,878	135	19	-	-	-	-
Venezuela	-	-	-	-	-	-	-	24
Vietnam	-	-	26	-	-	-	-	-
Other countries	8	58	2	104	9	-	2	1
Total all countries	**8,403**	**8,245**	**1,151**	**801**	**596**	**3**	**389**	**349**
Value of imports (cif)/export (fob) (£m) (3)	248.1	319.4	68.7	59.1	24.4	97.8	26.4	26.2
Value per tonne (£)	29.52	38.73	59.73	73.79	40.97	32.59	67.79	75.10

Thousand tonnes

	Imports (1)				Exports			
1997	Steam coal	Coking coal	Anthracite	Other solid fuel	Steam coal	Coking coal	Anthracite	Other solid fuel
European Union								
Austria	-	-	-	-	10	-	-	-
Belgium/Luxembourg	7	-	38	9	13	-	75	2
Denmark	-	-	-	-	-	-	2	1
Finland	-	-	-	-	-	-	-	22
France	3	-	6	14	101	-	59	30
Germany	3	-	83	16	133	-	19	-
Irish Republic	50	-	14	1	317	-	137	7
Italy	-	-	-	-	3	-	1	12
Netherlands (2)	66	-	117	15	45	-	6	2
Spain	-	-	-	6	31	-	11	-
Sweden	-	-	-	4	6	1	5	43
Total European Union	**129**	**-**	**258**	**65**	**659**	**1**	**314**	**119**
Australia	505	3,857	-	1	-	-	-	-
Canada	-	1,632	-	18	-	-	-	-
Colombia	2,763	-	47	-	-	-	-	-
Indonesia	78	-	6	-	-	-	-	-
Norway	100	-	2	1	64	-	91	221
People's Republic of China	3	-	75	616	-	-	-	-
Poland	532	51	117	24	-	-	-	-
Republic of South Africa	2,262	13	197	3	-	-	-	-
Russia	-	-	36	68	-	-	-	-
United States of America	4,281	2,519	150	-	-	-	-	1
Venezuela	97	-	-	-	-	-	-	-r
Vietnam	-	-	15	-	-	-	-	-
Other countries	6	-	28	90	7	-	7	4
Total all countries	**10,756**	**8,072**	**931**	**886**	**731**	**1**	**414**	**345r**
Value of imports (cif)/export (fob) (£m) (3)	289.8	298.3	80.5	55.6	31.5	0.1	25.7	26.0
Value per tonne (£)	27.85	36.95	62.93	62.79	43.0	44.33	62.03	71.82

	Imports (1)				Exports			
1998	Steam coal	Coking coal	Anthracite	Other solid fuel	Steam coal	Coking coal	Anthracite	Other solid fuel
European Union								
Austria	-	-	-	-	-	-	-	-
Belgium/Luxembourg	31	-	-	14	-	-	49	1
Denmark	-	-	-	-	-	-	8	-
Finland	-	-	-	-	-	-	-	16
France	-	-	-	20	78	-	23r	19
Germany	19	-	6	13	141	-	2	-
Irish Republic	28	-	4	1	234r	-	54	9
Italy	-	-	-	-	-	-	-	8
Netherlands (2)	150r	-	19r	1	41	-	11	1
Spain	3	-	-	-	78	-	21	-
Sweden	-	-	-	-	16	-	3	64
Total European Union	**231r**	**-**	**29r**	**49**	**587r**	**-**	**170r**	**118**
Australia	1,115	3,495	-	-	-	-	-	-
Canada	-	1,552	-	52	-	-	-	-
Colombia	3,819r	-	-	-	2	-	-	-
Indonesia	24	-	-	-	-	-	-	-
Norway	117	-	3	-	94r	-	107r	225r
People's Republic of China	-	-	132	602	-	-	-	-
Poland	875	-	75	20	-	-	1	-
Republic of South Africa	2,302	-	162r	-	1	-	1	-
Russia	-r	-	20r	49	-	-	-	-
United States of America	3,142	3,598	6	18	-	-	-	-
Venezuela	399	-	-	-	-	-	-	-
Vietnam	-	-	80	-	-	-	-	-
Other countries	55	-	12	51	5	-	3	2
Total all countries	**12,079r**	**8,646**	**519r**	**841**	**689r**	**-**	**282r**	**345r**
Value of imports (cif)/export (fob) (£m) (3)	314.1	298.2	28.3r	46.3	25.5r	-	18.0r	28.8r
Value per tonne (£)	25.96r	34.49	56.34r	54.96	37.01r	-	63.78r	66.00r

8.5 Imports and exports of solid fuel (continued)

Thousand tonnes

1999	Imports (1)				Exports			
	Steam coal	Coking coal	Anthracite	Other solid fuel	Steam coal	Coking coal	Anthracite	Other solid fuel
European Union								
Austria	-	-	-	-	-	-	-	-
Belgium/Luxembourg	59	-	11	14	-	-	69	-
Denmark	-	-	-	-	3	-	3	5
Finland	-	-	-	-	-	-	-	22
France	1	-	1	6	32	-	65	25
Germany	3	-	5	4	27	-	11	-
Irish Republic	34	-	10	8	200	-	77	8
Italy	-	-	-	-		-	-	-
Netherlands (2)	193	-	-	-	10	-	9	1
Portugal	-	-	-	-	3	-	-	-
Spain	-	-	-	-	25	-	43	-
Sweden	-	-	-	-	1	-	4	84
Total European Union	**291**	**-**	**27**	**31**	**303**	**-**	**282**	**147**
Australia	1,688	4,262	-	-	-	-	-	-
Canada	-	1,427	-	18	-	-	-	-
Colombia	4,592	-	-	-	-	-	-	-
Indonesia	82	-	-	-	-	-	-	-
Norway	43	-	-	-	113	-	42	149
People's Republic of China	81	24	144	322	-	-	-	-
Poland	1,286	-	13	14	1	-	-	-
Republic of South Africa	2,673	38	210	-	1	-	-	-
Russia	139	-	33	12	-	-	-	-
United States of America	746	2,268	2	-	-	-	-	-
Venezuela	472	-	-	-	-	-	-	-
Vietnam	-	-	167	-	-	-	-	-
Other countries	47	-	3	37	17	-	4	1
Total all countries	**12,139**	**8,020**	**599**	**436**	**434**	**-**	**328**	**297**
Value of imports (cif)/export (fob) (£m) (3)	302.5	244.3	30.7	19.2	19.8	-	18.2	19.0
Value per tonne (£)	24.95	30.47	51.25	43.98	45.74	-	55.49	63.97

Source : H.M. Customs and Excise

(1) *Country of origin basis.*
(2) *Includes extra-EU coal routed through the Netherlands.*
(3) *Value of imports are "cif" (cost, insurance and freight) and value of exports are "fob" (free on board). See technical note for fuller definition.*

Chapter 9
Prices and values

Introduction

9.1 This chapter covers energy prices and expenditure on energy products. Data in the tables are mainly in cash prices. However, price comparisons given in this chapter (unless otherwise stated) refer to movements in annual average data in real terms. These are prices from which the effects of inflation[1], as measured by the Gross Domestic Product (GDP) market prices deflator, have been removed. The GDP deflator provides an index of inflation in the whole economy and therefore is applicable consistently to domestic and industrial prices. The main domestic comparisons include Value Added Tax (VAT), which has applied to domestic sector prices from 1 April 1994, reflecting the price domestic consumers actually pay, although some price movements excluding VAT are also included.

9.2 For most fuels there is a difference in the prices paid by smaller consumers, typically households, and those paid by larger consumers, usually those in the industrial sector. Indeed there are differences in prices between large and small industrial users. An important reason for the differences is the presence of economies of scale in areas such as energy. In a competitive energy market, larger customers can secure the benefit of these economies by negotiating lower prices. Equally important is the fact that a household's energy demands may be more variable through the day and year (and therefore higher in peak price times) than those of industrial customers who use energy for continuous processes or can load manage. For these reasons the tables show prices separately for domestic and industrial consumers. Although no prices are given for commercial consumers, prices for the domestic sector should be fairly close to those for smaller commercial consumers and industrial prices should provide a reasonable proxy for larger customers in the commercial sector.

9.3 Domestic sector expenditure figures and typical bills are given first, followed by tables showing the prices of fuels purchased by the electricity and gas supply industries, prices paid by manufacturing industry and a table showing industrial price indices. Tables showing the prices of petrol and other oil products and the duties and taxes on them are followed by comparative domestic and industrial electricity and gas prices in EU and G7 countries, both including and excluding tax and a table of comparative petrol and diesel prices across the EU.

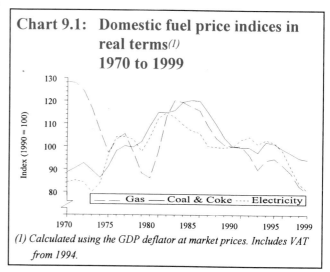

Chart 9.1: Domestic fuel price indices in real terms[1] 1970 to 1999

(1) Calculated using the GDP deflator at market prices. Includes VAT from 1994.

Domestic prices (Tables 9.1 to 9.9)

Retail prices of fuels (Table 9.1)

9.4 Table 9.1 shows the energy components of the Retail Prices Index (a good general indicator of domestic prices). Overall domestic prices fell in cash terms between 1998 and 1999, chiefly driven by large falls in heating oil prices, linked to the sharp falls in crude oil prices, regulation driven falls in electricity prices and the full impact of the reduction in VAT. Data for earlier years for fuel and light in total and for petrol (which includes diesel) and lubricating oil are given in Table 9.4.

9.5 The tax on domestic fuels changed in 1997 with the announcement in the 1 July Budget that VAT on domestic fuel would be cut from 8 to 5 per cent from 1 September 1997. This reduction in VAT is estimated to have reduced electricity prices by around 2 per cent in 1998 (following a 1 per cent fall in 1997) with similar downward impact on other fuels.

---

[1] Inflation is measured using the following values of the GDP deflator rescaled to show 1990=100;

Year	GDP deflator at market prices 1990=100
1990	100.0
1991	106.8
1992	111.0
1993	114.1
1994	115.8
1995	118.8
1996	122.7
1997	126.2
1998	130.2
1999	134.0

9.6   Charts 9.1 and 9.2 show real term movements in prices of fuels used in the home and in motor vehicles. Between 1988 and 1992, domestic electricity prices rose in each year resulting in a 6 per cent rise for the whole period (although prior to 1988 prices had fallen each year since 1982). Post 1992 prices have fallen, with the exception of 1994 when VAT was introduced. The fall accelerated in 1996 when measures such as price regulation on distribution charges and the first round of the reduction in the Fossil Fuel Levy (from 10 per cent to 3.7 per cent in November 1996) were major factors in lowering prices by 3½ per cent on 1995. However, the full impact of these changes and subsequent reductions in the Levy to 2.2 per cent in 1997 and 0.9 per cent in 1998 was clearly seen with a 7 per cent fall in prices in both years. The Levy was further reduced down to 0.7 per cent in January 1999 then 0.3 per cent in October 1999. 1999 saw a further fall of 4 per cent.

9.7   Overall domestic electricity prices in 1999 were 19 per cent lower than in 1990, the year the industry was restructured and privatised (23 per cent lower excluding VAT), although much of the fall has occurred in the past 3 years with prices falling by 18 per cent since 1996. Domestic competition is yet to have a significant impact on average electricity prices, the introduction of which is explained in more detail in paragraphs 9.26 to 9.30.

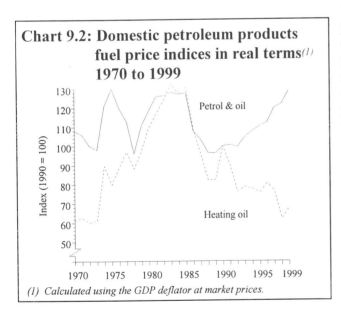

Chart 9.2: Domestic petroleum products fuel price indices in real terms[1] **1970 to 1999**

(1) Calculated using the GDP deflator at market prices.

9.8   Domestic gas prices moved broadly in line with oil prices until 1983, when a general de-coupling began in many UK fuel prices. Prices then fell fairly steadily over the next 10 years, until by 1993 they were about 25 per cent lower than in 1983, in real terms. The trend was broken in 1994 and 1995 when the introduction of VAT and tariff changes caused prices to rise by around 5 per cent. However, the advent of competition (see paragraphs 9.25 to 9.26 for more

details), the reduction in VAT and three years of unchanged and then reduced British Gas standard tariffs have led to real term falls in 1996 to 1999. In 1999 the average price of domestic gas was 3½ per cent lower than in 1998, 20 per cent lower than in 1990 in real terms (24 per cent lower excluding VAT), and down 13 per cent on 1996. However, consumers who have moved to new suppliers will have seen larger falls (see paragraph 9.31).

9.9   Heating oil prices typically follow crude oil prices, e.g. rising rapidly in 1990 due to the Gulf crisis, before falling back to a post 1973 low in 1995. During 1996 prices rose by 7 per cent, as crude oil prices climbed and international demand was generally higher, before falling back by 5 per cent in 1997 as oil prices began to weaken. 1998 saw a sharp reduction in crude prices (described in more detail in paragraph 9.68 to 9.69) to levels not seen since the early 1970s. As such heating oil prices also fell to real prices not seen for 20 years. Following the agreement of key oil producers to meet production targets crude oil prices began to rise again early in 1999 and have continued to do so since. Overall average heating oil prices rose by 9 per cent between 1998 and 1999 but were 32½ per cent lower than in 1990.

9.10 Since 1988 prices of fuels used in motor vehicles have generally increased in real terms year on year and in 1999 were 29 per cent higher than in 1990. The increases in petrol prices from 1993 through to 1999 resulted chiefly from Budget increases in the duty payable on petrol and diesel following the previous administration's commitment to raise duty by at least 5 per cent a year in real terms and the current Chancellor of the Exchequer's hardening of the policy to a 6 per cent real rise a year. However in the November 1999 pre-budget statement the Chancellor announced that in future Budgets there would be no commitment to increase duty rates in real terms. In the March 2000 Budget duty on motor spirits and diesel was increased by 3½ per cent. More detail on petrol and diesel prices are given in paragraphs 9.56 to 9.64. Overall since 1963 both domestic fuel and light, and petrol and oil prices have seen more than a 10 fold increase in real terms (see Table 9.4).

Expenditure on energy by households (Tables 9.2, 9.3 and 9.4)

9.11 These tables give figures for the amounts spent on energy by households. Table 9.2 shows total expenditure for all households in the United Kingdom and Table 9.3 shows weekly expenditure per consuming household (i.e. only those households which use each fuel).

9.12 Data in the first of these tables are the fuel components of household final consumption expenditure published by the Office for National

Statistics (ONS) in the national accounts. The figures are at market prices, i.e. the prices paid by purchasers, inclusive of taxes (i.e. VAT since 1994) and duty. They are shown on two price bases, firstly at current prices, which are the prices prevailing in the year to which they refer, and secondly at constant 1995 prices, the national accounts base year. The data show that after rising by 1½ per cent between 1995 and 1997 real expenditure on energy fell marginally in 1998 before falling by 1½ per cent in 1999. However, there have been significant variations in real term expenditure since 1990, notably a 47½ per cent fall in expenditure on coal and coke compared to a 9½ per cent increase in expenditure on electricity.

9.13   The overall pattern of expenditure on individual fuels as a percentage of the total expenditure on fuels is shown in Chart 9.3 for 1999. This shows that expenditure on motor fuels accounts for 53 per cent of consumers' expenditure on energy products (up from 50 per cent in 1998) with gas accounting for 18 per cent of consumers' expenditure, electricity 26 per cent whilst coal, coke and petroleum products (other than motor fuels) account for 3 per cent between them. Table 9.2 also shows the impact of falling energy prices relative to the prices of other goods and services. In 1999 expenditure on energy products represented 5 per cent of total consumers' expenditure, measured at current prices, compared to 6 per cent in 1995, 9 per cent in 1985 and 8 per cent in 1980.

After dipping in 1989 and 1990, there was a rise in 1991 as consumption rose by 10 per cent, in part due to demand for heating fuels rising in the colder weather. Between 1991 and 1995 expenditure remained fairly steady at its 1991 level, despite the introduction of VAT on domestic fuels in April 1994, before rising sharply in 1996 to a level 4 per cent higher than in 1991. Factors described above such as the large fall in electricity prices, reduction in VAT and the fact that 1997 was on average over 1 degree warmer than 1996, help to explain the fall in expenditure of 4 per cent in 1997. Real expenditure fell slightly in 1999 by around 2½ per cent to the lowest level since 1990 reflecting falling gas and electricity prices. Overall, real consumers' expenditure on fuel and light in 1999 was 4½ per cent lower than in 1991 and 9 per cent higher than in 1981.

9.15   Consumers' expenditure on petrol, diesel and lubricating oil (referred to as petrol and oil for simplicity) at 1995 prices has risen steadily over the last 30 years as car ownership increased (Table 9.4). Since 1990, despite increases in petrol prices, principally as a result of budget tax changes, real expenditure in 1995 prices has fallen by 9 per cent, between 1990 and 1999, falling by ½ per cent between 1998 and 1999. The downward movement in the 1990s is a combined result of fuel switching (i.e. the move to unleaded) and methodology which is described in paragraph 9.81.

Chart 9.3: Consumers' expenditure on energy 1999

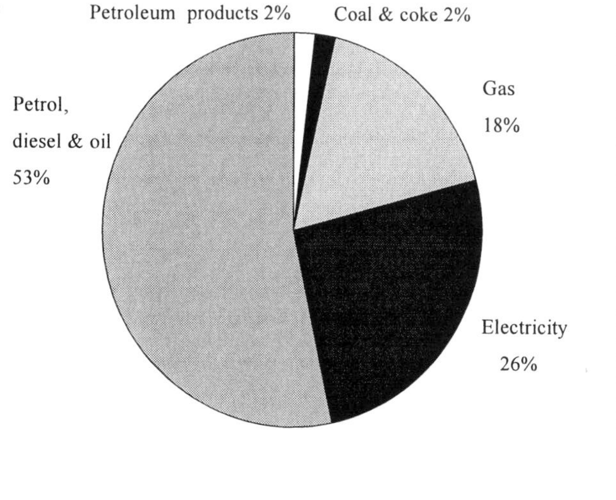

Petroleum products 2%   Coal & coke 2%

Petrol, diesel & oil 53%

Gas 18%

Electricity 26%

Total £28,634 million

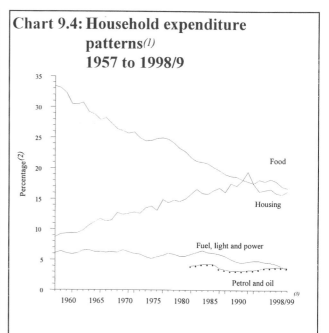

Chart 9.4: Household expenditure patterns[1] 1957 to 1998/9

Food

Housing

Fuel, light and power

Petrol and oil [3]

Percentage[2]

1960   1965   1970   1975   1980   1985   1990   1998/99

(1) Data are taken from Family Expenditure Survey Reports, 1957 to 1998/9 produced by the Office for National Statistics.
(2) Percentages are expenditure on commodity or service as a percentage of total household expenditure.
(3) The survey changed to a financial year basis in 1994/5.

9.14   Expenditure on fuel and light at 1995 prices (i.e. re-valued to a constant price base) from 1963 onwards are shown in Table 9.4. Expenditure increased until 1974 then fluctuated for about ten years with a peak in 1979 mainly caused by the effects of crude oil price rises. Between 1984 and 1988, expenditure rose again, as consumers increased consumption significantly in line with increases in their overall standard of living.

9.16   Expenditure on individual fuels, in households using each fuel, are shown in Table 9.3 based on data from the Family Expenditure Survey run by the ONS. Average expenditure on all fuels in cash terms rose by 35 per cent between 1990 and 1998/99 with expenditure on motor fuel seeing the largest rise (54 per cent) reflecting the rises in petrol prices. As stated above, electricity prices for domestic consumers have fallen faster than gas prices in the past three years.

9.17   Data in the table are collected via a household survey and as such can be affected by temperature differences year on year and sampling variations. The latter is evident in the 1996/7 data where the number of coal users in the survey increased, but nearly all were households without solid fuel central heating. These households purchase coal less regularly and in smaller quantities (at higher prices) pushing up average expenditure. A truer picture emerges from coal purchasers with solid fuel heating, their expenditure has fallen from £13.31 in 1995/6 to £13.10 in 1996/7.

9.18   Chart 9.4 shows average household expenditure on fuel, light and power and on petrol and oil (data only available from 1981) as a proportion of average total household expenditure. The corresponding data for expenditure on food and housing are also shown. It is clear that whilst expenditure on fuel and light as a percentage of the total has remained fairly stable, with a gradual fall since 1985, the proportions spent on housing and food have changed dramatically. Since 1970 expenditure on fuel has fallen from 6½ per cent of total household expenditure to 3½ per cent in 1998/99, whilst housing has risen from 12½ to 16 per cent and food has fallen from 25½ to 16½ per cent over the same period.

Table 9A: Household expenditure as a percentage of total expenditure by selected gross income decile 1998/9

	Lowest 10 per cent	Third decile	Sixth decile	Highest 10 per cent
Fuel and power	7.0	5.6	3.5	2.1
Housing	16.5	16.1	16.5	16.0
Food	22.6	21.0	16.9	13.6

Source: Family Expenditure Survey, ONS

9.19   The data in Chart 9.4 are for the average household. However, there are large variations in the percentages of income spent on the items by different income decile groups as Table 9A above shows. In terms of shares of total expenditure, households in the lowest decile (i.e. the 10 per cent of households with the lowest incomes) spend over twice as much on fuel as the average household. For these households expenditure on fuel and power typically accounted for 7 per cent of total expenditure in 1998/99 compared to 2 per cent for the highest decile i.e. over 3 times higher. Similar comparisons for food and housing show the percentage accounted for by the lowest decile to be around 1½ times that of the highest for food but around the same for housing. A time

series of these data show that although the actual percentage expenditure on fuel is falling for the lower income households the relative gap to other income groups remains broadly constant.

9.20   Households in the top decile (the top 10 per cent of incomes) typically spend twice as much on fuel and power as those in the bottom decile, whilst their overall expenditure is around 7 times greater. Again these ratios have remained fairly similar since 1993, although the percentage spent on fuels for most deciles has fallen as a result of generally warmer winters combined with falling real underlying energy prices. The differences between expenditure in the income deciles show that in absolute terms expenditure on fuel varies less by income than other items such as food and housing (i.e. there are not cheaper fuel substitutes).

9.21   Many factors influence household expenditure on fuel ranging from regional variation in weather to the number of people in the household. One important factor is the type of dwelling. Naturally the larger the property the more space that needs heating. Therefore the data in Table 9B are as expected, with fuel expenditure in the typically larger detached properties nearly double that spent in flats. When fuel expenditure as a percentage of total expenditure is considered, it is 20 per cent more for people living in flats than those living in detached houses. Therefore, the type of dwelling is a major factor in the debate on fuel poverty along with the main causes of low income and poor energy efficiency.

Table 9B: Household expenditure (£ per week) on fuel and power by type of dwelling in 1998/9

	Detached house	Semi detached	Terrace house	Flat
Fuel and power	15.00	12.20	11.10	8.10
Total expenditure	503.50	356.40	314.80	224.80
Fuel as percentage of total	3.0	3.4	3.5	3.6

Source: Family Expenditure Survey, ONS

Domestic fuels prices and bills and the impact of competition
(Tables 9.5 to 9.9)

9.22   Table 9.5 shows retail prices for gas and electricity purchased by domestic consumers in certain large towns/cities in the United Kingdom. The places have been selected to include one that is in each of the authorised areas of the Public Electricity Supply (PES) companies and in each of the Local Distribution Zones (LDZ) used for gas (see paragraph 9.84 for further details). Levels of consumption have been chosen to be representative of domestic consumers. Table 9.6 shows average bills for a range of different consumption levels.

9.23 British Gas stopped regional pricing in 1990. However, the advent of competition (described below) has meant that consumers across the country can obtain different prices, as such a regional breakdown for gas is again applicable. However, differences in the average bills will often reflect more British Gas' main competitors in each area rather than distinct regional pricing. For example, there is a £10 difference in average credit gas bills between Southampton and Cardiff, which mainly reflects market share lost by British Gas in each region (22 per cent in the Southern LDZ compared to 35 per cent in Wales, in Q3 1999) but in part reflects the market shares of individual new supply companies and their pricing structures.

9.24 Regional pricing has always been a feature of the electricity market as transmission and distribution costs vary from region to region. This is clearly shown in the bills where in 1999 an average credit consumer in the South East will pay £46 less per year than a similar consumer in Cardiff and £75 less than consumers in Northern Ireland, with a similar picture true for direct debit consumers. For pre-payment consumers, those in Birmingham had the lowest average costs in 1999, paying on average £53 less than consumers in Cardiff.

9.25 Until April 1996 all domestic consumers received their electricity from their local regional electricity company and their gas from British Gas. However, by May 1999 full competition in both markets had been introduced.

9.26 The first trial in competitive gas supply started in the South West of England in April 1996 with originally 11 new licensed suppliers competing with British Gas to supply 500,000 consumers in Cornwall, Devon and Somerset. In February and March 1997, competition was extended to a further 1.5 million customers in the South West and South East (the former County of Avon, Dorset, Kent and East and West Sussex). Competition was further extended to Scotland and other regions in England and Wales in late 1997 and the first half of 1998. Competition in domestic electricity supply began on 14 September 1998 with 750,000 consumers in Eastern, MANWEB, Scottish Power and Yorkshire's supply areas. They were allowed to compete in each other's areas and initially faced competition only from British Gas, although 4 other non-REC suppliers now have domestic licences.

9.27 Competition was further extended to a proportion of Midland and SEEBOARD regions on 28 October 1998. Sections of Scottish Hydro's, Northern's and NORWEB and SWEB's areas opened on 4, 11 and 30 November respectively. The first round was completed in East Midlands in late December,

with SWALEC and Southern in mid and late January 1999.

9.28 At the end of January 1999, 50 per cent of the domestic market was opened in the following PES areas: Yorkshire, Scottish Power, MANWEB, Eastern, Midlands and SEEBOARD, with Northern and Hydro-Electric opening in late January/February. In February, competition was rolled out to all consumers in Eastern, MANWEB, Scottish Power and Yorkshire's supply areas and further extended in 9 others. Competition was extended to all consumers in Great Britain on 24 May 1999.

9.29 By March 2000 approximately 5½ million gas consumers (27½ per cent) were no longer supplied by British Gas Trading. Table 9C gives market penetration in more detail, by LDZ, showing that it is the markets in the North and Wales that new suppliers have had most success in and that the proportion of consumers switching supplier does vary by payment type. At the end of Q3 1999 British Gas had lost around 24 per cent of the credit and 30 per cent of the direct debit market compared to 13 per cent of the pre-payment market, although it should be noted that British Gas's pre-payment prices are below the average of new suppliers.

Table 9C: Domestic gas market penetration (in terms of percentage of customers supplied) by local distribution zone and payment type, quarter 3 1999

Region	British Gas Trading			Non-British Gas		
	Credit	Direct Debit	Prepay-ment	Credit	Direct Debit	Prepay-ment
East Midlands	73	69	85	27	31	15
Eastern	77	73	89	23	27	11
North East	76	69	83	24	31	17
North Thames	77	73	90	23	27	10
North West	74	76	86	26	24	14
Northern	70	60	85	30	40	15
Scotland	79	66	94	21	34	6
South East	79	66	89	21	34	11
South West	80	71	87	20	29	13
Southern	78	68	87	22	32	13
Wales	65	72	76	35	28	24
West Midlands	79	76	87	21	24	13

9.30 By March 2000 just over 4 million electricity consumers (16½ per cent) were no longer with their home supplier. As previously stated, competition in electricity was not fully rolled out until May 1999. Table 9D gives early indications of market penetration from Q3 1999, showing that it is the markets in East Midlands and Midlands that new suppliers had the most success in. At the end of Q3 1999, the regional electricity companies had lost around 9 per cent of the credit and 15 per cent of the direct debit market compared to 4 per cent of the pre-payment market.

Table 9D: Domestic electricity market penetration (in terms of percentage of customers supplied) by Public Electricity Supply area and payment type, quarter 3 1999

Region	Home Supplier			Non-Home Supplier		
	Credit	Direct Debit	Prepay-ment	Credit	Direct Debit	Prepay-Ment
Eastern	91	83	96	9	17	4
East Midlands	86	85	92	14	15	8
London	90	85	95	10	15	5
Midlands	89	84	89	11	16	11
North East	90	80	94	10	20	6
North Scotland	95	86	99	5	14	1
Merseyside and North Wales	90	79	97	10	21	3
North West	92	92	86	8	8	14
South	91	91	87	9	9	13
South East	89	89	85	11	11	15
South Scotland	93	93	85	7	7	15
South Wales	90	90	83	10	10	17
South West	97	97	90	3	3	10
Yorkshire	91	91	87	9	9	13

9.31 Competition has in general meant lower prices and therefore lower bills to consumers switching supplier. For example, in 1999 a gas consumer could save £53 by switching from British Gas to a new supplier on credit terms or £24 on direct debit terms. In 1999 a combined saving of £66 (21 per cent) was possible by switching company and payment type. Savings for electricity are less, reflecting the historical position that supply costs (and it is electricity and gas supply that have been opened up to competition) make up a small proportion of the domestic electricity bill. Average savings of around £20 per year, rising to over £40 in some areas, are possible on credit and direct debit terms.

9.32 For pre-payment customers, competition is not directly bringing as large savings. British Gas cut their pre-payment prices 3 times during 1998 and 1999 and as a result charged on average £10 less than the average of new suppliers in 1999 (compared to around £5 more in 1998). Whilst in the electricity market new suppliers are on average charging £10 less than the incumbent supplier. In some areas prices offered by a new supplier can be over £60 more per year, although in all areas there are companies offering slightly lower prices.

9.33 The savings on offer from new suppliers are reflected in average gas and electricity bills presented in this chapter. It is however, important to remember that the reduction gained from switching supplier will depend on when in the year a change takes place and to whom the consumer changes. These variables combined with the fact that approximately 74½ per cent of gas consumers were still with British Gas at the end of Q3 1999, mean that average all supplier bills do not show the kind of reductions described in paragraph 9.31 above.

9.34 Tables 9.6 and 9.7 show the way standing charges and unit prices (the price for each kWh of energy consumed) contribute to domestic energy bills. Whilst the total unit cost columns show how much each unit costs overall.

9.35 Electricity standing charges vary a great deal by company and payment type. The weighted average UK standing charge in 1999 for credit consumers from home suppliers was £8.46 per quarter, but ranges from zero to £11.76. Direct debit and pre-payment average and ranges are £7.70 (-£1.99 to £11.36) and £13.67 (£3.37 to £16.58) respectively. Data shown in Table 9.6 include VAT. A similar story is true for gas in 1999 with standing charges of new suppliers being typically higher than British Gas' for direct debit and pre-payment customers, by 1 - 3 p/day. Whilst British Gas credit customers paid on average over 4p/day more than customers of new suppliers.

9.36 For low volume electricity consumers standing charges can account for over 20 per cent of a credit or direct debit electricity bill rising to nearly 30 per cent of a pre-payment bill. The nature of gas standing charges where British Gas charges are lower for pre-payment customers than other types means that standing changes account for around 11 per cent of a low volume pre-payment gas bill compared to 23 per cent of an equivalent direct debit bill and 32 per cent of a credit bill.

9.37 However, a low standing charge does not necessarily mean a low total bill (although for a low volume electricity user it will in most cases). For gas, although standing charges are lower for pre-payment consumers, the associated higher unit rate means that in 1999 the average unit cost was around 1.9 p/kWh for a low volume pre-payment customer compared to 1.6 and 1.8 for direct debit and credit consumers respectively. Therefore, the average unit cost (total bill/volume consumed) gives a complete picture of cost of electricity. Tables 9.6 and 9.7 give UK average unit costs for specific consumption, but once again variation is magnified on a company basis. A low volume consumer on a pre-payment meter can be paying over 10¾ pence per unit (including VAT) in Northern Ireland or South Wales, whilst a high volume direct debit consumer will pay less than 6¾ pence in East Anglia.

9.38 Average household electricity and gas bills including rebates and VAT are given in Tables 9.8 and 9.9. The bills are calculated from published price information and unpublished customer numbers provided by the Regional Electricity Companies, British Gas and their competitor companies as part of the DTI's domestic competition inquiry. They are calculated for typical annual consumption levels of 18,000 kWh for gas and 3,300 kWh for electricity. The tables show bills for the three main payment methods, as in the last few years there has been an expansion in the use of direct debit tariffs and the number of customers on these tariffs. For example between Q4 1998 and Q3 1999 the percentage of consumers on direct debit tariffs rose from 31 to 33 per cent for electricity and from 37 to 39 per cent for gas, a considerable growth from their general universal introduction in 1995.

9.39   Electricity and gas bills are clearly falling in real and cash terms, as the full impact of the reduction in the Fossil Fuel Levy, competition, regulation and the reduction in VAT and other changes, described above, feed through to bills. A typical standard credit tariff UK electricity bill was £264 in 1999 compared to £299 in 1995 with an equivalent gas bill standing at £305 compared to £327 in 1995.

on other payment methods (19 per cent for pre-payment and 17 per cent for standard credit). Equivalent figures for electricity are savings of 24 per cent for direct debit, compared to 22 per cent for pre-payment and for standard credit customers.

9.41   The growth of competition is also leading to expansion and innovation in the methods available to pay for electricity and gas. Already companies are introducing tariffs with no standing charges and are linking up with other non-energy companies to offer deals linked to air miles or points on shop's loyalty schemes. One company, Equigas, has introduced a gas tariff where all consumers pay the same unit rate and no standing charge regardless of payment type.

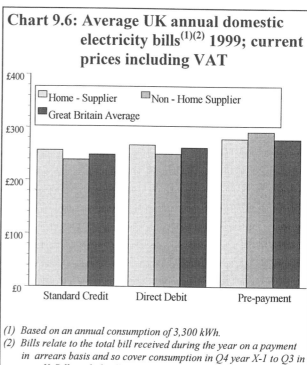

Chart 9.6: Average UK annual domestic electricity bills[1][2] 1999; current prices including VAT

(1) Based on an annual consumption of 3,300 kWh.
(2) Bills relate to the total bill received during the year on a payment in arrears basis and so cover consumption in Q4 year X-1 to Q3 in year X. Bills include all tariff changes and rebates but exclude the value of incentives offered by certain suppliers.

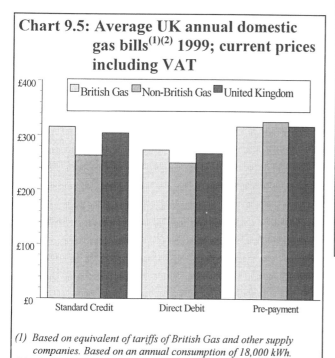

Chart 9.5: Average UK annual domestic gas bills[1][2] 1999; current prices including VAT

(1) Based on equivalent of tariffs of British Gas and other supply companies. Based on an annual consumption of 18,000 kWh.
(2) Bills relate to the total bill received during the year on a payment in arrears basis and so cover consumption in Q4 year X-1 to Q3 in year X. Bills include all tariff changes and rebates but exclude the value of incentives offered by certain suppliers.

Industrial prices (Tables 9.10 to 9.13)

Fuel inputs for electricity generation (Table 9.10)

9.40   Customers on direct debit pay less than customers on other payment methods, as the tables demonstrate. In 1999, an electricity direct debit customer will pay on average £28 less than a pre-payment customer and £11 less than a credit customer, with equivalent average differences for gas of £50 and £37. What is also clear is that, certainly for gas, customers paying by direct debit have seen larger falls (e.g. 24 per cent in real terms 1995 to 1999) than those

9.42   Table 9.10 shows prices paid by electricity companies for the fuels they use to generate electricity and the average beach price of gas.

9.43   The price of coal purchased by the electricity generators fell by about £1 per tonne in 1999 following a fall of about £3.60 per tonne in 1998 and in cash terms was 31½ per cent lower than in 1993 (42 per cent in real terms). New coal supply contracts were negotiated in early 1998 following the ending of the previous 5 year deals established prior to coal privatisation, resulting in a fall in prices. Over the same period the price for oil rose by 53½ per cent (30½ per cent in real terms), but fell by 20 per cent in 1998 on the back of sharp falls in the price of crude oil. In 1999 the price of oil for generation rose sharply back to the level last seen in 1996. Oil purchased for generation, like all generation fuels, is more likely to be purchased on longer-term contracts. This, coupled with the mix of oils purchased, means that oil for generation is less closely related to spot prices than other industrial users' contracts. The use of gas for electricity generation has increased significantly since 1993, (see Chapter 5 paragraph 5.26). Since then its price has fallen by 13 per cent in cash terms (26 per cent in real terms).

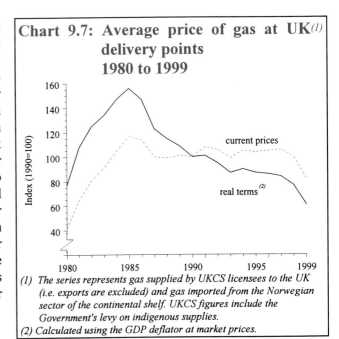

Chart 9.7: Average price of gas at UK[(1)] delivery points 1980 to 1999

(1) The series represents gas supplied by UKCS licensees to the UK (i.e. exports are excluded) and gas imported from the Norwegian sector of the continental shelf. UKCS figures include the Government's levy on indigenous supplies.
(2) Calculated using the GDP deflator at market prices.

9.44 When converted to a common pence per kWh (p/kWh) basis, oil was approximately 76½ per cent more expensive than coal in 1999 a sharp increase from the 42 per cent differential seen in 1998, when oil prices had fallen substantially. However, comparison of fuel input prices in common units (p/kWh) does not necessarily reflect differences in the cost of generating electricity using different fuels. As well as fuel input costs, generation costs are also affected by non fuel costs and by the efficiency with which fuel inputs are converted into electricity. For example, combined cycle gas stations have higher efficiencies than conventional steam stations. Therefore, just comparing the fuel input costs per kWh, which show gas to be more expensive, does not provide a picture of full costs. Data on station efficiencies are shown in Table 5.9.

9.45 Chart 9.7 and the lower part of Table 9.10 show the movement in the "beach price" of gas in real and cash terms. The series is derived from gas sales by licensees in the UKCS to delivery points in the UK. It excludes exported gas and is adjusted to include imported gas. The trade adjusted beach price averaged 0.462 pence per kWh in 1999. Over supply of gas, principally from 1995 onwards, has led to the development of a spot gas market where some producer's un-contracted gas and initial purchases, in excess of their own needs, are sold. As a result the spot price fell sharply from about 0.69 pence per kWh to below 0.35 pence per kWh (10 pence a therm) in 1995. Since then the spot and new International Petroleum Exchange futures markets have grown in volume terms and the price of gas on these markets has developed a seasonal trend peaking at around 0.5 p/kWh (15p /therm) in December, but falling to around 0.35 p/kWh in the summer.

9.46 The beach price series includes the gas levy which was introduced in 1980/81. The Chancellor announced in his 1998 Budget that the levy would be abolished from 1 April 1998 and reduced by 25 per cent to around 0.1p/kWh for the year 1997/8. These changes are reflected in the table with further details given in paragraph 9.92.

Prices of fuels purchased by manufacturing industry (Tables 9.11 and 9.12)

9.47 These tables show fuel prices paid by manufacturing industry in cash terms. As fuel prices vary considerably according to the amount purchased, data are compiled separately for different sizes of consumers (the definitions of the size bands are given in Table 9G on page 219). Average annual prices in original units are shown in Table 9.11 whilst annual and quarterly prices, for the latest two years, in common pence per kWh form are shown in Table 9.12. (Quarterly prices in original units are published in Energy Trends.)

9.48 The prices aim to represent average 'market' prices, i.e. prices which are available to the majority of purchasers in each size band. Only fuels used for energy purposes are included in the price estimates, fuels used as raw materials in manufacturing processes (including blast furnace supplies) are excluded.

9.49 Chart 9.8, shows that the average prices of coal, heavy fuel oil and gas in 1999 were between 0.47 and 0.74 pence per kWh. Electricity prices were considerably higher averaging 3.6 pence per kWh in 1999, reflecting the costs incurred in converting other fuels and electricity's greater efficiency in use.

9.50 Real price changes by size of user have varied somewhat for each fuel. This reflects the bargaining position of the larger users and factors such as: the timing of the introduction of competition and previous pricing arrangements; length of contracts; underlying factors such as abundance of fuel supply in gas; and the relative (to size) impact of crude prices on fuel oils.

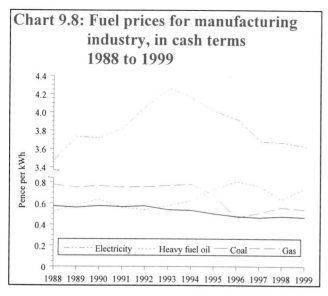

Chart 9.8: Fuel prices for manufacturing industry, in cash terms 1988 to 1999

9.51 A clear example of these differences is shown in gas prices. Large and small gas users have seen real prices fall by around 46 per cent between 1990 and 1999. However, between 1996 and 1998 real prices for large users rose by an average of 15½ per cent, compared to a fall of 7½ per cent for small users. The differences largely reflect how the market has developed; more small users are moving to cheaper contracts from their previous tariff terms, whilst large users, who have had to negotiate new contracts in 1997, 1998, and 1999 have found prices higher as the fierce price competition evident throughout 1995 and 1996 receded.

9.52 The largest price changes in 1999 were for large consumers of oil. The very largest heavy fuel oil consumers have seen an increase of 17 per cent during 1999 compared to around 5½ per cent for small consumers. Equivalent changes for gas oil consumers are a rise of 6½ and a fall of 2½ per cent respectively. The rises are of course linked to crude oil prices and the strong pound.

9.53 Between 1990 and 1999 the largest electricity users have seen price falls of 22½ per cent compared to 34½ per cent for the smallest users. The larger fall for small users (who are a mixture of franchise and non-franchise customers) reflects the fact that customers with maximum demand over 100 kW could negotiate cheaper contracts from 1994 onwards, whilst a sizeable number of larger users had price advantages prior to the establishment of competition to over 1 MW customers in 1990 and these took some time to be fully unwound and new contracts established. However, the largest users have faired better since 1994 recording falls of 26 per cent. Small users have seen significantly larger gains in 1999 than the largest consumers with an average real fall of 6½ per cent compared to 3½ per cent for the largest users. Since 1990 coal prices have fallen by 40 per cent for large users and 33 per cent for small.

Industrial fuel price indices (Table 9.13)

9.54 This table shows the long term index series for industrial fuel prices, whilst Chart 9.9 shows the movement in industries average fuel prices since 1970 in real terms. The series for electricity and gas are averages of all industrial users calculated from energy supply company data. Coal and heavy fuel oil data are based on the average data as published in Table 9.11. Real term electricity and coal prices are at their lowest levels since 1970. The same is true for gas, with the exception of 1996, as average real gas prices have risen in recent years following the dramatic falls driven by excess supply and strong market share competition seen during 1995 and 1996. Heavy fuel oil prices are lower than in all years since 1970 except 1991 to 1994 and 1998.

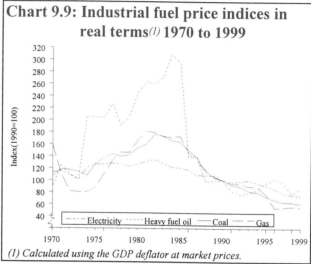

Chart 9.9: Industrial fuel price indices in real terms[1] 1970 to 1999

(1) Calculated using the GDP deflator at market prices.

9.55 Industrial electricity prices fell again in 1999, by an average of 2½ per cent. Real electricity prices have fallen by 19½ per cent between 1995 and 1999. Industrial gas prices fell in 1999 by an average of 2½ per cent. Real gas prices have fallen by 27½ per cent since 1995. UK heavy fuel oil prices moved broadly in line with international markets (and crude oil prices) falling by 19½ per cent in 1998 then rising by 10½ per cent in 1999 meaning they have fallen by 11 per cent since 1995. Average coal prices recorded a fall of 4½ per cent in 1999, equivalent to a fall of 17 per cent since 1995.

Petroleum products: prices and rates of duty (Tables 9.14 to 9.16)

9.56 Table 9.14 gives national average mid month retail prices for motor fuels and other petroleum products, inclusive of taxes, for the last two years and annual average prices for the past twenty. A price index for crude oil is presented in Table 9.4 to show the relationship between oil prices and petroleum product prices. A further run of product prices, back to 1954, is given in Table 9.15. In Table 9.16 the rates of duty applicable in the period are given, with the VAT rates shown in the footnotes. These duties are included in the prices of petrol and other products shown in

Tables 9.14 and 9.15. The March 2000 Budget changes are included in Table 9.16.

9.57 The data published are national average prices calculated from prices supplied by all major motor fuel marketing companies. Prior to 1977 price data were collated from a variety of sources notably the published scheduled wholesale prices of the oil companies to which retailers' margins were added. The results of various consumers' surveys were also taken into consideration in arriving at a typical price. Users of this table should bear in mind that, because of the multiplicity of petroleum marketing companies operating in the United Kingdom and the diversity of their pricing policies, prices differ from dealer to dealer and from area to area. From January 1995, sales by super/hypermarkets, which now make up around 25 per cent of the retail petrol market (see Chapter 3 paragraph 3.61 for more details) are included in the price estimates.

9.58 A key feature of Table 9.14 and Chart 9.10, below, is that since the Gulf crisis in 1990/91 motor fuel prices have increased at a steady rate, chiefly caused by Budget tax changes as listed in Table 9.16. Strong competition in late 1995 and early 1996 with promotions such as "Price Watch" led to sharp price falls and at the peak of the price competition in April 1996 prices had fallen to below December 1995, pre Budget, levels. To achieve these price cuts many operators cut margins on retail sales from a traditional 4 to 5 pence per litre down to as low as one pence per litre in some cases in mid 1996. During 1997 and 1998 as price competition stabilised somewhat (although remaining strong at a local scale) and crude oil prices moved sharply downward, margins moved back towards their previous levels. However, during the sharp increases in crude oil prices in 1999 and 2000 the margins reduced again.

9.59 In April 2000 motor fuel prices were 7 to 10 pence per litre higher than in April 1999. The Budget in March 2000 increased duty levels on petrol and diesel by a significantly smaller amount than in recent years. The duty on 4-star was increased by (1.80 p/litre before VAT) 2.11p including VAT with similar, VAT inclusive, rises of 1.89 p/litre for unleaded, diesel, and ultra low sulphur diesel and 1.97 p/litre for lead replacement petrol. Prices at the pump rose by about 1½ pence per litre after the Budget.

9.60 Duty is not the only factor that influences the prices of petrol. Between April 1998 and February 1999 the prices of crude oil fell by around 20 per cent in cash terms. Linked to this the prices of motor fuel excluding duty also fell by around 19 per cent (around 2 p/litre), meaning for example unleaded petrol excluding tax was less than 10 pence/litre and the tax inclusive price of unleaded in February 1999 was less

than that seen in January 1998 (a budget change earlier). However, crude oil prices rose by 173 per cent between February 1999 (when prices troughed) and February 2000 (when prices peaked). Over this period there were increases of around 11-14 pence per litre in the price of motor spirit and DERV.

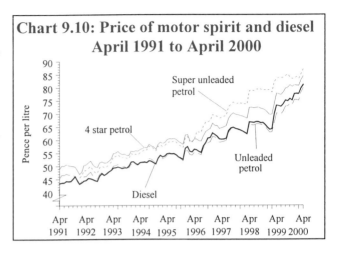

Chart 9.10: Price of motor spirit and diesel April 1991 to April 2000

9.61 A historical perspective of petrol and diesel prices is given in Table 9.15. This shows that petrol prices increased only gradually from 1954 to 1973, when they were affected by the sharp rise in crude oil prices and then controlled by Government Order during most of 1974. The next big increase was during 1979 as a result of the second oil price shock. Prices then rose until 1985 before falling during 1986. The Gulf crisis of 1990/91 had only a temporary effect on prices (shown in the January 1991 figures) with the prices of motor fuels in recent years being driven upwards by tax changes but offset to some extent by strong competition in the retail sales market and as described above the recent changes in crude oil prices.

9.62 A reduced rate of duty for unleaded petrol, initially one pence lower than 4 star, was introduced in the 1987 Budget. This duty difference was increased in stages to reach nearly 5 pence per litre after the March 1993 Budget. Since then the duty differential has been held at or slightly above that level in subsequent changes of duty. In April 2000 the duty differential was 5.9 pence (or 6.9 pence including VAT), however the sales of leaded petrol ceased in January 2000, except for a limited number of special cases. 4 star was replaced by lead replacement petrol (LRP) which has the same duty rate as super unleaded petrol, 2.1 pence per litre above that for premium unleaded.

9.63 Prior to 1994 diesel was cheaper than premium unleaded by as much as 3 pence per litre in some months. Since then, duty rates were first equalised, in monetary terms, in 1994 and then raised more for diesel than unleaded in the March 1998 Budget. In this period the differential in prices has fluctuated but the position in 1996 to 1998 was typically one of diesel being around a penny per litre more expensive. The further tax differential in favour of unleaded in the

1999 Budget meant that in early 1999 unleaded was up to 3 pence per litre cheaper than diesel. In the 1998 Budget, a lower duty rate was introduced for Ultra Low Sulphur diesel (ULSD), (further details are given in Chapter 3 paragraph 3.57). The rate was originally set 2p/litre lower than for standard diesel, but was put at a discount of 3p/litre in the 1999 Budget and has remained at this discount since. During the summer of 1999 the majority of diesel being sold changed to ULSD and prices quoted for July 1999 onwards are for ULSD. The increasing sales of ULSD during mid 1999, combined with the discounted duty rate narrowed the differential in pump prices of deisel and unleaded petrol during the latter half of 1999. Prices for super-unleaded motor spirit are available only from January 1991. Sales of the fuel have been falling partly as it became the most expensive fuel following the duty rate being increased above that for premium unleaded in 1996. However, the duty rate of super-unleaded motor spirit was reduced from 1 October 1999 to facilitate the introduction of Lead Replacement Petrol (LRP). LRP is treated as higher octane (super) unleaded petrol for duty purposes.

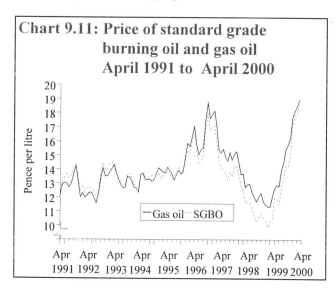

Chart 9.11: Price of standard grade burning oil and gas oil April 1991 to April 2000

9.64 Standard grade burning oil and gas oil prices generally move in line with crude oil prices (see below). This means that events such as the Gulf crisis in 1990/91 caused the price of these fuels to rise initially but then fall back, as crude oil prices rose and fell. Annual average prices rose sharply during 1996, to near 1990 levels, before gradually easing throughout 1997. 1998 saw a dramatic fall in prices to new real term lows since data has been available in 1977. This was followed by dramatic increases in the price during 1999 and early 2000, mirroring the movements in crude oil prices. In real terms, standard grade burning oil and gas oil prices rose by 10 and 8 per cent respectively in 1999 and fell by 39 and 29 per cent respectively since 1990. These movements are illustrated in Chart 9.11 above with a longer run of data in Table 9.15.

Crude oil prices (Table 9.4)

9.65 Throughout this chapter movements in crude oil prices have been discussed and cited as reasons for changes in various domestic and industrial fuel prices as well as, quite naturally, petroleum products. The price of crude oil can change for a variety of reasons, but a common feature is that they are all global events. Examples include: oil shortages (1973); political uncertainty (1990/1); and general over supply coupled with weaker Far East demand (1998).

9.66 Crude oil prices are shown in Table 9.4 as an index based on a "basket" of crude oil prices that are used as an input, along with other fuel prices, for the Producer Prices Index (produced by ONS). The overall index, which is also given, aims to measure changes in the cost of inputs to manufacturing and therefore is a leading indicator of inflation. The crude oil data are shown graphically in Chart 9.12.

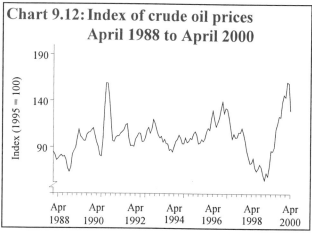

Chart 9.12: Index of crude oil prices April 1988 to April 2000

9.67 The recent history of crude oil prices is of them peaking, in cash terms, in 1984 before falling by 60 per cent to a low point in 1988. Prices then rose in 1989 and 1990, the latter as a result of the Gulf crisis before falling back again in early 1991 then remaining fairly stable (for crude) at about their 1986 level until mid of 1995, although the withdrawal of Sterling from the ERM produced a small peak in 1993 as the pound was devalued relative to the dollar.

9.68 Crude oil prices rose sharply in late 1995 and continued to do so through 1996 as a result of factors such as the uncertainty over the re-introduction of Iraqi exports, low oil stocks and lower than expected non-OPEC production. The upward pressures peaked in October 1996 when the price was higher than at any time since December 1990, during the Gulf crisis. Crude prices fluctuated throughout 1997 and by October were around the same level as in January 1996. From October 1997 crude prices fell virtually every month until December 1998 chiefly on the basis of excess supply made worse by economic uncertainty in the Far East and milder winter weather. During 1998 the average price of crude oil fell by 35 per cent in cash terms making crude prices over 50 per cent lower in real terms in 1998 than in 1990 and 5 per cent lower than in 1970.

9.69 Average crude oil prices have been increasing since late 1998 when production cuts were agreed upon by key oil producers to tackle the general over supply and reduced Far East demand which precipitated the rapid fall in prices in 1998. By April 2000 crude prices were more than twice the December 1998 low.

International comparisons (Tables 9.17 to 9.19)

9.70 One way of looking at fuel price movements in the United Kingdom over the past few years is to consider them in the context of prices in similar countries. A summary of prices derived from IEA data up to 1998, the latest full year for which data are available, are presented in Tables 9.17 and 9.18. More detailed international comparisons of fuel prices are included in publications of the International Energy Agency (IEA), and of the Statistical Office of the European Communities (Eurostat), and the European Commission as follows:

IEA quarterly *Energy Prices and Taxes*.
Eurostat *Electricity Prices 1990-1999*.
Eurostat *Gas Prices 1990-1999*, both of which are updated annually.
European Commission *Oil Bulletin*, published weekly by the Directorate General for Energy.

9.71 It is important when comparing international prices to keep in mind the impact of exchange rates (as the data are presented in a common pound sterling basis, the changing level of the pound will cause some changes in relative prices) and inflation rates in individual countries. The relative strength of the pound in 1997, 1998 and 1999 (e.g. sterling appreciated by 21 per cent against the German Mark between 1996 and 1999) to some extent will have had an adverse effect on comparisons of UK data.

were the lowest. Some countries have enjoyed larger price falls, especially during 1997 and into 1998 when UK prices have risen but prices have fallen across Europe, where gas prices are still linked to crude oil prices.

9.73 The average prices paid for electricity by industrial consumers in the UK were the 10th lowest in the EU in 1998, maintaining the position held in 1997. However, the UK has seen the sixth largest fall in prices in the EU since 1990. In late 1998 and early 1999 the largest electricity users in the UK were faring less well in the UK, however since the end of 1999 the largest industrial electricity users in the UK are thought to have seen significant reductions in prices in advance of the introduction of the New Electricity Trading Arrangements (NETA) (see paragraph 5.4).

9.74 Domestic gas and electricity prices in 1998 including all taxes were the third and fourth lowest in the EU respectively, representing a slight improvement since 1990 for electricity.

9.75 Table 9.18 shows prices including and excluding tax (apart from the USA and Canada where full data including and excluding tax are not available) and how UK prices compare to the median of EU prices. Excluding tax makes little difference to the ranking of UK industrial prices (although as many countries are looking at energy taxes this could change in the future). Excluding taxes does, however, worsen the standing of UK domestic prices which fall to 5th and 9th lowest for gas and electricity respectively. Overall UK gas prices excluding tax, are 4 and 25 per cent below the EU median for domestic and industrial consumers, with domestic electricity consumers 1½ per cent above the median and industrial consumers paying 12 per cent (0.4p/kWh) more.

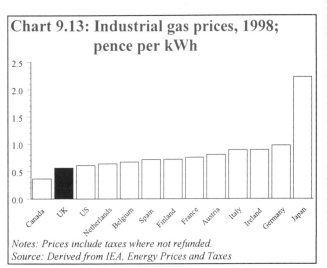

Chart 9.13: Industrial gas prices, 1998; pence per kWh

Notes: Prices include taxes where not refunded.
Source: Derived from IEA, Energy Prices and Taxes

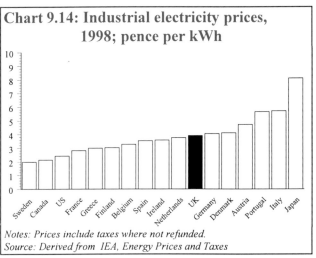

Chart 9.14: Industrial electricity prices, 1998; pence per kWh

Notes: Prices include taxes where not refunded.
Source: Derived from IEA, Energy Prices and Taxes

9.72 In general average UK industrial gas prices in 1998 are the lowest in Europe and the relative position has improved or maintained an improvement over the past few years. For example, in 1990 UK industrial gas prices were the seventh lowest in the EU, but by 1996

9.76 International comparisons can also be informative when looking at motor fuel prices and Table 9.19 presents unleaded petrol and diesel prices including and excluding tax and duty for the three most recent years. The table clearly shows that, as a result of tax policy, UK prices at the pump are the highest in Europe, despite having amongst the lowest prices before tax. It should, however, be remembered that the strong pound in recent years relative to the Euro and other EU currencies will magnify the differences. In April 2000 a litre of unleaded cost 11½ pence more in the UK than in the second highest country, the Netherlands, and was over 31 pence higher than in Spain, Luxembourg and Greece. Diesel prices in the UK were more than 30 p/litre more expensive than in all other EU countries and double the price in Greece, Luxembourg and Portugal.

9.77   In the UK, 76 per cent of the price of a litre of unleaded petrol is tax and duty compared to 73 per cent in Germany, 71 per cent in France and 52 per cent in Portugal. For diesel the tax impact is even starker with

the UK tax percentage of 75 per cent of pump price compared to the next highest of 67½ per cent in Germany and the lowest of 53 per cent in Luxembourg.

Contact: *Lesley Petrie (Statistician)*
020 7215 2720
Domestic energy market
Sharon Young
 020 7215 6531
Industrial, international and oil prices
Sara Atkins
020 7215 6532

Technical notes and definitions

Table 9.1

9.78 The source of the prices in this table is the Retail Prices Index (RPI), published by the Office for National Statistics. The fuel components within the RPI are published, together with the all items RPI. Also given (below in Table 9E) are the weights within the total index, in parts per 1,000, of the fuel components. RPI is calculated using prices collected on a day near the middle of the month.

9.79 The following notes apply to Table 9.1:

Coal and smokeless fuel (coal and solid fuels) - Retail prices of one standard grade of household coal and of the boiler/room heater grade of smokeless fuel sold by the retailer, obtained from local retailers in up to 146 areas throughout the United Kingdom.

Gas and electricity - The indices are calculated using published tariff information from British Gas (and since April 1996 other suppliers), the Public Electricity Supply Companies and Northern Ireland Electricity (NIE). When prices change in an area (including discounts and lump sum rebates), an index is re-calculated for a selection of the tariffs in use in that area at typical levels of consumption at each tariff. Electricity area indices are weighted together using the total receipts of each Public Electricity Supply Company and NIE from their sales to domestic consumers under each tariff. Gas companies are weighted by customer numbers, which currently gives most weight to British Gas. Both indices are calculated using mainly credit tariffs only.

Other fuel and light (oil and other fuels) - This comprises bottled gas and paraffin until January 1986, and domestic heating oils. Prices of heating oil are provided by the main suppliers.

Motor spirit and lubricating oil - Retail prices of the different grades of motor spirit and engine oil are obtained from garages in more than 180 areas throughout the United Kingdom.

Table 9.2

9.80 Household final consumption expenditure comprises household expenditure in the United Kingdom on the fuels specified and fuel purchases by foreign tourists. It excludes expenditure on fuels by businesses. VAT was levied on domestic fuels at 8 per cent in April 1994, reduced to 5 per cent in September 1998, and is included in the table from 1994 onwards. For coal, coke and petroleum products it was assumed that all consumers paid VAT from the date of its introduction. For electricity and gas an estimate was made that 5 per cent of electricity sales and 4 per cent of gas sales were covered by customers pre-paying their bills to avoid VAT in 1994 and 1995. Figures for total consumers' expenditure are also shown for comparison.

9.81 The following notes apply to Table 9.2:

Coal and coke – Household final consumption expenditure on these fuels is based on estimates of inland sales of solid fuels to domestic consumers. Expenditure in Northern Ireland is estimated based on values of colliery despatches of house coal to Northern Ireland.

Gas - Personal consumption in the United Kingdom is taken as sales to domestic premises. The average price used is the average revenue per kWh for public supply sales of gas to domestic consumers.

Electricity - Sales from the public electricity supply system to domestic consumers in the United Kingdom plus estimates of the domestic element included in sales to dual use premises. Sales are valued at the average revenue per unit for electricity sold to domestic consumers, which takes into account discounts and lump sum rebates.

Petroleum products, motor spirit, diesel and lubricating oil - Estimates of the quantity and value of paraffin, liquid gases and lubricating oil purchased by domestic consumers are provided by the petroleum industry. For fuel oil and heating oils, information is available from the petroleum industry on quantities delivered to domestic consumers. For motor spirit and diesel, estimates of business purchases of the fuels are made and deducted from total deliveries to arrive at purchases by domestic consumers. The figures for domestic consumption for all four of these fuels are then valued using monthly prices collected by the Department from oil companies.

Table 9E: Retail prices index, fuel component weights

	All items	Fuel and light	Coal and solid fuels	Coal	Solid fuels	Gas	Electricity	Oil and other fuels	Petrol and lubricating oil
1975	1,000	53	11	9	2	12	25	5	47
1980	1,000	59	9	8	1	16	29	4	43
1985	1,000	65	8	6	2	24	29	4	50
1990	1,000	50	4	3	1	19	24	3	33
1992	1,000	47	3	2	1	18	24	2	33
1994	1,000	45	2	1	1	18	23	2	37
1996	1,000	43	2	1	1	18	22	1	37
1997	1,000	41	1	0.5	0.5	17	21	2	39
1998	1,000	36	1	0.5	0.5	16	18	1	39
1999	1,000	34	1	0.5	0.5	15	17	1	38

Revaluation at constant prices - The estimates of expenditure on gas and electricity have been obtained by deflating the current price estimates by movements in the gas and electricity components of the index of retail prices. For other fuels, consumers' expenditure at 1995 average prices is calculated by applying average 1995 prices of these fuels to the current quantities used for calculating expenditure at current prices. In times of motor spirit fuel switching as seen in the 1990 this can lead to counter intuitive results with real term expenditure falling simply as a result of people moving to cheaper unleaded fuel (i.e. greater volume at fixed lower cost), when all petrol prices have indeed been rising. The methodology underlying calculations of expenditure on diesel has recently been reviewed by DTI. ONS are considering how the new series can be introduced.

Table 9.3
9.82 Figures for Table 9.3 are taken from the Family Expenditure Survey (FES) conducted by the ONS. The figures are estimates based upon a representative sample of households. The averages in the table have been calculated on the basis of consuming households, i.e. only those households who consumed the particular fuel in question are included in the calculation of the average expenditure. These estimates therefore differ from those published by the ONS in the report, "Family Spending", where the total of all households is used to calculate average fuel expenditure. After the publication of data for 1993 the FES moved to a financial year basis meaning 1998/9 is the latest year for which information is available. The data presented on expenditure on fuel as a proportion of total expenditure in Chart 9.4 and Tables 9A and 9B are based on all households, not just those consuming the fuel or other commodity, for ease of comparison.

Table 9.4, Chart 9.11, Crude oil
9.83 The index in Table 9.4 is based on price data collected from oil companies covering both imported oil and indigenous crude oil. The imported oil component is valued at the c.i.f. prices. The index is used to represent the average cost of crude oil within the producers price index, produced by ONS, to measure manufacturing industries input and output prices. DTI also calculate series for gas, electricity,

coal, and petroleum products which feed into the index.

Tables 9.5 to 9.9
9.84 Table 9.5 shows representative gas and electricity bills by payment type in each of the 15 Public Electricity Supply (PES) areas in the UK and 12 gas Local Distribution Zones (LDZ) in Great Britain. The unit cost (UC) represents the total cost to the consumer per unit consumed and is calculated by dividing the bill shown by the number of units consumed (18,000 kWh for gas, 3,300 kWh for electricity). The electricity PES areas and gas LDZ associated with each of the towns and cities in Table 9.5 are shown in Table 9F.

Table 9F: Towns and Cities in Table 9.5 by LDZ and PES area

	Gas LDZ	Electricity PES area
Aberdeen	Scotland	Northern Scotland
Belfast	n/a	Northern Ireland
Birmingham	West Midlands	West Midlands
Canterbury	South East	South East
Cardiff	Wales	South Wales
Edinburgh	Scotland	Southern Scotland
Ipswich	Eastern	Eastern
Leeds	Northern	Yorkshire
Liverpool	North Western	Merseyside & North Wales
London	London	London
Manchester	North Western	North West
Newcastle	North East	North East
Nottingham	East Midlands	East Midlands
Plymouth	South West	South West
Southampton	Southern	Southern

9.85 Bills and unit costs   are based on published prices and include standing charges. No allowances are made for introductory offers or non-cash benefits that may be available from new suppliers. Both electricity and gas bills and costs reflect the prices of all suppliers. This basis is used for all the domestic bills and cost data used in Tables 9.5 to 9.9.

9.86 The bills shown in Tables 9.5, 9.8 and 9.9 relate to the total bill including VAT in cash terms received during the calendar year, for the tariff type shown, including all tariff changes and rebates. Averages are weighted by the number of domestic customers. For electricity an annual consumption of 3,300 kWh is used whilst the equivalent figure for gas is 18,000 kWh.

9.87 The weighted average all supplier gas bills are based on equivalent tariffs of British Gas and other supply companies. As the estimate (like all the bills in the table) is based on bills received during the calendar year, that is consumption in Q4 of year X-1 and Q1 - Q3 of year X, customers of new gas suppliers will have received some of their gas in a year from British Gas prior to switching. This, coupled with the fact that British Gas in 1999 still supplied around 70 per cent of the domestic market, means that especially in the early years of competition the all supplier average is not substantially lower than the British Gas figure, despite the large savings available, as shown by the average non British Gas bill.

9.88 Table 9.6 includes data for 'Economy 7' tariffs, where a lower unit cost is applied to off-peak (night) consumption. For total consumption of 6,600 off-peak consumption has been taken as 3,600 with the equivalent figure for 10,000 kWh total consumption taken as 5,500. The 'white meter' tariffs are used to calculate values in Scotland, where 'Economy 7' tariffs are not available.

9.89 Tables 9.5, 9.6 and 9.7 attempt to give an indication of variation in pricing by showing the largest and smallest charges, calculated costs and component percentages of total bills. In Tables 9.6 and 9.7 the figures given for the standing charge are a weighted (by customer numbers) average of all companies standing charges. All other calculations are also based on weighted averages.

Table 9.10

9.90 The prices for fuels used in electricity generation are collected via a quarterly inquiry of electricity generators in the United Kingdom. The data reported are the value and volume of fuel purchased during the quarter and may not always reflect the fuel actually used (i.e. there can be stocking and destocking especially of coal). The prices reported are typically for long-term contracts, with price escalator factors, some of which may have been entered into some time ago. As such the prices can be higher than those paid by large industrial users who typically negotiate contracts each year.

9.91 The gas beach price series is calculated as follows:

$$\frac{\text{Value of (UKCS gas sales + gas imports - gas exports)}}{\text{Volume of (UKCS gas sales + gas imports - gas exports)}}$$

where the UKCS sales value and volume data are derived from the DTI's quarterly statistical inquiry into oil and gas extraction (PQ1100). Returns from the inquiry give the value and volume of gas sold by each licensee from a particular field (or group of fields). Data from the inquiry on sales and expenditure by licensees are covered in Table C9 and further explained in Annex C, paragraph C43. Trade data are supplied by Customs and Excise and published in Chapter 8, Table 8.1 and in *Energy Trends*.

9.92 The gas levy applied to gas purchased under certain contracts originally entered into before July 1975. The cost of gas under these pre-July 1975 contracts had historically been substantially less than the prevailing market price. Gas sold under these contracts was not subject to Petroleum Revenue Tax (PRT) because the contracts were classified as "tax-exempt" when PRT was introduced in 1975. Instead, under the Gas Levy Act 1981, the purchaser of gas subject to the relevant contracts had to pay a levy on every therm of such gas that they purchased. The purpose of the gas levy was to capture for the Exchequer the bulk of the economic rent which would otherwise accrue to the purchaser from purchasing this gas at below market prices. However, current and expected future gas market prices are now below the average cost of this gas (even before adding the cost of the levy). The gas levy was abolished from 1 April 1998.

Tables 9.11 and 9.12

9.93 Prices are derived from information collected via the Quarterly Fuels Inquiry on fuel purchases from a panel of about 1,200 establishments within manufacturing industry (which excludes electricity generation). The panel consists of companies purchasing fuels in small and large quantities. To maximise the coverage of each fuel type and minimise the burden on business, larger users are surveyed proportionally more than smaller users.

9.94 For each size of consumer the average price for a fuel (exclusive of VAT) is calculated by dividing the total quantity of purchases into their total value. The "all consumers-average" price uses base weighting and weights the prices for each size band according to purchases by businesses in the size band recorded in the 1984 Purchases Inquiry. (This is a large scale survey conducted every 5 years until 1989, and conducted annually for a rotating selection of industries from 1994; it provides information on purchases of materials and fuels by the whole of UK industry.) The weights will be reviewed when comprehensive up-to-date purchases data are available. The size bands are defined, for each fuel individually, according to the approximate range of annual purchases covered. (See Table 9G below).

9.95 As described above the prices given are representative market prices. This means trades that, because of their size or dominance of total consumption would produce an unrepresentative price, are excluded. For example, coal and coke purchased by the iron and steel sector are excluded as is gas purchased for electricity generation.

9.96   For some fuels the relative size in volume terms of the largest users can have the effect of moving the weighted average more towards the large user price. This is true for gas where because of the growth in consumption the weights provided by the 1984 purchases survey may be out of date. Therefore, for some fuels (e.g. gas and gas oil) the median price (the price at which 50 per cent of the prices paid are higher and 50 per cent lower) may be another useful guide to average prices.

9.97   There is no sub-division into size bands of the prices for medium fuel oil, liquefied petroleum gases and hard coke owing to the small number of sites purchasing each of these fuels. The small sample sizes reflect the small overall consumption, relative to the major fuels covered, which means that although the prices are still representative, they can be subject to more sample effects than the other fuels (e.g. if a relatively large purchaser switches fuel).

9.98   To enable coal prices to be calculated in common units, companies record the calorific value of the coal they purchase. Conversion factors for fuel oil (both heavy and medium), gas oil, liquefied petroleum gas and hard coke are given in Annex A.

9.99   Table 9.12 also shows the 10 per cent and 90 per cent deciles and the median price for each fuel in addition to the prices for each size band. The 10 per cent decile is the point within the complete range of prices below which the lowest 10 per cent of those prices fall. Similarly the 90 per cent decile is the point above which the highest 10 per cent of the prices occur. Thus, these values give some indication of the spread of prices paid by purchasers. The deciles and the median are calculated by giving equal "weight" to each purchaser, but are scaled to represent the mix of

fuel users by size in the industrial population that the panel represents.

Tables 9.17 to 9.19

9.100 International comparisons are based on data published by international organisations. For electricity and gas the data used are collated and published by the International Energy Agency in *Energy Prices and Taxes*. Individual countries supply data to the IEA so methodology can vary from country to country. Motor fuel prices are taken from the European Commission's Oil Bulletin which contains weekly and mid-month data. Mid-month data is generally more representative in that it covers a greater proportion of sales. Again collection methodologies vary between countries, but these tend to be more consistent than with other fuels.

Table A9.1

9.101 This table provides data previously contained in the electricity and gas chapters on the total value of gas and electricity sold to final consumers. The data are collected from the energy supply companies. The data are useful in indicating relative total expenditure between sectors, but the quality of data provided in terms of industrial classification has been worsening in recent years. Net selling values provide some indication of typical prices paid in broad sectors and can be of use to supplement more detailed and accurate information contained in the rest of this chapter.

Monthly and quarterly data

9.102 Monthly data on petroleum product prices and the crude oil index are provided in Table 30 of the DTI's monthly statistical bulletin *Energy Trends*. Also provided are: quarterly data for fuel prices paid by manufacturing sector (Table 26); fuel costs for electricity generation (Table 27); and industrial and domestic fuel price indices (Tables 28 and 29).

Table 9G: Range of annual purchases for the Quarterly Fuels Inquiry

Fuel	Large Greater than	Of which: Extra large Greater than	Of which: Moderately large	Medium	Small Less than
Coal (tonnes)	7,600	..	..	760 to 7,600	760
Heavy fuel oil (tonnes)	4,900	15,000	4,900 to 15,000	490 to 4,900	490
Gas oil (tonnes)	175	..	..	35 to 175	35
Electricity (thousand kWh)	8,800	150,000	8,800 to 150,000	880 to 8,800	880
Gas *(1)* (thousand kWh)	8,800	..	..	1,500 to 8,800	1,500

(1) Respondents purchasing more than one type of supply (tariff, firm contract and interruptible contract) are treated as separate entities in respect of each type of supply.

9.1 Retail prices index : fuel components[1][2]

United Kingdom

1990 = 100

Indices (5)		All items	Fuel and light							Petrol and lubricating oil
			Total (3)	Coal and smokeless fuels	Coal	Smokeless fuels	Gas	Electricity	Heating oils (4)	
	1975	27.1	25.1	24.6	25.2	22.0	26.2	26.0	21.6	35.1
	1980	53.0	53.4	58.9	60.7	50.9	46.6	56.2	59.1	64.0
	1985	75.0	85.1	90.5	90.9	88.8	88.3	80.5	98.5	96.2
	1990	100.0	100.0	100.0	100.0	100.0	100.0	100.0	100.0	100.0
	1995	118.2	116.1	120.2	119.5	124.6	112.5	120.8	89.9	131.2
	1998	129.1	107.8	123.6	123.1	127.5	107.8	109.3	80.8	159.1
	1999	131.2	107.4	126.0	125.7	129.7	107.1	108.0	90.5	172.5
1999	January	129.6	107.3	126.7	126.3	130.7	107.7	108.4	75.9	155.9
	February	129.8	107.2	126.9	126.4	130.7	107.7	108.3	75.4	155.2
	March	130.1	107.4	126.8	126.3	130.7	108.0	108.3	78.1	164.1
	April	131.0	107.2	127.2	..	..	106.8	108.3	84.1	172.7
	May	131.3	107.0	123.9	..	..	106.8	108.1	83.5	172.6
	June	131.3	106.9	123.6	..	..	106.8	108.0	83.5	172.1
	July	130.9	107.2	123.8	..	..	106.8	107.8	91.9	174.4
	August	131.2	107.2	123.7	..	..	106.8	107.7	93.3	179.2
	September	131.8	107.4	124.8	..	..	106.8	107.7	98.6	179.0
	October	132.0	107.5	128.0	..	..	106.8	107.7	99.9	180.5
	November	132.2	107.8	128.3	..	..	106.8	107.7	104.4	179.7
	December	132.6	108.3	128.6	..	..	106.8	107.7	116.8	184.4
2000	January	132.1	108.2	128.8	..	..	106.8	107.7	114.7	183.9
	February	132.8	108.2	128.7	..	..	106.8	107.7	115.2	183.8
	March	133.5	108.3	129.1	..	..	106.8	107.7	117.6	190.5

Index numbers relative to the GDP deflator (6)

		All items	Total (3)	Coal and smokeless fuels	Coal	Smokeless fuels	Gas	Electricity	Heating oils (4)	Petrol and lubricating oil
	1975	100.4r	93.0	91.2r	93.3	81.5	97.2r	96.3	80.0	130.0
	1980	98.1	98.8r	109.1	112.4	94.3	86.2r	104.1	109.5r	118.6r
	1985	99.7	113.2	120.4r	120.9	118.1	117.5r	107.0	131.0	127.9
	1990	100.0	100.0	100.0	100.0	100.0	100.0	100.0	100.0	100.0
	1995	99.5	97.7	101.2	100.6	104.8	94.7	101.7	75.7	110.4
	1998	99.2r	82.8r	94.9r	94.6r	97.9r	82.8r	84.0r	62.1r	122.2r
	1999	97.9	80.1	94.1	..	..	79.9	80.6	67.5	128.7
1999	January	97.9r	81.1r	95.8r	95.4r	98.8r	81.4r	81.9r	57.4r	117.8r
	February	98.1r	81.0r	95.9r	95.6r	98.8r	81.4r	81.9r	57.0r	117.3r
	March	98.3r	81.2r	95.9r	95.5r	98.8r	81.6r	81.9r	59.1r	124.0r
	April	98.0	80.2	95.1	..	..	79.9	81.0	62.9	129.2
	May	98.2	80.0	92.6	..	..	79.9	80.8	62.4	129.1
	June	98.2	80.0	92.4	..	..	79.9	80.8	62.4	128.7
	July	97.2	79.6	91.9	..	..	79.3	80.0	68.2	129.5
	August	97.4	79.6	91.8	..	..	79.3	79.9	69.3	133.0
	September	97.8	79.8	92.7	..	..	79.3	79.9	73.2	132.9
	October	97.5	79.4	94.6	..	..	78.9	79.5	73.8	133.3
	November	97.6	79.6	94.8	..	..	78.9	79.5	77.1	132.7
	December	98.0	80.0	95.0	..	..	78.9	79.5	86.3	136.2
2000	January	97.2	79.6	94.8	..	..	78.6	79.2	84.4	135.3
	February	97.7	79.6	94.7	..	..	78.6	79.2	84.8	135.3
	March	98.2	79.7	95.0	..	..	78.6	79.2	86.6	140.2

Source: Office for National Statistics

(1) Quarterly data are published in Table 29 of Energy Trends.
(2) Rebased to 1990 by DTI from original ONS indices. Indices are given to one decimal place to provide as much information as is available but precision is greater at higher levels of aggregation i.e. at sub-group and group levels.
(3) Including oil for domestic heating and light but excluding motor spirit and lubricating oil.
(4) Including domestic heating oils, and bottled gas for use in the home, but excluding paraffin from February 1986.
(5) From April 1994 VAT at 8% has been included for coal and solid fuels, gas, electricity and heating oils, decreasing to 5% in September 1997.
(6) Deflated using seasonally adjusted GDP deflator at market prices with base year of 1995 but rescaled to 1990=100.

9.2 Household expenditure on energy[1]
United Kingdom

£ million

	1980	1985	1990	1995	1998	1999
At current market prices						
Coal & coke	677	1016	660	530	459r	474
Gas	1,852	4034	4,864	5,909	5,145r	5,030
Electricity	3,370	4910	6,278	8,195	7,715r	7,392
Petroleum products (2)	456	600	448	483	494r	488
All fuel and power	6,355	10,560	12,250	15,117	13,813r	13,384
Motor spirit, Derv fuel and lubricating oil	4,646	8,018	10,172	12,001	14,241r	15,250
Total energy products (3)	**11,001**	**18,578**	**22,422**	**27,118**	**28,054r**	**28,634**
Total consumers' expenditure	**132,663**	**206,600**	**336,492**	**438,453**	**527,149r**	**560,924**
Revalued at 1995 prices						
Coal & coke	1,638	1,575	930	530	488r	489
Gas	4,502	5,138	5,478	5,909	5,507r	5,419
Electricity	7,184	7,334	7,541	8,195	8,491r	8,241
Petroleum products (2)	624	448	417	483	592r	530
All fuel and power	13,948	14,495	14,366	15,117	15,078r	14,679
Motor spirit, Derv fuel and lubricating oil	9,414	10,775	13,176	12,001	12,036r	11,969
Total energy products (3)	**23,362**	**25,270**	**27,542**	**27,118**	**27,114r**	**26,648**
Total consumers' expenditure	**297,256**	**331,404**	**415,788**	**438,453**	**487,479r**	**506,742**

Source: Office for National Statistics

(1) These figures are based on the March 2000 edition of Consumers Trends as published by the Office for National Statistics. The full title of the survey conducted by ONS is 'Household final consumption expenditure'. Some early data has been revalued to allow for full aggregation. The figures exclude business expenditure. All data may be subject to change by ONS.

(2) Includes only heating fuels e.g. burning oil and LPG.

(3) Quarterly data on energy expenditure are shown in the Office for National Statistics Monthly Digest of Statistics.

9.3 Average expenditure on fuel per consuming household[1]
United Kingdom

£ per week

	1980	1985	1990	1995/96	1997/98	1998/99
Electricity						
All households	2.95	4.56	5.62	6.72	6.58	6.49
With electric central heating	4.73	7.28	8.65	9.33	9.48	9.62
Without electric central heating	2.71	4.28	5.26	6.44	6.27	6.14
Heating oils and other fuels	3.02	4.46	5.88	9.42	11.05	9.26
Solid fuel	8.07	10.77	11.18	8.65	9.16	10.20
Gas						
All households	2.47	5.17	5.53	6.70	7.16	6.49
With gas central heating	3.19	6.11	6.21	7.19	7.62	6.87
Without gas central heating	1.72	3.49	3.49	4.20	4.28	3.96
All fuels (excluding motor fuel)	**6.15**	**11.66**	**11.07**	**12.74**	**12.98**	**12.36**
Motor fuel	7.72	12.64	13.02	16.96	19.86	20.09
Average expenditure all fuels (2)	**13.87**	**24.30**	**24.09**	**29.70**	**32.84**	**32.44**

Source: Family Expenditure Survey, ONS

(1) Data is based on a survey and therefore samples sizes will vary from year to year which can give misleading results, especially for the lesser used fuels which will have a greater sample error rate. Data shows average expenditure recorded in households consuming the specified fuel. See text for further details.

(2) Arithmetic sum of average fuel expenditure.

9.4 Price indices and consumers' expenditure 1963 to 1999

United Kingdom

	Retail Prices Index		Producer Price Index		Household expenditure on energy (£ million) (1)			
	Fuel and light (2)	Petrol and oil	Fuels purchased by manufacturing industry (3)	Crude oil acquired by refineries (4)	Fuel and light	Petrol and oil	Fuel and light	Petrol and oil
	13 January 1987 = 100		1995 = 100		At current market prices		Revalued at 1995 prices	
1963	11.1	13.8	..	..	1,000	309	11,425	3,436
1964	11.5	14.0	..	..	992	382	10,930	4,194
1965	12.0	15.1	..	..	1,076	470	11,321	4,830
1966	12.7	15.5	..	..	1,149	539	11,454	5,412
1967	13.0	15.8	..	..	1,196	608	11,511	5,707
1968	14.0	16.8	..	..	1,328	714	11,843	6,336
1969	14.5	18.5	..	..	1,416	809	12,273	6,625
1970	15.3	18.8	..	8.4	1,480	859	12,362	6,919
1971	16.9	20.1	..	10.1	1,605	942	12,233	7,163
1972	18.2	20.6	..	10.3	1,781	1,093	12,686	7,987
1973	18.7	21.5	..	13.7	1,880	1,249	13,035	8,623
1974	21.9	30.8	23.1	42.5	2,249	1,717	13,268	8,358
1975	29.1	41.9	29.2	47.2	2,887	2,232	12,952	8,190
1976	36.0	44.6	35.1	50.1	3,557	2,500	12,842	8,696
1977	41.8	47.8	43.3	69.0	4,219	2,676	13,157	8,781
1978	45.0	45.4	46.2	62.9	4,613	2,610	13,391	9,050
1979	49.5	59.7	53.0	80.6	5,292	3,554	14,049	9,171
1980	61.9	76.5	67.2	124.7	6,355	4,646	13,948	9,414
1981	75.1	91.4	80.5	160.7	7,727	5,695	13,434	9,540
1982	85.6	97.6	87.6	170.4	8,696	6,331	13,174	9,906
1983	92.0	104.4	93.6	177.4	9,348	6,872	13,017	10,085
1984	94.6	108.0	104.4	179.0	9,492	7,481	12,934	10,622
1985	98.7	114.9	109.2	173.3	10,560	8,018	14,495	10,775
1986	100.0	100.0	95.4	97.7	10,865	7,354	14,699	11,439
1987	99.1	100.9	91.6	98.9	10,855	7,773	14,795	11,915
1988	101.6	99.8	93.8	76.2	11,064	8,227	14,751	12,757
1989	107.3	106.9	98.0	98.1	11,401	9,059	14,355	13,006
1990	115.9	119.5	98.2	113.5	12,250	10,1 2	14,366	13,176
1991	125.1	128.4	100.0	105.5	14,197	10,758	15,381	12,924
1992	127.8	132.1	104.6	99.6	14,380	11,019	15,129	12,865
1993	126.2	142.6	106.1	103.7	14,618	11 518	15,598	12,514
1994	131.7	149.1	104.6	94.3	14,885	' ',972	15,311	12,443
1995	134.5	156.8	100.0	100.0	15,117	2,001	15,117	12,001
1996	134.8	164.7	91.7	122.5	16,074	12,748	15,966	12,244
1997	130.6	181.1	88.3	107.8	14,879r	13,714r	15,315r	12,112r
1998	125.0	190.1	89.2	70.4	13,813r	14,241r	15,078r	12,036r
1999	124.4	206.1	89.1	100.9	13,384	15,250	14,679	11,969
1998Q1	125.9	183.5	95.0	79.4	4,361r	3,398r	4,715r	2,969r
1998Q2	125.4	193.0	85.8	73.2	2,981r	3,539r	3,242r	2,947r
1998Q3	124.2	193.4	83.7	67.8	2,440r	3,669r	2,696	3,049r
1998Q4	124.4	190.3	92.3	61.3	4,031r	3,635r	4,425r	3,071r
1999Q1	124.3	189.3	94.1	63.7	4,257	3,410	4,674	2,910
1999Q2	124.0	206.1	85.5	87.3	2,755	3,759	3,031	2,950
1999Q3	124.3	212.1	84.8	116.3	2,422	3,964	2,664	3,028
1999Q4	125.0	216.9	91.9	136.2	3,950	4,117	4,310	3,081

Source: Office for National Statistics & DTI

(1) ONS data may be subject to revision.
(2) Includes coal and other solid fuels, gas, electricity, heating oils and other petroleum products, but excludes petrol and motor oil.
(3) Includes coal, gas and electricity. Prior to 1986 also includes petroleum products.
(4) Prior to 1974 includes only imported oil.

9.5 Annual 1999 domestic gas & electricity bills[1][2] and average unit costs (UC)[3]

Pence per kWh and pounds

Payment type		Credit				Direct debit [5]				Pre-payment			
		Gas		Electricity		Gas		Electricity		Gas		Electricity	
Town/city[4]	Bill range	UC	Bill	UC	Bill	UC	Bill	UC	Bill	UC	Bill	UC	Bill
Aberdeen	Largest [6]	1.76	316	9.02	298	1.58	284	8.17	269	1.99	359	9.37	309
	Average	1.70	307	8.38	277	1.49	268	8.04	265	1.77	318	8.30	274
	Smallest [6]	1.37	247	6.96	230	1.31	237	6.64	219	1.76	317	7.85	259
Belfast	Average	..	..	9.89	326	..	..	9.60	317	..	..	10.47	345
Birmingham	Largest	1.78	320	7.84	259	1.58	284	7.44	245	1.86	335	9.13	301
	Average	1.70	307	7.64	252	1.51	271	7.40	244	1.77	318	7.91	261
	Smallest	1.40	251	6.96	230	1.35	243	6.67	220	1.76	317	7.61	251
Canterbury	Largest	1.78	320	7.62	251	1.58	284	7.51	248	1.99	358	9.02	298
	Average	1.69	305	7.59	251	1.49	269	7.33	242	1.76	317	7.96	263
	Smallest	1.41	253	6.87	227	1.35	242	6.64	219	1.70	306	7.44	246
Cardiff	Largest	1.81	326	9.32	307	1.58	284	9.04	298	1.94	350	10.55	348
	Average	1.65	297	9.01	297	1.48	266	8.66	286	1.81	325	9.51	314
	Smallest	1.37	247	8.12	268	1.34	241	7.37	243	1.76	317	8.55	282
Edinburgh	Largest	1.76	316	9.01	297	1.58	284	8.20	270	1.99	359	9.56	315
	Average	1.70	307	8.25	272	1.49	268	7.99	264	1.77	318	8.74	288
	Smallest	1.37	247	7.30	241	1.31	237	6.97	230	1.76	317	8.62	284
Ipswich	Largest	1.78	321	7.64	252	1.58	284	7.44	245	1.94	350	8.49	280
	Average	1.70	305	7.61	251	1.48	266	7.36	243	1.76	317	7.92	261
	Smallest	1.40	251	6.82	225	1.35	243	6.51	215	1.67	301	7.38	243
Leeds	Largest	1.77	319	7.64	252	1.58	284	7.57	250	1.94	350	10.55	348
	Average	1.69	304	7.46	246	1.49	268	7.33	242	1.77	319	8.51	281
	Smallest	1.39	251	6.92	228	1.33	239	6.59	217	1.76	317	7.72	255
Liverpool	Largest	1.77	319	8.47	279	1.58	284	8.28	273	1.85	333	10.04	331
	Average	1.69	304	8.43	278	1.49	267	8.18	270	1.77	318	8.97	296
	Smallest	1.42	256	7.54	249	1.33	240	7.19	237	1.76	317	8.17	270
London	Largest	1.80	323	8.03	265	1.58	284	7.92	261	2.00	359	9.46	312
	Average	1.70	305	7.81	258	1.49	268	7.66	253	1.77	318	8.12	268
	Smallest	1.42	255	6.86	226	1.33	240	6.82	225	1.76	317	8.00	264
Manchester	Largest	1.77	319	8.25	272	1.58	284	7.68	253	1.85	333	9.38	310
	Average	1.69	304	7.78	257	1.49	267	7.50	247	1.77	318	8.32	274
	Smallest	1.42	256	7.07	233	1.33	240	6.80	224	1.76	317	8.03	265
Newcastle	Largest	1.77	319	9.36	309	1.58	284	8.74	288	1.99	359	9.90	327
	Average	1.67	301	8.87	293	1.45	262	8.52	281	1.78	320	9.18	303
	Smallest	1.40	252	7.45	246	1.33	240	7.16	236	1.76	317	7.93	262
Nottingham	Largest	1.78	321	8.02	265	1.58	284	7.77	256	1.94	348	10.03	331
	Average	1.68	303	7.71	254	1.48	266	7.39	244	1.76	316	8.43	278
	Smallest	1.39	250	7.01	231	1.34	242	6.73	222	1.58	284	8.05	266
Plymouth	Largest	1.77	319	8.78	290	1.58	284	8.44	279	1.99	358	10.58	349
	Average	1.70	306	8.30	274	1.49	268	8.06	266	1.77	318	8.66	286
	Smallest	1.40	252	7.77	256	1.33	239	7.42	245	1.72	309	8.34	275
Southampton	Largest	1.78	320	7.95	262	1.58	284	7.61	251	1.99	357	10.00	330
	Average	1.71	307	7.73	255	1.50	270	7.54	249	1.78	320	8.17	270
	Smallest	1.41	254	6.98	230	1.35	242	6.75	223	1.76	317	7.61	251
UK	Largest	1.81	326	9.89	326	1.58	284	9.60	317	2.00	359	10.58	349
	Average	1.69	305	8.00	264	1.49	268	7.67	253	1.77	318	8.53	282
	Smallest	1.37	247	6.82	225	1.31	237	6.51	215	1.58	284	7.38	243

(1) All bills are calculated assuming an annual consumption of 18,000 kWh for gas and 3,300 kWh for electricity. They are calculated as weighted (by average customer numbers in 1998) averages of individual tariff bills. All values include VAT.

(2) Bills relate to the total bill received during the year, which covers consumption from quarter four 1998 to quarter three 1999.

(3) Unit costs (UC) are calculated by dividing the bills shown by the relevant consumption levels.

(4) The towns/cities specified indicate which electricity/gas region these bills apply to. See Table 9F.

(5) Includes monthly and quarterly direct debit payments (although vast majority are monthly direct debit payments).

(6) The largest/smallest annual all tariff average company gas/electricity bills and unit costs.

9.6 Average[1] UK electricity costs by payment type and consumption level, 1999, including VAT

Pounds per quarter and pence per kWh

	Standard Domestic							Economy 7				
	Rate [2]	**as % of the bill**			**Unit costs** [3]			**Rate**	**as % of the bill**		**Unit costs**	
		2,000	3,300	6,600	2,000	3,300	6,600		6,600	10,000	6,600	10,000
Credit												
Largest [4]	11.76	21.9	14.4	7.7	10.75	9.89	9.23	20.08	18.5	13.2	6.59	6.10
Average [5]	8.46	19.4	12.8	6.9	8.75	8.00	7.41	10.34	11.4	7.9	5.48	5.20
Smallest [4]	0.00	0.0	0.0	0.0	7.42	6.82	6.06	0.35	0.5	0.3	4.68	4.42
Direct Debit [6]												
Largest	11.36	22.0	14.3	7.6	10.35	9.60	9.08	19.56	18.0	12.6	6.60	6.21
Average	7.70	18.4	12.2	6.6	8.36	7.67	7.12	9.21	10.5	7.2	5.33	5.08
Smallest [7]	-1.99	-5.7	-3.7	-2.0	7.00	6.51	6.05	-1.69	-2.3	-1.6	4.48	4.28
Pre-Payment												
Largest	16.58	25.5	19.0	10.5	12.98	10.58	9.54	21.89	17.9	12.9	7.43	6.80
Average	13.67	28.3	19.4	10.8	9.67	8.53	7.65	15.46	16.1	11.4	5.81	5.44
Smallest	3.37	8.5	5.5	3.1	7.92	7.38	6.69	3.37	4.2	3.0	4.84	4.49

(1) All averages are weighted using customer numbers, and include all tariffs offered by each company.

(2) Averages are weighted averages of all tariffs.

(3) Unit costs are calculated by dividing the bills shown by the given consumption levels, i.e. they include the contribution of the standing charge to the overall bill.
    A companies largest (smallest) standing charge for an individual modal tariff; the largest (smallest) % of a companies average bill accounted for by a companies average standing charge; the highest (lowest) company annual all tariff bill divided by given consumption.

(4) Average standing charges (weighted by customer numbers) as a % of average bills (weighted as in Table 9.5).

(5) Includes monthly and quarterly direct debit payments (although vast majority are monthly direct debit payments).

(6) When a company does not have a separate standing charge, the application of any fixed rate discount (for prompt payment or direct debit for instance) means the standing charge in this table may show as a negative value.

9.7 Average[1] gas costs, in GB by payment type and consumption level, 1999, including VAT

Pence per day and pence per kWh

		Standing charge				**Total unit costs** [3]			
	Rate [2]	**as a % of the bill**							
Consumption levels (kWh)		6,000	12,000	18,000	24,000	6,000	12,000	18,000	24,000
Credit									
British Gas	13.39	33.4	20.4	14.7	11.5	2.33	1.90	1.76	1.68
Largest [4]	13.16	32.4	19.6	14.0	10.9	2.35	1.95	1.81	1.74
New suppliers Average [5]	8.91	27.8	16.5	11.8	9.1	1.85	1.56	1.46	1.42
Smallest [4]	0.00	0.0	0.0	0.0	0.0	1.37	1.33	1.37	1.35
All company average	12.45	32.4	19.8	14.2	11.1	2.22	1.82	1.69	1.63
Direct Debit [6]									
British Gas	6.66	21.4	12.1	8.5	6.5	1.81	1.59	1.52	1.49
Largest	10.59	31.6	18.4	12.9	10.0	1.94	1.67	1.58	1.54
New suppliers Average	7.81	25.9	15.3	10.8	8.4	1.74	1.48	1.39	1.35
Smallest	0.00	0.0	0.0	0.0	0.0	1.31	1.24	1.31	1.28
All company average	7.02	22.7	13.0	9.1	7.0	1.79	1.56	1.49	1.45
Pre-payment									
British Gas	4.01	10.0	6.1	4.4	3.4	2.33	1.90	1.76	1.69
Largest	9.00	20.7	12.4	8.7	6.7	2.51	2.10	2.00	1.94
New suppliers Average	7.46	19.3	11.2	7.9	6.1	2.24	1.92	1.82	1.76
Smallest	0.00	0.00	0.0	0.0	0.0	2.04	1.75	1.58	1.49
All company average	4.43	11.1	6.7	4.8	3.8	2.32	1.91	1.77	1.70

(1) All averages are weighted using customer numbers, and include all tariffs offered by each company.

(2) Averages are weighted averages of all tariffs.

(3) Unit costs are calculated by dividing the bills shown by the given consumption levels, i.e. they include the contribution of the standing charge to the overall bill.

(4) A company's largest/smallest standing charge for an individual modal tariff; the largest/smallest % of a company's average bill accounted for by a company's average standing charge; the highest (lowest) company annual all tariff bill divided by given consumption.

(5) Average standing charges (weighted by customer numbers) as a % of average bills (weighted as in Table 9.5)

(6) Includes monthly and quarterly direct debit payments (although vast majority are monthly direct debit payments).

9.8 Average domestic annual standard[1][2] electricity bill, for 3,300 kWh annual consumption, including VAT

Pounds

| | | Year | | | | % Change | | | | |
| | | Cash terms | | | | Cash terms | | | Real terms [3] | |
		1990	1995	1998	1999	1990 -1999	1995 -1999	1998 -1999	1995 -1999	1998 -1999
Standard credit	England & Wales	246	299	265	261	6	-13	-2	-23	-5
	Scotland	230	293	275	273	19	-7	..	-17	-3
	Northern Ireland	261	346	326	326	25	-6	..	-16	-3
	UK	245	299	268	264	8	-12	-2	-22	-4
Direct debit	England & Wales	..	294	256	251	..	-15	-2	-24	-5
	Scotland	..	290	270	264	..	-9	-2	-19	-5
	Northern Ireland	..	346	317	317	..	-8	..	-19	-3
	UK	..	295	258	253	..	-14	-2	-24	-5
Pre-payment [4]	England & Wales	265	319	283	279	5	-13	-2	-22	-4
	Scotland	253	309	288	285	12	-8	-1	-18	-4
	Northern Ireland	275	373	345	345	25	-7	..	-18	-3
	UK	264	319	285	282	6	-12	-1	-22	-4
Pre-payment less credit [5]		19	20	17	17	..	..	..	..	..
Pre-payment less direct debit [5]		..	24	27	28	..	..	..	..	..

(1) For an Economy 7 user with an annual consumption of 6,600 kWh, 1999 England & Wales, Scotland, Northern Ireland and UK bills are £357, £396, £435 and £361 respectively.
(2) Bills relate to total bill received in the year, e.g. covering consumption from Q4 1998 to Q3 1999. Bills in 1998 for home supplier only.
(3) 1995 prices are calculated using the GDP (market prices) deflator.
(4) For a pre-payment user with an annual consumption of 2,000 kWh 1999 England & Wales, Scotland, Northern Ireland and UK bills are £192, £194, £234 and £193 respectively.
(5) Based on UK bills.

9.9 Average GB domestic annual gas bill, for 18,000[1] kWh annual consumption

Pounds

| | | Year | | | | | % Change | | | | |
| | | Cash terms | | | | | Cash terms | | | Real terms [2] | |
		1986	1990	1995	1998	1999	1990 -1999	1995 -1999	1998 -1999	1995 -1999	1998 -1999
Standard Credit	British Gas	261	285	327	320	316	11	-3	-1	-14	-4
	Non British Gas	..	..	..	263	263	..	..	..	..	-3
	All suppliers	261	285	327	315	305	7	-7	-3	-17	-6
Direct Debit	British Gas	..	..	311	281	274	..	-12	-3	-22	-5
	Non British Gas	..	..	..	249	250	..	..	1	..	-2
	All suppliers	..	..	311	277	268	..	-14	-3	-24	-6
Pre-Payment [3]	British Gas	277	303	347	331	317	5	-9	-4	-19	-7
	Non British Gas	..	..	..	326	327	..	..	..	..	-3
	All suppliers	277	303	347	331	318	5	-8	-4	-19	-7
Pre-payment less credit [4]		16	18	20	16	13	..	..	..	..	..
Pre-payment less direct debit		..	..	36	54	50	..	..	..	..	..

(1) Figures for 1998 and 1999 are based on equivalent tariffs of British Gas and other supply companies but exclude the value of incentives offered by certain suppliers. Prior to 1997 bills based on British Gas tariffs only. Based on an annual consumption of 18,000 kWh, however prior to February 1997, kWh's were calculated differently and a figure of 17,600 is used. Bills relate to total bill received in the year, covering consumption from Q4 1998 to Q3 1999.
(2) 1995 prices are calculated using the GDP (market prices) deflator.
(3) For a pre-payment user with an annual consumption of 12,000 kWh 1999 British Gas, non-British Gas and all company bills are £228, £187 and £219 respectively.
(4) Based on all supplier bills.

9.10 Average prices of fuels purchased by the major UK power producers[1] for electricity generation and of gas at UK delivery points[2][3]

	Unit	1993	1994	1995	1996	1997	1998	1999
Electricity companies								
Coal *(4)*	£ per tonne	42.44	36.35	35.11	35.22	33.74	30.17	29.01
	pence per kWh *(5)*	0.611	0.528	0.500	0.507	0.474	0.421	0.405
Oil *(6)(7)*	£ per tonne	55.91	67.90	81.12	84.15	89.75	71.87	85.84
	pence per kWh *(5)*	0.472	0.526	0.684	0.709	0.746	0.599	0.715
Natural gas	pence per kWh	0.706	0.667	0.643	0.628	0.647	0.656	0.613
Gas at UK delivery points								
Natural gas	pence per kWh	0.556	0.588	0.584	0.592	0.593	0.560	0.462
Natural gas excluding levy *(8)*	pence per kWh	0.523	0.564	0.561	0.571	0.576	0.560	0.462

(1) See paragraph 5.58 for information on companies covered.
(2) The series represents gas supplied by UKCS licensees to the UK (i.e. exports are excluded) and gas imported from the Norwegian Sector of the continental shelf. UKCS figures include the Government's levy on indigenous supplies.
(3) Quarterly data are published in Table 27 of Energy Trends
(4) Excludes slurry.

(5) Converted from original units using factors on page 247.
(6) Excludes orimulsion.
(7) Excludes any natural gas liquids burnt at Peterhead Power Station.
(8) These average prices exclude the gas levy that was introduced in 1981and abolished on 1 April 1998.

9.11 Industrial fuel prices[1][2]

Great Britain

In original units

	Size of consumer	1994	1995	1996	1997	1998	1999
Coal [3]	Small	63.95	60.43	57.62	56.91	60.35	59.58
(£ per tonne)	Medium	56.14	52.43	50.43	45.46	46.55	45.79
	Large	37.30	34.52	32.69	32.27	32.89	32.56
All consumers -	average	40.19	37.27	35.41	34.42	35.16	34.77
Heavy fuel oil [4]	Small	78.89	94.83	104.64	101.73	88.16	95.41
(£ per tonne)	Medium	76.12	90.90	98.82	95.24	82.60	91.62
	Large	71.75	81.75	91.35	86.52	69.10	82.46
Of which:	Extra large	70.16	78.30	89.11	84.81	64.62	77.76
	Moderately large	74.66	88.04	95.45	89.63	77.28	91.01
All consumers -	average	74.21	86.66	95.70	91.55	76.31	87.36
Gas oil [4]	Small	154.76	154.27	172.88	174.74	152.47	152.93
(£ per tonne)	Medium	143.44	145.15	164.31	166.00	139.23	143.22
	Large	127.79	131.02	157.88	155.06	123.09	135.15
All consumers -	average	130.81	133.72	159.22	157.19	126.25	136.82
Electricity	Small	6.612	6.196	6.058	5.725	5.575	5.377
(Pence per kWh)	Medium	4.711	4.671	4.573	4.289	4.247	4.193
	Large	3.739	3.581	3.496	3.296	3.295	3.267
Of which:	Extra large	3.388	3.083	3.046	2.899	2.922	2.903
	Moderately large	4.010	3.966	3.844	3.603	3.584	3.548
All consumers -	average	4.150	4.007	3.916	3.687	3.667	3.623
Gas	Small	1.227	1.105	0.937	0.896	0.920	0.884
(Pence per kWh)	Medium	0.940	0.865	0.661	0.696	0.746	0.729
	Large	0.738	0.638	0.433	0.478	0.530	0.513
All consumers -	average	0.780	0.677	0.464	0.509	0.560	0.546
-	firm	0.894	0.791	0.512	0.564	0.633	0.605
-	interruptible	0.666	0.569	0.416	0.452	0.498	0.487
-	tariff	1.373	1.322	1.346	1.280	..	..
Medium fuel oil [4]							
(£ per tonne)							
All consumers -	average	84.03	92.93	98.33	90.86	87.23	104.93
Liquefied petroleum gases [5]							
(£ per tonne)							
All consumers -	average	139.48	146.62	156.69	177.62	151.33	144.38
Hard coke [6]							
(£ per tonne)							
All consumers -	average	101.34	111.99	126.31	119.05	112.93	101.13

(1) Quarterly data are published in Table 26 of Energy Trends.
(2) Prices of fuels purchased by manufacturing industry. For more details see explanatory notes in paragraphs 9.93 to 9.99.
(3) Excludes blast furnace supplies.
(4) Oil product prices include hydrocarbon oil duty.
(5) The median for liquefied petroleum gas is £209.9 per tonne.
(6) Excludes breeze and supplies to blast furnaces.

9.12 Industrial fuel prices [1][2]

Great Britain

Pence per kW

	Size of consumer	1991	1992	1993	1994	1995	1996	1997	1998	1999
Coal [3]	Small	0.862	0.905	0.904	0.842	0.796	0.763	0.752	0.776	0.778
	Medium	0.720	0.743	0.753	0.738	0.686	0.657	0.593	0.616	0.609
	Large	0.541	0.544	0.501	0.498	0.468	0.444	0.441	0.449	0.442
All consumers -	**average**	**0.569**	**0.576**	**0.541**	**0.535**	**0.502**	**0.477**	**0.466**	**0.477**	**0.469**
-	10% decile [5]	0.544	0.559	0.564	0.526	0.535	0.519	0.514	0.521	0.545
-	median [5]	0.812	0.847	0.857	0.791	0.690	0.667	0.656	0.687	0.689
-	90% decile [5]	0.931	0.990	0.992	0.967	0.954	0.943	0.838	0.916	0.971
Heavy fuel oil [4]	Small	0.622	0.577	0.654	0.665	0.799	0.882	0.846	0.735	0.800
	Medium	0.583	0.556	0.579	0.642	0.766	0.833	0.792	0.688	0.770
	Large	0.534	0.521	0.558	0.605	0.689	0.770	0.719	0.576	0.690
Of which:	Extra large	0.557	0.534	0.554	0.591	0.660	0.751	0.705	0.538	0.650
	Moderately	0.521	0.515	0.564	0.629	0.742	0.805	0.745	0.644	0.760
All consumers -	**average**	**0.562**	**0.541**	**0.578**	**0.626**	**0.731**	**0.807**	**0.761**	**0.636**	**0.740**
-	10% decile [5]	0.507	0.506	0.524	0.574	0.690	0.754	0.692	0.604	0.634
-	median [5]	0.587	0.561	0.597	0.653	0.785	0.859	0.803	0.716	0.790
-	90% decile [5]	0.764	0.719	0.729	0.764	0.895	1.023	0.957	0.926	1.059
Gas oil [4]	Small	1.343	1.176	1.252	1.219	1.213	1.359	1.385	1.206	1.211
	Medium	1.291	1.106	1.197	1.130	1.141	1.291	1.316	1.101	1.134
	Large	1.172	1.032	1.102	1.007	1.030	1.241	1.229	0.974	1.070
All consumers -	**average**	**1.194**	**1.051**	**1.116**	**1.030**	**1.051**	**1.251**	**1.246**	**0.999**	**1.083**
-	10% decile [5]	1.043	0.955	1.040	0.984	1.012	1.112	1.138	0.914	0.900
-	median [5]	1.224	1.095	1.175	1.104	1.121	1.309	1.301	1.086	1.126
-	90% decile [5]	1.530	1.303	1.375	1.291	1.290	1.528	1.498	1.344	1.480
Electricity	Small	6.736	7.058	6.879	6.612	6.196	6.058	5.725	5.575	5.377
	Medium	4.515	4.711	4.903	4.711	4.671	4.573	4.289	4.247	4.193
	Large	3.328	3.572	3.810	3.739	3.581	3.496	3.296	3.295	3.267
Of which:	Extra large	3.030	3.226	3.458	3.388	3.083	3.046	2.899	2.922	2.903
	Moderately	3.559	3.839	4.082	4.010	3.966	3.844	3.603	3.584	3.548
All consumers -	**average**	**3.825**	**4.061**	**4.264**	**4.150**	**4.007**	**3.916**	**3.687**	**3.667**	**3.623**
-	10% decile [5]	3.821	4.095	4.313	4.237	4.175	4.099	3.816	3.820	3.778
-	median [5]	6.100	6.398	6.153	5.987	5.778	5.529	5.250	5.172	5.110
-	90% decile [5]	8.131	8.448	8.331	8.020	7.883	7.451	7.102	6.947	6.810
Gas	Small	1.360	1.382	1.279	1.227	1.105	0.937	0.896	0.920	0.884
	Medium	1.002	0.987	0.970	0.940	0.865	0.661	0.696	0.746	0.729
	Large	0.693	0.701	0.713	0.738	0.638	0.433	0.478	0.530	0.513
All consumers -	**average**	**0.753**	**0.756**	**0.769**	**0.780**	**0.677**	**0.464**	**0.509**	**0.560**	**0.546**
-	firm	0.933	0.945	0.931	0.894	0.791	0.512	0.564	0.633	0.605
-	interruptible	0.612	0.628	0.640	0.666	0.569	0.416	0.452	0.498	0.487
-	tariff	1.471	1.460	1.388	1.373	1.322r	1.346	1.280	..	..
-	10% decile [5]	0.844	0.866	0.885	0.860	0.716	0.513	0.537	0.592	0.606
-	median [5]	1.344	1.385	1.331	1.193	1.051	0.812	0.828	0.861	0.851
-	90% decile [5]	1.593	1.592	1.530	1.501	1.495	1.441	1.323	1.177	1.175
Medium fuel oil [4]										
All consumers -	average	0.704	0.644	0.664	0.708	0.783	0.829	0.755	0.727	0.882
Liquefied petroleum gases [6]										
All consumers -	average	1.138	1.059	1.116	1.013	1.064	1.137	1.294	1.103	1.048
Hard coke [7]										
All consumers -	average	1.343	1.386	1.503	1.300	1.437	1.620	1.525	1.364	1.297

(1) Quarterly data are published in Table 26 of Energy Trends.
(2) Prices of fuels purchased by manufacturing industry. For more details see explanatory notes in paragraphs 9.93 to 9.99.
(3) Excludes blast furnace supplies.
(4) Oil product prices include hydrocarbon oil duty.
(5) The 10% decile is the point within the complete range of prices below which the bottom 10% of those prices fall. Similarly the 90% decile is the point above which the top 10% of the prices occur. The median is the midway point.
(6) The median for liquefied petroleum gas is 1.524 pence per kWh
(7) Excludes breeze and supplies to blast furnaces.

9.12 Industrial fuel prices[1][2] (continued)

Great Britain

Pence per kWh

	Size of consumer	1998				1999			
		First quarter	Second quarter	Third quarter	Fourth quarter	First quarter	Second quarter	Third quarter	Fourth quarter
Coal *(3)*	Small	0.744	0.770	0.779	0.806	0.805	0.757	0.765	0.763
	Medium	0.594	0.608	0.626	0.644	0.626	0.607	0.592	0.607
	Large	0.439	0.454	0.451	0.452	0.446	0.438	0.467	0.424
All consumers -	**average**	**0.464**	**0.479**	**0.480**	**0.483**	**0.475**	**0.466**	**0.489**	**0.454**
-	10% decile *(5)*	0.508	0.522	0.515	0.515	0.545	0.546	0.528	0.545
-	median *(5)*	0.677	0.691	0.698	0.709	0.709	0.693	0.657	0.682
-	90% decile *(5)*	0.857	0.864	0.904	0.943	0.971	0.969	0.985	0.973
Heavy fuel oil *(4)*	Small	0.789	0.729	0.707	0.704	0.685	0.767	0.879	0.985
	Medium	0.734	0.685	0.672	0.651	0.665	0.755	0.843	0.851
	Large	0.605	0.578	0.562	0.557	0.559	0.660	0.744	0.937
Of which:	Extra large	0.573	0.543	0.518	0.517	0.523	0.643	0.693	0.924
	Moderately large	0.664	0.643	0.643	0.629	0.623	0.691	0.837	0.961
All consumers -	**average**	**0.674**	**0.635**	**0.619**	**0.609**	**0.613**	**0.707**	**0.796**	**0.913**
-	10% decile *(5)*	0.607	0.598	0.597	0.597	0.578	0.681	0.716	0.778
-	median *(5)*	0.762	0.729	0.697	0.696	0.679	0.765	0.637	0.983
-	90% decile *(5)*	0.900	0.851	0.851	0.834	0.850	0.934	1.104	1.168
Gas oil *(4)*	Small	1.269	1.203	1.175	1.141	1.093	1.127	1.229	1.475
	Medium	1.175	1.104	1.062	1.028	0.984	1.060	1.178	1.382
	Large	1.046	1.000	0.944	0.900	0.900	0.992	1.112	1.275
All consumers -	**average**	**1.071**	**1.021**	**0.967**	**0.925**	**0.917**	**1.005**	**1.125**	**1.295**
-	10% decile *(5)*	1.010	0.972	0.904	0.862	0.842	0.926	0.959	1.185
-	median *(5)*	1.164	1.104	1.049	1.016	1.002	1.155	1.186	1.396
-	90% decile *(5)*	1.381	1.344	1.306	1.297	1.307	1.283	1.404	1.581
Electricity	Small	5.756	5.382	5.441	5.679	5.445	5.414	5.305	5.317
	Medium	4.416	4.139	4.122	4.289	4.330	4.140	4.141	4.152
	Large	3.575	3.131	3.059	3.427	3.561	3.038	3.188	3.310
Of which:	Extra large	3.318	2.742	2.599	3.064	3.323	2.559	2.875	2.926
	Moderately large	3.774	3.431	3.414	3.707	3.744	3.407	3.432	3.607
All consumers -	**average**	**3.913**	**3.515**	**3.465**	**3.775**	**3.863**	3.452	**3.552**	**3.638**
-	10% decile *(5)*	3.940	3.691	3.692	3.858	3.874	3.716	3.695	3.760
-	median *(5)*	5.460	5.101	5.091	5.350	5.325	5.108	5.000	4.999
-	90% decile *(5)*	7.034	6.437	6.571	6.875	7.069	6.632	6.575	6.880
Gas	Small	0.920	0.919	0.920	0.920	0.885	0.901	0.884	0.868
	Medium	0.748	0.736	0.741	0.754	0.741	0.731	0.716	0.726
	Large	0.530	0.525	0.526	0.538	0.537	0.516	0.501	0.497
All consumers -	**average**	**0.568**	**0.552**	**0.546**	**0.571**	**0.577**	**0.547**	**0.520**	**0.531**
-	firm	0.639	0.625	0.615	0.647	0.641	0.613	0.568	0.595
-	interruptible	0.501	0.493	0.492	0.503	0.509	0.484	0.483	0.474
-	10% decile *(5)*	0.599	0.582	0.588	0.605	0.608	0.610	0.597	0.605
-	median *(5)*	0.868	0.854	0.855	0.861	0.863	0.853	0.857	0.839
-	90% decile *(5)*	1.176	1.160	1.177	1.179	1.177	1.142	1.198	1.160
Medium fuel oil *(4)*									
All consumers -	average	0.727	0.726	0.724	0.729	0.743	0.801	0.949	1.022
Liquefied petroleum gases									
All consumers -	average	1.173	1.099	1.063	1.052	1.034	1.026	1.043	1.101
	median *(6)*	1.827	1.790	1.771	1.679	1.409	1.273	1.313	1.530
Hard coke *(7)*									
All consumers -	average	1.415	1.378	1.345	1.308	1.363	1.327	1.272	1.214

9.13 Fuel price indices for the industrial sector[1][2]
1970 to 1999
United Kingdom

1990 = 100

	Coal (3)	Heavy fuel oil (3)	Gas (4)	Electricity (4)	Total fuel (5)	Coal (3)	Heavy fuel oil (3)	Gas (4)	Electricity (4)	Total fuel (5)
	Current fuel price index numbers					Fuel price index numbers relative to the GDP deflator (6)				
1970	15.7	13.2	23.0	16.5	15.9	107.4	90.5	157.8	113.3	108.7
1971	18.7	19.8	17.6	18.3	18.8	117.8	124.6	110.4	115.1	118.1
1972	20.2	18.9	14.0	18.7	18.4	117.2	110.0	81.5	108.7	107.0
1973	21.1	18.5	14.8	18.8	18.5	114.0	100.0	80.1	101.7	99.7
1974	22.9	43.8	17.1	25.7	32.6	107.7	205.8	80.2	120.5	153.2
1975	34.4	55.2	23.6	34.1	41.3	127.4	204.4	87.4	126.3	152.8
1976	42.6	63.2	34.4	39.4	47.7	137.0	203.3	110.6	126.8	153.4
1977	50.8	80.2	46.9	45.4	59.3	143.6	226.5	132.6	128.3	167.6
1978	55.2	75.2	58.0	50.0	59.9	139.7	190.3	146.7	126.5	151.7
1979	64.2	93.4	66.1	55.5	70.8	142.1	206.6	146.3	122.8	156.6
1980	83.3	132.3	89.9	68.5	94.0	154.2	245.0	166.5	126.8	174.1
1981	96.2	158.5	109.0	79.3	112.0	160.1	263.7	181.4	131.9	186.3
1982	114.2	167.4	115.9	86.9	119.4	176.8	259.2	179.4	134.5	184.8
1983	118.2	184.4	117.9	87.2	124.0	173.6	270.7	173.1	128.1	182.1
1984	118.3	219.2	121.8	87.1	134.5	166.4	308.3	171.3	122.4	189.2
1985	123.0	222.3	130.0	90.5	137.4	163.6	295.6	172.9	120.3	182.6
1986	116.9	106.9	113.4	91.4	101.6	150.6	137.8	146.1	117.7	131.0
1987	112.1	114.4	106.8	88.7	98.1	137.4	140.2	130.9	108.7	120.2
1988	99.6	84.5	101.7	93.5	94.4	115.1	97.6	117.6	108.1	109.1
1989	97.3	92.0	98.3	100.2	98.6	104.7	99.0	105.8	107.9	106.1
1990	100.0	100.0	100.0	100.0	100.0	100.0	100.0	100.0	100.0	100.0
1991	98.5	87.8	101.0	103.3	100.5	92.2	82.3	94.5	96.7	94.1
1992	99.8	84.5	104.5	109.0	104.8	89.9	76.1	94.1	98.2	94.4
1993	93.6	90.1	102.7	114.2	107.9	82.1	78.9	90.0	100.1	94.6
1994	92.5	97.4	103.6	110.1	106.4	79.9	84.1	89.5	95.1	91.9
1995	86.8	113.8	90.4	109.1	105.6	73.1	95.8	76.1	91.8	88.9
1996	82.6	125.7	66.1	105.3	102.6r	67.3	102.4	53.9	85.8	83.6r
1997	80.6	120.2	69.7	99.3	97.5r	63.9r	95.3r	55.3r	78.7r	77.3r
1998	82.5	100.2	73.6	98.4	93.6r	63.4r	77.0r	56.5r	75.6r	71.9r
1999	81.2	114.2	73.7	99.0	95.8	60.6	85.2	55.0	73.8	71.5
1997Q3	79.9	114.6	65.9	90.4	90.2r	63.2r	90.7r	52.2r	71.5r	71.3r
1997Q4	82.8	121.9	71.2	104.4	101.4r	64.7r	95.2r	55.6r	81.6r	79.2r
1998Q1	80.4	106.3	73.2	107.1	100.0r	62.7r	82.8r	57.0r	83.5r	78.0r
1998Q2	83.0	100.1	70.3	91.3	88.4r	64.0r	77.2r	54.2r	70.4r	68.2r
1998Q3	83.1	97.6	70.3	90.9	87.8r	63.5r	74.6r	53.7r	69.5r	67.1r
1998Q4	83.6	96.0	74.4	103.6	96.7r	63.4r	72.8r	56.4r	78.5r	73.3r
1999Q1	82.3	87.0	74.9	105.4	96.8	62.2	65.7	56.6	79.7	73.1
1999Q2	80.7	103.1	70.5	93.7	90.3	60.4	77.1	52.7	70.1	67.5
1999Q3	84.6	116.2	68.2	95.3	92.7	62.8	86.3	50.6	70.7	68.8
1999Q4	78.5	145.1	74.3	100.9	101.2	58.0	107.2	54.8	74.5	74.7

(1) Quarterly data are published in Table 28 of Energy Trends.
(2) Index numbers shown represent the average for the period specified. VAT is excluded, as it is generally refundable.
(3) Indices based on the survey of the prices of fuels purchased by manufacturing industry customers in Great Britain only, as shown in Table 9.11.
(4) Indices based on the average unit value of sales to industrial customers.
(5) Total fuel indices are annually weighted.
(6) Deflated using seasonally adjusted GDP deflator at market prices with base year of 1995 but rescaled to 1990=100.

9.14 Typical retail prices of petroleum products[1] 1977 to 2000

United Kingdom

Pence per litre

	Motor spirit (2)			DERV fuel (2)(3)	Standard grade burning oil (4)	Gas oil (4)
	4 star(5)	Super unleaded	Premium unleaded			
Annual average price						
1977	17.64	..	..	18.21	8.40	8.37
1978	16.77	..	..	18.46	8.39	8.42
1979	22.66	..	..	23.65	10.89	10.90
1980	28.32	..	..	29.67	14.78	14.77
1981	34.29	..	..	34.01	18.01	17.51
1982	36.62	..	..	35.86	20.75	20.11
1983	39.28	..	..	37.30	21.19	20.71
1984	40.62	..	..	38.33	19.67	20.44
1985	43.14	..	..	41.94	21.12	21.58
1986	37.35	..	..	35.60	13.95	13.77
1987	37.90	..	..	34.58	12.55	13.16
1988	37.38	..	..	34.00	10.65	10.88
1989	40.39	..	38.29	36.18	12.04	11.64
1990	44.87	..	42.03	40.48	15.56	14.64
1991	48.47	47.31	45.06	43.81	14.11	13.65
1992	50.28	48.38	46.11	45.01	12.89	12.49
1993	54.12	52.91	49.44	49.20	13.64	13.42
1994	56.87	55.98	51.56	51.53	13.37	13.27
1995	59.70	58.55	53.77	54.24	13.80	13.87
1996	61.63	63.67	56.52	57.71	15.93	16.53
1997	67.22	71.31	61.82	62.47	14.36	15.45
1998	71.11	77.80	64.80	65.50	11.25	12.47
1999	77.20	82.92	70.16	72.49	12.73	13.89
Monthly price						
1999 January	69.61	79.23	62.87	63.95	9.89	11.36
February	69.78	78.26	63.02	64.17	10.22	11.33
March	73.85	82.24	66.51	69.94	10.52	12.06
April	77.83	83.39	70.20	73.23	12.00	12.64
May	77.61	83.82	70.04	73.09	11.89	12.90
June	77.32	83.74	69.80	72.78	11.54	12.79
July	78.26	83.87	70.98	73.81	12.74	13.96
August	79.76	84.57	72.87	75.21	13.31	14.48
September	80.05	85.11	73.02	74.96	14.31	15.45
October	80.99	83.90	73.85	75.81	14.27	15.74
November	80.35	83.40	73.36	75.23	14.95	16.27
December	81.02	83.54	75.42	77.65	17.11	17.73
2000 January	80.84	84.15	75.38	77.75	17.84	18.15
February	80.75	83.42	75.14	77.68	17.92	18.50
March	82.99	85.24	78.32	79.82	18.63	19.06
April	84.45	87.18	79.96	81.07	18.33	18.61

(1) Latest monthly data for petroleum products prices are published in Table 30 of Energy Trends.

(2) The estimates are generally representative of prices paid (inclusive of taxes) at the pump on or about the 15th of the month. Estimates are based on information provided by oil companies until December 1994. From January 1995 data from super/hypermarket chains have been included.

(3) From July 1999 diesel prices represent average prices for Ultra Low Sulphur Diesel which now accounts for virtually all diesel sold. Prices for the period March to June 1999 represent a mixture of both types of diesel as companies switched to only selling ULSD. Pump prices for both diesels are broadly the same.

(4) These estimates are based on information provided by marketing companies and are for deliveries of up to 1,000 litres for standard grade burning oil and for deliveries of 2,000 to 5,000 litres for gas oil. These prices include Valued Added Tax which was introduced from 1 April 1994 at a rate of 8% until 1 September 1997 when this rate was reduced to 5%.

(5) From October 1999 Four Star prices represent 'Lead Replacement Petrol' (LRP) which had replaced Four Star at 95% of outlets at that time. Leaded petrol has now been phased out. Pump prices for both petrols are broadly the same.

9.15 Typical retail prices of petroleum products[1] 1954 to 2000

Pence per litre

		Motor spirit (2)				DERV fuel (2)(3)	Standard grade burning oil (4)	Gas oil (4)
		2 star	4 star(5)	Super unleaded	Premium unleaded			
1954	January	4.47	4.90	..	..	..	..	..
1955	January	4.49	4.90	..	..	4.38	..	..
1956	January	4.49	5.04	..	..	4.47	..	..
1957	January	6.09	6.64	..	..	6.07	..	..
1958	January	4.65	5.13	..	..	4.69	..	..
1959	January	4.67	5.13	..	..	4.70	..	..
1960	January	4.67	5.18	..	..	4.79	1.55	1.55
1961	January	4.58	5.09	..	..	4.65	1.65	1.52
1962	January	4.86	5.36	..	..	4.93	1.89	1.73
1963	January	4.77	5.36	..	..	4.93	1.86	1.73
1964	January	4.77	5.22	..	..	4.97	1.86	1.73
1965	January	5.36	5.82	..	..	5.52	1.86	1.73
1966	January	5.36	5.68	..	..	5.52	1.86	1.73
1967	January	5.73	6.05	..	..	5.89	1.86	1.73
1968	January	5.77	6.14	..	..	6.07	2.06	1.95
1969	January	6.57	6.94	..	..	6.92	2.09	1.95
1970	January	6.78	7.15	..	..	6.99	2.17	2.02
1971	January	7.06	7.42	..	..	7.22	2.34	2.14
1972	January	7.26	7.70	..	..	7.48	2.45	2.21
1973	January	7.48	7.70	..	..	7.48	2.51	2.27
1974	January	8.91	9.24	..	..	9.13	3.39	3.39
1975	January	15.62	15.95	..	..	12.21	5.06	5.26
1976	January	16.50	16.83	..	..	13.53	6.49	6.47
1977	January	17.05	17.49	..	..	17.16	8.15	7.93
1978	January	16.37	16.76	..	..	18.57	8.43	8.48
1979	January	17.09	17.50	..	..	18.42	8.37	8.36
1980	January	25.98	26.39	..	..	27.80	13.07	13.03
1981	January	28.64	29.05	..	..	30.70	15.90	15.80
1982	January	34.28	35.02	..	..	34.89	20.33	19.68
1983	January	35.85	36.70	..	..	37.64	22.71	22.52
1984	January	39.44	40.35	..	..	36.78	19.84	20.31
1985	January	40.71	41.54	..	..	40.59	21.60	22.62
1986	January	40.81	41.63	..	..	41.13	19.48	19.47
1987	January	37.57	38.42	..	..	35.00	13.52	14.70
1988	January	35.98	36.79	..	..	33.94	11.97	12.29
1989	January	36.36	37.14	..	36.02	34.17	11.41	11.15
1990	January	..	40.92	..	38.37	39.21	15.45	15.46
1991	January	..	45.13	44.38	42.14	43.31	17.52	17.13
1992	January	..	46.93	45.57	43.43	43.19	12.47	12.02
1993	January	..	51.27	49.76	47.13	47.05	14.10	13.52
1994	January	..	55.50	54.48	50.83	51.72	12.94	12.72
1995	January	..	59.11	58.00	53.44	54.13	13.32	13.93
1996	January	..	61.97	61.26	55.93	57.43	15.38	15.86
1997	January	..	65.46	69.24	61.09	62.02	17.13	18.14
1998	January	..	69.03	73.96	63.13	63.34	12.92	13.67
1999	January	..	69.61	79.23	62.87	63.95	9.89	11.36
2000	January	..	80.84	84.15	75.38	77.75	17.84	18.15

(1) The estimates are generally representative of prices paid (inclusive of taxes) at the pump on or about the 15th of the month. Estimates are based on information provided by oil companies from 1977 until 1994. From January 1995 data from super/hypermarket chains have been included.

(2) Maximum retail prices imposed by Order during the period 15 December 1973 to 20 December 1974.

(3) From January 2000 diesel prices represent average prices for Ultra Low Sulphur Diesel which now accounts for virtually all diesel sold. Pump prices for both diesels are broadly the same.

(4) Typical prices for deliveries of up to 1,000 litres of standard grade burning oil and between 2,000 and 5,000 litres of gas oil. Prior to 1977, prices were for deliveries of 900 litres of standard grade burning oil and 2,275 litres of gas oil. Since January 1995 prices include VAT at a rate of 8% until January 1998 when the applicable rate was 5%.

(5) From October 1999 Four Star prices represent 'Lead Replacement Petrol' (LRP) which had replaced Four Star at 95% of outlets at that time. Leaded petrol has now been phased out. Pump prices for both petrols are broadly the same.

9.16 Effective rates of duty on principal hydrocarbon oils, 1964 to 2000[1]

Pence per litre

Date from which duty effective		Aviation gasoline (2)	Gas for use as road fuel (2)	Motor spirit (2)(3)		DERV fuel (2)	Fuel oil (6)	Gas oil (6)(7)	Kerosene (6)
				Unleaded (4)	Leaded (5)				
11 November	1964	3.575			3.575	3.575			
21 July	1966	3.932			3.932	3.932	0.202	0.202	0.202
11 April	1967	3.941			3.941	3.941			
19 March	1968	4.308			4.308	4.308			
22 November	1968	4.739			4.739	4.739	0.222	0.222	0.222
15 April	1969	4.949			4.949	4.949	0.220	0.220	0.220
3 July	1972		2.475						
10 April	1976	6.599	3.300		6.599	6.599			
30 March	1977	7.699	3.849		7.699	7.699	0.550	0.550	
8 August	1977	6.599	3.300		6.599				
13 June	1979	8.100	4.050		8.100	9.200	0.660	0.660	
26 March	1980	10.000	5.000		10.000	10.000	0.770	0.770	
10 March	1981	13.820	6.910		13.820	13.820			
2 July	1981					11.910			
9 March	1982	7.770	7.770		15.540	13.250			
15 March	1983	8.150	8.150		16.300	13.820			
13 March	1984	8.580	8.580		17.160	14.480			zero
19 March	1985	8.970	8.970		17.940	15.150			
19 March	1986	9.690	9.690		19.380	16.390		1.100	
17 March	1987			18.420					
15 March	1988	10.220	10.220		20.440	17.290			
14 March	1989			17.720					
20 March	1990	11.240	11.240	19.490	22.480	19.020	0.830	1.180	
19 March	1991	12.930	12.930	22.410	25.850	21.870	0.910	1.290	
10 March	1992	13.900	13.900	23.420	27.790	22.850	0.950	1.350	
16 March	1993	15.290	15.290	25.760	30.580	25.140	1.050	1.490	
30 November	1993	16.570	16.570	28.320	33.140	27.700	1.160	1.640	
29 November	1994	17.630	33.140 (8)	30.440	35.260	30.440	1.660	2.140	
1 January	1995	18.070		31.320	36.140	31.320			
28 November	1995	19.560	28.170 (8)	34.300	39.120	34.300	1.810	2.330	
26 November	1996	20.840	21.130 (8)	36.860	41.680	36.860	1.940	2.500	
2 July	1997	22.550		40.280	45.100	40.280	2.000	2.580	
17 March	1998	24.630		43.990	49.260	44.990(9)	2.180	2.820	
9 March	1999	26.440	15.000 (8)	47.210	52.880	50.210(9)	2.650	3.030	
21 March	2000	27.340		48.820	54.680	51.820(9)	2.740	3.130	

(1) Duty rates remain the same unless otherwise stated.
(2) These fuels became liable to Value Added Tax as follows:-
   (i) 10% with effect from 1 April 1974
   (ii) 8% with effect from 29 July 1974
   (iii) For motor spirit 25% with effect from 18 November 1974
   (iv) For motor spirit 12.5% with effect from 12 April 1976
   (v) 15% with effect from 18 June 1979
   (vi) 17.5% with effect from 1 April 1991
(3) With effect from 14 March 1989 until 20 March 1990, the rate of duty for 2-star and 3-star leaded motor spirit was 21.220 pence per litre.
(4) From 15 May 1996 a different rate of duty from unleaded fuel was introduced for super unleaded fuel as follows:
   15 May 1996   37.62 pence per litre
   26 November 1996   40.18   ..
   2 July 1997   43.60   ..
   17 March 1998   48.76   ..
   9 March 1999   52.33   ..
   1 October 1999   49.21   ..
   21 March 2000   50.89   ..

(5) From 1 October 1999 a different rate of duty from leaded fuel was introduced for Lead Replacement Petrol as follows:
   1 October 1999   49.21 pence per litre
   21 March 2000   50.89   ..
(6) For industrial and commercial consumers these fuels became liable to the standard rate of Value Added Tax on 1 July 1990 (at 15% to 31 March 1991 and at 17.5% from 1 April 1991), recoverable by the majority of such consumers. These fuels attracted Value Added Tax for domestic consumers from 1 April 1994 at an intial rate of 8%. This was reduced to 5% from 1 September 1997.
(7) AVTUR (aviation turbine fuel) attracted the gas oil rate until 18 March 1986 after which it was zero rated.
(8) From 29 November 1994 this duty is priced in pence per kilogram as the relative calorific values of the different types of road fuel gases are very similar when related to mass (kilogram).
(9) From 17 March 1998 a different rate of duty from derv fuel was introduced for Ultra Low Sulphur Diesel (ULSD) as follows:
   17 March 1998   42.99 pence per litre
   9 March 1999   47.21   ..
   21 March 2000   48.82   ..

9.17 Electricity and gas prices in the EU and the G7 countries

Electricity *(2)*

	Industrial					Domestic				
	1985	1990	1995	1997	1998	1985	1990	1995	1997	1998
EU										
Austria	3.05	3.67	5.14	4.93	4.73	6.58	8.75	12.15	10.33	10.07
Belgium	3.34	3.59	4.31	3.36	3.28	7.89	9.58	12.88	10.23	9.97
Denmark	3.58	3.50	4.39	3.92	4.12	6.69	9.24	13.22	11.92	12.86
Finland	3.15	3.55	3.81	3.16	3.03	4.06	5.78	6.89	6.13	5.91
France	2.65	3.17	3.81	2.98	2.82	6.75	8.43	10.56	8.18	7.79
Germany	3.63	5.13	6.33	4.38	4.06	6.36	9.20	12.88	9.73	9.59
Greece	3.35	3.64	3.92	3.30	3.00	4.77	6.64	7.21	6.28	5.96
Ireland	4.46	3.80	4.16	3.87	3.59	6.89	7.50	8.37	7.97	7.45
Italy	4.77	5.49	5.87	5.74	5.72	6.83	8.81	10.72	9.73	9.62
Luxembourg	3.24	..	..	..	..	5.21	6.97	9.27	7.56	7.40
Netherlands	3.09	2.94	4.73	3.84	3.77	6.80	6.58	8.56	7.91	7.73
Portugal	4.33	6.51	7.71	5.95	5.66	5.90	8.27	11.45	9.53	9.30
Spain	3.59	5.48	5.13	3.94	3.55	6.65	10.68	12.37	9.98	9.30
Sweden	2.17	2.80	2.49	2.09	1.98	3.06	4.93	5.98	6.18	5.87
UK	3.60	3.98	4.34	3.95	3.92	5.37	6.67	8.06	7.64	7.29
Rest of G7										
Canada*(3)*	1.97	2.01	2.41r	2.31	2.13	2.84	3.03	3.76r	3.59	3.31
Japan	7.39	6.88	11.74	9.18	8.13	9.80	9.95	17.12r	13.05	11.57
USA*(3)*	3.98	2.66	2.97	2.67	2.43	6.16	4.39	5.33	5.16	4.98

Pence per kWh *(1)*

Gas *(2)(4)*

	Industrial					Domestic				
	1985	1990	1995	1997	1998	1985	1990	1995	1997	1998
EU										
Austria	1.28	0.75	0.94	0.86	0.80	2.51	1.89	2.51	2.26	2.16
Belgium	1.18	0.70	0.77	0.69	0.67	2.22	2.01	2.66	2.17	2.13
Denmark	..	..	..	..	..	2.49	3.25	3.77	3.56	3.35
Finland	1.11	0.60	0.80	0.75	0.72	1.11	0.72	0.97	0.89	0.87
France	1.20	0.75	0.88	0.80	0.76	2.50	2.29	2.73	2.24	2.27
Germany	1.24	0.91	1.13r	1.00	0.97	2.01	1.92	2.60	2.19	2.13
Greece	..	..	..	..	..	..	..	..	..	..
Ireland	1.80	1.60	1.74	1.58	0.89	2.86	2.33	2.58	2.35	2.18
Italy	1.10	0.76	0.95	1.01	0.89	2.14	3.10	3.64	3.67	3.59
Luxembourg	..	..	..	..	..	1.56	1.19	1.63	1.56	1.43
Netherlands	1.05	0.61	0.81	0.70	0.64	1.64	1.62	1.97	1.89	1.86
Portugal	..	..	..	..	..	..	..	..	..	..
Spain	1.33	0.79	0.87	0.82	0.72	2.76	2.96	3.32	2.88	2.77
Sweden	..	..	..	..	..	..	..	..	..	..
UK	1.00	0.77	0.69	0.54	0.56	1.41	1.59	1.79	1.78	1.72
Rest of G7										
Canada*(3)*	0.69	0.40	0.38	0.38	0.37	1.07	0.89	0.88	0.90	0.88
Japan	2.80	2.00	2.68	2.51	2.22	4.22	4.58	7.69	6.97	6.18
USA*(3)*	1.01	0.54	0.55	0.71	0.62	1.56	1.13	1.33	1.40	1.36

Source: Derived from the International Energy Agency publication, Energy Prices and Taxes Q4 1999

(1) Prices converted to pounds sterling using annual average exchange rates.
(2) Prices include all taxes where not refundable on purchase.
(3) Prices shown exclude taxes as no figures inclusive of taxes are available (see table 9.18).
(4) Prices for natural gas have been converted from original data to kWh using gross calorific values.

9.18 Electricity and gas prices (excluding and including taxes) in the EU and G7 countries

Pence per kWh *(1)*

Excluding taxes	Electricity				Gas *(2)*			
	Industrial		Domestic		Industrial		Domestic	
	1990	1998	1990	1998	1990	1998	1990	1998
EU								
Austria	3.67	4.73	7.29	7.90	0.75	0.80	1.58	1.56
Belgium	3.59	3.28	8.18	8.15	0.70	0.67	1.72	1.68
Denmark	3.50	3.41	4.62	5.11	..	..	2.66	2.41
Finland	3.55	2.77	4.80	4.43	0.58	0.62	0.58	0.62
France	3.17	2.82	6.87	5.96	0.75	0.78	1.93	1.82
Germany	5.13	4.06	7.46	8.29	0.82	0.85	1.59	1.73
Greece	3.64	3.00	5.65	5.05	..	..	..	..
Ireland	3.80	3.59	6.76	6.62	1.60	0.89	2.12	1.94
Italy	4.15	4.50	6.40	6.83	0.72	0.80	2.10	2.01
Luxembourg	..	..	6.57	6.96	..	..	1.13	1.43
Netherlands	2.94	3.77	5.56	5.89	0.59	0.60	1.35	1.33
Portugal	6.51	5.66	7.03	8.86	..	..	..	..
Spain	5.48	3.38	9.53	7.63	0.79	0.72	2.60	2.35
Sweden	2.80	1.98	3.45	3.73	..	..	..	..
UK	3.98	3.92	6.67	6.94	0.77	0.56	1.59	1.64
Rest of G7								
Canada *(3)*	2.01	2.13	3.03	..	0.40	0.37	0.89	..
Japan	6.51	7.54	9.52	10.80	1.94	2.12	4.44	5.89
USA *(3)*	2.66	2.43	4.39	4.96	0.54	0.62	1.13	..
UK relative to:								
EU Median (%)	8.9	12.0	..	1.6	2.1	-24.8	-3.8	-4.0
EU rank	10	10	8	9	7	1	6	5
G7 rank *(4)*	4	5	4	4	5	2	4	1

Including taxes	Electricity *(5)*				Gas *(2)(5)*			
	Industrial		Domestic		Industrial		Domestic	
	1990	1998	1990	1998	1990	1998	1990	1998
EU (see table 9.17)								
G7 countries								
Canada *(3)*	2.01	2.13	3.03	3.31	0.40	0.37	0.89	0.88
France	3.17	2.82	8.43	7.79	0.75	0.76	2.29	2.27
Germany	5.13	4.06	9.20	9.59	0.91	0.97	1.92	2.13
Italy	5.49	5.72	8.81	9.62	0.76	0.89	3.10	3.59
Japan	6.88	8.13	9.95	11.57	2.00	2.22	4.58	6.18
USA *(3)*	2.66	2.43	4.39	4.98	0.54	0.62	1.13	1.36
UK	3.98	3.92	6.67	7.29	0.77	0.56	1.59	1.72
UK relative to:								
EU Median (%)	8.9	6.5	-19.3	-6.4	1.5	-23.4	-18.8	-19.9
EU rank	10	9	5	4	7	1	3	3
G7 rank	4	4	3	3	5	2	3	3

Source: Derived from the International Energy Agency publication, Energy Prices and Taxes Q4 1999

(1) Prices converted to pounds sterling using annual average exchange rates.
(2) Prices for natural gas have been converted from original data to kWh using gross calorific values.
(3) The energy tax situation in Canada and the US is unclear. For Canada, natural gas prices quoted do not include any federal or provincial taxes whilst electricity prices include provincial taxes but these are not available separately. For the US, natural gas prices include general sales taxes levied by the states but their national average is unknown while electricity prices shown exclude taxes.
(4) Assuming ranking of Canada and US remains the same whether taxes are included or excluded.
(5) Prices include all taxes where not refundable on purchase.

9.19 International petrol and diesel prices

European unleaded petrol *(2)* prices at mid April

	Price excluding tax and duty			Pump price			Tax component (%)		
	1998	1999	2000	1998	1999	2000	1998	1999	2000
Austria	17.78	15.21	21.62	53.37	51.00	56.10	66.7	70.2	61.5
Belgium	14.04	14.60	21.60	56.51	58.08	63.35	75.2	74.9	65.9
Denmark	14.63	15.01	19.88	54.68	60.52	64.23	73.2	75.2	69.0
Finland	14.39	13.84	21.66	61.57	61.85	67.83	76.6	77.6	68.1
France	11.61	11.64	18.83	59.83	60.85	65.26	80.6	80.9	71.1
Germany	13.50	14.34	15.76	53.12	57.26	57.84	74.6	74.9	72.8
Greece	13.53	15.00	20.39	41.01	42.38	45.29	67.0	64.6	55.0
Ireland	15.87	13.35	18.78	49.19	46.33	50.51	67.7,	71.2	62.8
Italy	14.83	15.79	20.96	58.66	61.77	63.01	74.7	74.4	66.7
Luxembourg	14.41	15.30	20.92	41.18	44.58	48.70	65.0	65.7	57.0
Netherlands	15.91	16.56	22.12	62.30	64.87	68.52	74.5	74.5	67.7
Portugal	14.34	12.91	25.75	52.40	53.54	53.84	72.6	75.9	52.2
Spain	13.88	13.87	19.34	43.44	44.49	48.58	68.0	68.8	60.2
Sweden	15.51	15.97	20.41	62.00	61.02	66.45	75.0	73.8	69.3
United Kingdom	11.99	12.53	19.23	65.77	70.19	79.96	81.8	82.2	76.0
UK Rank in EU	2	2	4	15	15	15	15	15	15

European diesel prices at mid April

	Price excluding tax and duty			Pump price			Tax component (%)		
	1998	1999	2000	1998	1999	2000	1998	1999	2000
Austria	16.02	13.54	19.37	41.62	39.14	44.33	61.5	65.4	56.3
Belgium	14.17	14.48	19.31	39.75	40.64	44.65	64.3	64.4	56.8
Denmark	13.60	13.66	19.46	41.72	42.44	50.48	67.4	67.8	61.5
Finland	15.52	14.71	22.17	42.89	42.42	49.59	63.8	65.3	55.3
France	11.01	11.39	16.90	42.10	44.03	48.63	73.8	74.1	65.3
Germany	12.94	13.27	13.78	38.71	41.95	42.60	66.6	68.4	67.6
Greece	10.93	11.91	16.96	30.28	33.94	37.99	63.9	64.9	55.4
Ireland	16.79	14.48	20.26	46.45	43.82	48.74	63.9	67.0	58.4
Italy	13.55	14.29	19.16	46.14	49.01	50.77	70.6	70.8	62.3
Luxembourg	12.72	13.58	18.13	33.36	34.77	38.48	61.9	60.9	52.9
Netherlands	14.45	15.21	19.04	42.70	44.63	47.44	66.2	65.9	59.9
Portugal	13.27	12.03	17.42	36.33	36.83	37.81	63.5	67.3	53.9
Spain	13.23	14.11	18.87	35.19	36.98	40.87	62.4	61.9	53.8
Sweden	14.25	18.32	19.39	48.27	47.45	50.99	70.5	61.4	62.0
United Kingdom	11.87	12.11	20.18	66.81	73.22	81.07	82.2	83.5	75.1
UK Rank in EU	3	4	13	15	15	15	15	15	15

Source: European Commission Oil Bulletin

(1) Prices converted to pounds sterling using mid April exchange rates.
(2) Premium unleaded petrol, 95RON.

A9.1 Sales of electricity and gas by sector

United Kingdom

Total selling value (£ million) (1)

	1995	1996	1997	1998	1999
Electricity generation - Gas	938	1,198	1,619r	1,749r	1,921
Industrial - Gas	1,163	891r	920r	1,053r	1,120
- Electricity	4,155	4,075	3,805	3,698	3,844
of which:					
Fuel industries	261	221	179	162	141
Industrial sector	3,894	3,854	3,626	3,536	3,703
Domestic sector - Gas	5,489	6,067r	5,725r	5,290r	4,928
- Electricity	7,614	7,759r	7,443r	7,335r	7,096
Other - Gas	1,124	998r	894r	910r	915
- Electricity	5,349	5,632r	5,422	5,275r	5,010
of which:					
Agricultural sector	251	260	246	219r	222
Commercial sector	3,510	3,702r	3,689	3,577r	3,363
Transport sector	301	309	297	301	306
Public lighting	146	145	146	122	100
Public admin. and other services	1,141	1,216	1,044	1,056	1,019
Total, all consumers	**25,832**	**26,620r**	**25,828r**	**25,310r**	**24,834**
of which gas	**8,714**	**9,154r**	**9,158r**	**9,002r**	**8,884**
of which electricity	**17,118**	**17,466r**	**16,670r**	**16,308r**	**15,950**

Average net selling value per kWh sold (pence) (1)

	1995	1996	1997	1998	1999
Electricity generation - Gas	0.643	0.628	0.647	0.656	0.613
Industrial - Gas	0.678	0.472	0.516	0.565	0.548
- Electricity	4.358	4.159	3.860	3.795	3.900
of which:					
Fuel industries	4.977	4.013	3.530	3.585	3.436
Industrial sector	4.322	4.167	3.878	3.806	3.920
Domestic sector - Gas	1.700	1.700r	1.657r	1.468r	1.399
- Electricity	7.280	7.172r	6.984r	6.583r	6.495
Other - Gas	1.002	0.816r	0.749r	0.770r	0.757
- Electricity	5.939	6.032r	5.519	5.377r	5.052
of which:					
Agricultural sector	6.619	6.803	6.463	5.711r	5.736
Commercial sector	6.094	6.138r	5.565	5.399r	5.061
Transport sector	4.478	4.615	4.316	4.430r	4.459
Public lighting	5.976	5.625	5.450	5.563r	5.097
Public admin. and other services	5.850	6.091	5.616	5.549r	5.090
Average, all consumers	**2.478r**	**2.298r**	**2.158r**	**2.044r**	**1.913**
of which gas	**1.158r**	**1.066r**	**1.025r**	**0.966r**	**0.897**
of which electricity	**5.903**	**5.831r**	**5.495r**	**5.313r**	**5.196**

(1) Excludes VAT where payable - see paragraph 9.101 for a definition of average net selling value and see paragraphs 1.74 for relevant technical definitions.

Annexes

Annex A: Energy and commodity balances, calorific values and conversion factors

Annex B: Energy and the environment

Annex C: UK oil and gas resources

Annex D: Glossary

Annex E: Major events in the energy industry

Annex F: Further sources

Department of Trade and Industry

Annex A
Energy and commodity balances, conversion factors and calorific values

Balance principles

A.1   This Annex outlines the principles behind the balance presentation of energy statistics.   It covers these in general terms. Fuel specific details are given in the appropriate chapters of this publication.

A.2   Balances are divided into two types which perform different functions.   The most basic and important is the *commodity balance* expressed for each energy commodity using the units usually associated with the commodity.   These are presented in the individual fuel chapters of this publication. By using a single column of figures, it shows the flow of the commodity from its sources of supply through to its final use.   However, because some commodities are manufactured from others, it is also useful to present the commodity balances in a common unit and place them alongside one another in a manner which shows the dependence of the supply of one commodity on another.   This format is an *energy balance*; figures are presented in this format in Chapter 1. The layout of the energy balance also differs slightly from the commodity balance.

A.3   Commodities which are made from others are termed secondary commodities, whilst primary energy commodities are those which are drawn (extracted or captured) from natural reserves or flows. Crude oil or coal are examples of primary commodities whilst petrol or coke are secondary commodities manufactured from them.   For balance purposes electricity may be considered to be both primary electricity (for example, hydro, wind) or secondary (produced from steam turbines using steam from the combustion of fuels).

A.4   Both commodity and energy balances show the flow of the commodity from its production or extraction or import through to its final use for combustion and its disappearance from the account.

A.5   A simplified model of the commodity flow underlying the balance structure is given in Chart A.1. It illustrates how primary commodities may be used directly and/or be transformed into secondary commodities.   The secondary fuels then enter final consumption or may also be transformed into another energy commodity (electricity produced from fuel oil, for example).   To keep the diagram simple these "second generation" flows have not been shown.

A.6   The arrows at the top of the figure represent flows to and from the "pools" of primary and secondary commodities from imports and exports and, in the case of the primary pool, extraction from reserves (e.g. the production of coal, gas and crude oil).

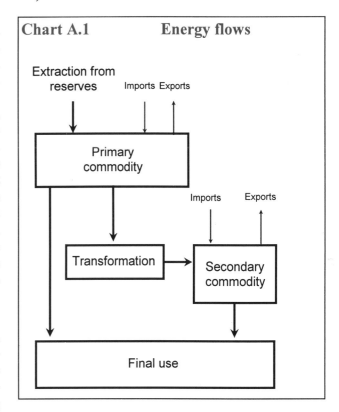

Chart A.1   Energy flows

Commodity balances (Tables 2.1 to 2.6, 3.1 to 3.6, 4.1, 5.1, and 7.1 to 7.3)

A.7   A commodity balance comprises a supply section giving the sources of supply (net in the sense that exports are subtracted to show national available supply) and a demand section.   The demand section is divided into a transformation section, a section showing uses in the energy industries (other than for transformation) and a section covering uses by final consumers for energy or non-energy purposes. Final consumption for energy purposes is divided into use by sector of economic activity.   The breakdowns used are described below.

Supply

Production
A.8   Production, within the commodity balance, covers indigenous production (extraction or capture of primary commodities) and generation or manufacture of secondary commodities.   Production is always

gross; that is it includes the quantities used during the extraction or manufacturing process.

Other sources

A.9   Production from other sources covers sources of supply which do not represent "new" supply. These may be recycled products, recovered fuels (slurry or waste coal), or electricity from pumped storage plants. The production of these quantities will have been reported in an earlier accounting period or have already been reported in the current period of account. Exceptionally, the *Other sources* row in the commodity balances for ethane, propane and butane is used to receive transfers of these hydrocarbons from gas stabilisation plants at North Sea terminals. In this manner the supplies of primary ethane, propane and butane from the North Sea are combined with the production of these gases in refineries so that the disposals may be presented together in the balances.

Imports and exports

A.10   The figures for imports and exports relate to energy commodities moving into or out of the United Kingdom as part of transactions involving United Kingdom companies. Exported commodities are produced in the United Kingdom and imported commodities are for use within the United Kingdom (although some may be re-exported before or after transformation). The figures thus exclude commodities either exported from or imported into HM Customs bonded areas or warehouses. These areas, although part of the United Kingdom, are regarded as being outside of the normal United Kingdom's customs boundary, and so goods entering into or leaving them are not counted as part of the statistics on trade used in the balances.

A.11   Similarly, commodities that only pass through the United Kingdom on their way to a final destination in another country are also excluded. However, for gas these transit flows are included because it is difficult to identify this quantity separately without detailed knowledge of the contract information covering the trade. This means that for gas there is some over statement of the level of imports and exports, but the net flows are correct.

A.12   The convention in these balances is that exports are shown with a negative sign.

Marine bunkers

A.13   These are deliveries of fuels (usually fuel oil or gas oil) to ships of any flag (including the United Kingdom) for consumption during the voyage to other countries. Marine bunkers are treated rather like exports and shown with a negative sign.

Stock changes

A.14   Additions to (- sign) and withdrawals from stocks (+ sign) held by producers and transformation industries correspond to withdrawals from and additions to supply, respectively.

Transfers

A.15   There are several reasons why quantities may be transferred from one commodity balance to another:

- a commodity may no longer meet the original specification and be reclassified;
- the name of the commodity may change through a change in use;
- to show quantities returned to supply from consumers. These may be by-products of the use of commodities as raw materials rather than fuels.

A.16   A quantity transferred from a balance is shown with a negative sign to represent a withdrawal from supply and with a positive sign in the receiving commodity balance representing an addition to its supply.

Total supply

A.17   The total supply available for national use is obtained by summing the flows above this entry in the balance.

Total demand

A.18   The various figures for the disposals and/or consumption of the commodities are summed to provide a measure of the demand for them. The main categories or sectors of demand are described in paragraphs A.20 to A.41.

Statistical difference

A.19   Any excess of supply over demand is shown as a statistical difference. A negative figure indicates that demand exceeds supply. Statistical differences arise when figures are gathered from a variety of independent sources and reflect differences in timing, in definition of coverage of the activity, or in commodity definition. Differences also arise for methodological reasons in the measurement of the flow of the commodity e.g. if there are differences between the volumes recorded by the gas producing companies and the gas transporting companies. A non-zero statistical difference is normal and, provided that it is not too large, is preferable to a statistical difference of zero as this suggests that a data provider has adjusted a figure to balance the account.

Transformation

A.20   The transformation sector of the balance covers those processes and activities which transform the original primary (and sometimes secondary) commodity into a form which is better suited for specific uses than the original form. Most of the transformation activities correspond to particular energy industries whose main business is to manufacture the product associated with them. Certain activities involving transformation take place to make products which are only partly used for energy needs (coke oven coke) or are by-products of other manufacturing processes (coke oven and blast furnace gases). However, as these products and by-products are then used, at least in part, for their energy content they are included in the balance system.

A.21 The figures given under the activity headings of this sector represent the quantities used for transformation. The production of the secondary commodities will be shown in the *Production* row of the corresponding commodity balances.

Electricity generation

A.22 The quantities of fuels burned for the generation of electricity are shown in their commodity balances under this heading. The activity is divided into two parts, covering the major power producers (for whom the main business is the generation of electricity for sale) and autogenerators (whose main business is not electricity generation but who produce electricity for their own needs and may also sell surplus quantities). The amounts of fuels shown in the balance represent the quantities consumed for the gross generation of electricity. Where a generator uses combined heat and power plant, the figures include only the part of the fuel use corresponding to the electricity generated.

A.23 Care is needed when seeking figures for energy consumption in industries which generate their own electricity. The figures for quantities of fuel used for electricity generation appear under the appropriate fuel headings in the *Transformation* sector heading for *Autogenerators,* whilst the electricity generated, appears in the *Electricity* column under *Production*. A breakdown of the information according to the branch of industry in which the generation occurs is not shown in the balance but is given in Chapter 1, Table 1.8. The figures for energy commodities consumed by the industry branches shown under final consumption include all use of electricity, but exclude the fuels combusted by the industry branches to generate the electricity.

Petroleum refineries

A.24 Crude oil, natural gas liquids and other oils needed by refineries for the manufacture of finished petroleum products are shown under this heading.

Coke manufacture and blast furnaces

A.25 Quantities of coal for coke ovens and all fuels used within blast furnaces are shown under this heading. The consumption of fuels for heating coke ovens and the blast air for blast furnaces are shown under "Energy industry use".

Patent fuel manufacture

A.26 The coals and other solid fuels used for the manufacture of solid patent fuels are reported under this heading.

Other

A.27 Any minor transformation activities not specified elsewhere are captured under this heading.

Energy industry use

A.28 Consumption by both extraction and transformation industries to support the transformation process (but not for transformation itself) are included here according to the energy industry concerned. Typical examples are the consumption of electricity in power plants (e.g. for lighting, compressors and cooling systems) and the use of extracted gases on oil and gas platforms for compressors, pumps and other uses. The headings in this sector are identical to those used in the transformation sector with the exception of *Pumped storage*. In this case the electricity used to pump the water to the reservoir is reported.

Losses

A.29 This heading covers the intrinsic losses which occur during the transmission and distribution of electricity and gas (including manufactured gases). Other metering and accounting differences for gas and electricity are within the statistical difference, as are undeclared losses in other commodities.

Final consumption

A.30 *Final consumption* covers both final energy consumption (by different consuming sectors) and the use of energy commodities for non-energy purposes, that is *Non energy use*. Consumption is considered to be final in the sense that the commodities used are not for transformation into secondary commodities. The energy concerned disappears from the account after use. Any fuel use for electricity generation by final consumers is identified and reported separately within the transformation sector. It is important to appreciate that when an enterprise generates electricity, the figure for final consumption of the industrial sector to which the enterprise belongs includes its use of the electricity it generates itself (as well as supplies of electricity it purchases from others) but does not include the fuel used to generate that electricity.

A.31 The classification of consumers according to their main business follows, as far as practicable, the *Standard Industrial Classification (SIC1992)*. The qualifications to, and constraints on, the classification are described in the technical notes to Chapter 1, paragraphs 1.74 to 1.78. Table 1E in Chapter 1 shows the breakdown of final consumers used, and how this corresponds to the SIC1992.

Industry

A.32 Two sectors of industry (iron and steel and chemicals) require special mention because the activities they undertake fall across the transformation, final consumption and non-energy classifications used for the balances. Also, the data permitting an accurate allocation of fuel use within each of these major divisions are not readily available.

Iron and steel

A.33 The iron and steel industry is a heavy energy user for transformation and final consumption activities. Figures shown under final consumption for this industry branch reflect the amounts which remain after quantities used for transformation and energy sector own use have been subtracted from the industry's total energy requirements. Use of fuels for transformation by the industry may be identified within the transformation sector of the commodity balances.

A.34 The amounts of coal used for coke manufacture by the iron and steel industry are in the transformation sector of the coal balance. Included in this figure is the amount of coal used for coke manufacture by the companies outside of the iron and steel industry, i.e. solid fuel manufacturers. The corresponding production of coke and coke oven gas may be found in the commodity balances for these products. The use of coke in blast furnaces is shown in the commodity balance for coke, and the gases produced from blast furnaces and the associated basic oxygen steel furnaces are shown in the production row of the commodity balance for blast furnace gas.

A.35 Fuels used for electricity generation by the industry are included in the figures for electricity generation by autogenerators and are not distinguishable as being used by the iron and steel sector in the balances. Electricity generation and fuel used for this by broad industry group are given in Table 1.8.

A.36 Fuels used to support coke manufacture and blast furnace gas production are included in the quantities shown under *Energy industry use*. These gases and other fuels do not enter coke ovens or blast furnaces but are used to heat the ovens and the blast air supplied to furnaces.

Chemicals

A.37 The petro-chemical industry uses hydrocarbon fuels (mostly oil products and gases) as feedstock for the manufacture of its products. Distinguishing the energy use of delivered fuels from their non-energy use is complicated by the absence of detailed information. The procedures adopted to estimate the use are described in paragraphs A.40 and A.41 under *Non energy use*.

Transport

A.38 Figures under this heading are almost entirely quantities used strictly for transport purposes. However, the figures recorded against road transport usually include some fuel that is actually consumed in some "off-road" activities. Similarly, figures for railway fuels include some amounts of burning oil not used directly for transport purposes. Transport sector use of electricity includes all electricity used in industries classified to SIC1992 Groups 60 to 63. Fuels supplied to cargo and passenger ships undertaking international voyages are reported as *Marine bunkers* (see paragraph A.13). Supplies to fishing vessels are included under "agriculture".

Other sectors

A.39 The classification of all consumers groups under this heading, except *domestic*, follows *SIC1992* and is described in Table 1E in Chapter 1. The consistency of the classification across different commodities cannot be guaranteed because the figures reported are dependent on what the data suppliers can provide.

Non energy use

A.40 The non energy use of fuels may be divided into two types. They may be used directly for their physical properties e.g. lubricants or bitument used for road surfaces, or by the petro-chemical industry as raw materials for the manufacture of goods such as plastics. In their use by the petro-chemical industry, relatively little combustion of the fuels takes place and the carbon and/or hydrogen they contain are largely transferred into the finished product. However, in some cases heat from the manufacturing process or from combustion of by-products may be used. Data for this energy use are rarely available. Depending on the feedstock, non energy consumption is either estimated or taken to be the deliveries to the chemicals sector.

A.41 Both types of non energy use are shown under the *Non energy use* heading at the foot of the balances.

The energy balance (Tables 1.1 to 1.3)

Principles

A.42   The energy balance conveniently presents:

- an overall view of the United Kingdom's energy supplies;
- the relative importance of each energy commodity;
- dependence on imports;
- the contribution of our own fossil and renewable resources;
- the interdependence of one commodity on another.

A.43   The energy balance is constructed directly from the commodity balances by expressing the data in a common unit, placing them beside one another and adding appropriate totals. However, some rearrangement of the commodity balance format is required to show transformation of primary into secondary commodities in an easily understood manner.

A.44   Energy units are widely used as the common unit, and the current practice for the United Kingdom and the international organisations which prepare balances is to use the tonne of oil equivalent or a larger multiple of this unit, commonly thousands. One tonne of oil equivalent is defined as 10^7 kilocalories (41.87 gigajoules). The tonne of oil equivalent is another unit of energy like the gigajoule, kilocalorie or kilowatt hour, rather than a physical quantity. It has been chosen as it is easier to visualise than the other units. Due to the natural variations in heating value of primary fuels such as crude oil, it is rare that one tonne of oil has an energy content equivalent to one tonne of oil equivalent, however it is generally within a few per cent of the heating value of a tonne of oil equivalent. The energy figures are calculated from the natural units of the commodity balances by multiplying by factors representing the calorific (heating) value of the fuel. The gross calorific values of fuels are used for this purpose. When the natural unit of the commodity is already an energy unit (electricity in kilowatt hours, for example) the factors are just constants converting one energy unit to another.

A.45   Most of the underlying definitions and ideas of commodity balances can be taken directly over into the energy balance. However, production of secondary commodities and, in particular, electricity are treated differently and need some explanation. The components of the energy balance are described in turn below, drawing out the differences of treatment compared with the commodity balances.

Primary supply

A.46   Within the energy balance, the production row covers only extraction of primary fuels and the generation of primary energy (hydro, nuclear, wind). Note the change of row heading from *Production* in the commodity balances to *Indigenous production* in the energy balance. Production of secondary fuels and secondary electricity are shown in the transformation sector and not in the indigenous production row at the top of the balance.

A.47   For fossil fuels, indigenous production represents the marketable quantity extracted from the reserves. Indigenous production of *Primary electricity* comprises hydro-electricity, wind and nuclear energy. The energy value for hydro-electricity is taken to be the energy content of the electricity produced from the hydro power plant and not the energy available in the water driving the turbines. A similar approach is adopted for electricity from wind generators. The electricity is regarded as the primary energy form because there are currently no other uses of the energy resource "upstream" of the generation. The energy value attached to nuclear electricity is discussed in paragraph A.51.

A.48   The other elements of the supply part of the balance are identical to those appearing in the commodity balances. In particular, the sign convention is identical so that figures for exports and international marine bunkers carry negative signs. A stock build carries a negative sign to denote it as a withdrawal from supply whilst a stock draw carries a positive sign to show it as an addition to supply.

A.49   The *Primary supply* is the sum of the figures above it in the table, taking account of the signs, and expresses the national requirement for primary energy commodities from all sources and foreign supplies of secondary commodities. It is an indicator of the use of indigenous resources and external energy supplies. Both the amount and mixture of fuels in final consumption of energy commodities in the United Kingdom will differ from the primary supply. The "mix" of commodities in final consumption will be much more dependent on the manufacture of secondary commodities, in particular electricity.

Transformation

A.50   Within an energy balance the presentation of the inputs to and outputs from transformation activities requires special mention as it is carried out using a compact format. The transformation sector also plays a key role in moving primary electricity from its own column in the balance into the electricity column so that it can be combined with electricity from fossil fuelled power stations and the total disposals shown. A discussion of the transformation sector also provides a convenient place to explain the treatment of nuclear electricity.

A.51 Nuclear electricity is obtained by passing steam from nuclear reactors through conventional steam turbine sets. The heat in the steam is considered to be the primary energy available and its value is calculated from the electricity generated using the average thermal efficiency of nuclear stations, currently 36.8 per cent in the United Kingdom. Indigenous production of primary electricity comprises nuclear electricity, hydro electricity and electricity from wind generation. The electrical energy from hydro and wind is transferred from the *Primary electricity* column to the *Electricity* column using the *transfers* row because electricity is the form of primary energy and no transformation takes place. However, because the form of the nuclear energy is the steam from the nuclear reactors, the energy it contains is shown entering electricity generation and the corresponding electricity produced is included with all electricity generation in the figure, in the same row, under the *Electricity* column.

A.52 Quantities of fuels entering transformation activities (fuels into electricity generation, crude oil into petroleum product manufacture (refineries), or coal into coke ovens) are shown with a negative sign to represent the input and the resulting production is shown as a positive number.

A.53 For electricity generated by Major power producers, the inputs are shown in the *Major power producers* row of the *coal, manufactured fuel, primary oils, petroleum products, gas, renewables* and *primary electricity* columns. The total energy input to electricity generation is the sum of the values in these first seven columns. The *Electricity* column shows total electricity generated from these inputs and the transformation loss is the sum of these two figures, given in the *Total* column.

A.54 Within the transformation sector, the negative figures in the *Total* column represent the losses in the various transformation activities. This is a convenient consequence of the sign convention chosen for the inputs and outputs from transformation. Any positive figures represent a transformation gain and, as such, are an indication of incorrect data.

A.55 In the energy balance the columns containing the input commodities, for both electricity generation and oil refining, are separate from the columns for the outputs. However, for the transformation activities involving solid fuels this is only partly the case. Coal used for the manufacture of coke is shown in the coke manufacture row of the transformation section in the coal column, but the related coke and coke oven gas production are shown combined in the *Manufactured fuels* column. Similarly, the input of coke to blast furnaces and the resulting production of blast furnace gas are not identifiable and have been combined in the *Manufactured fuels* column in the *Blast furnace* row. As a result, only the net loss from blast furnace transformation activity appears in the column.

A.56 The share of each commodity or commodity group in primary supply can be calculated from the table. This table also shows the demand for primary as well as foreign supplies. Shares of primary supplies may be taken from the *Primary supply* row of the balance. Shares of fuels in final consumption may be calculated from the final consumption row.

Energy industry use and final consumption
A.57 The figures for final consumption and energy industry use follow, in general, the principles and definitions described under commodity balances in paragraphs A.28 to A.41.

Standard conversion factors

1 tonne of oil equivalent (toe)	= 10^7 kilocalories
	= 396.8 therms
	= 41.87 GJ
	= 11,630 kWh

1 therm = 100,000 British thermal units (Btu)

The following prefixes are used for multiples of joules, watts and watt hours:

kilo (k)	= 1,000	or 10^3
mega (M)	= 1,000,000	or 10^6
giga (G)	= 1,000,000,000	or 10^9
tera (T)	= 1,000,000,000,000	or 10^{12}
peta (P)	= 1,000,000,000,000,000	or 10^{15}

WEIGHT 1 kilogramme (kg) = 2.2046 pounds (lb)
1 pound (lb) = 0.4536 kg
1 tonne (t) = 1,000 kg
= 0.9842 long ton
= 1.102 short ton (sh tn)
1 Statute or long ton = 2,240 lb
= 1.016 t
= 1.120 sh tn

VOLUME 1 cubic metre (cu m) = 35.31 cu ft
1 cubic foot (cu ft) = 0.02832 cu m
1 litre = 0.22 Imperial gallon (UK gal.)
1 UK gallon = 8 UK pints
= 1.201 U.S. gallons (US gal)
= 4.54609 litres
1 barrel = 159.0 litres
= 34.97 UK gal
= 42 US gal

LENGTH 1 mile = 1.6093 kilometres
1 kilometre (km) = 0.62137 miles

TEMPERATURE 1 scale degree Celsius (C) = 1.8 scale degrees Fahrenheit (F)
For conversion of temperatures: °C = 5/9 (°F - 32); °F = 9/5 °C + 32

Average conversion factors for petroleum

	Imperial gallons per tonne	Litres per tonne		Imperial gallons per tonne	Litres per tonne
Crude oil:			Gas/diesel oil:		
Indigenous	264	1,199	Gas oil	257	1,170
Imported	260	1,181	Marine diesel oil	253	1,150
Average of refining throughput	262	1,192			
			Fuel oil:		
Ethane	601	2,730	All grades	222	1,011
Propane	433	1,967	Light fuel oil:		
Butane	381	1,733	1% or less sulphur	237	1,079
Naphtha (l.d.f.)	322	1,465	>1% sulphur	232	1,055
			Medium fuel oil:		
Aviation gasoline	308	1,395	1% or less sulphur	237	1,078
			>1% sulphur	225	1,023
Motor spirit:			Heavy fuel oil:		
All grades	299	1,361	1% or less sulphur	226	1,027
Unleaded Super	292	1,326	>1% sulphur	222	1,010
Premium	299	1,360			
Ultra low sulphur petrol	297	1,348			
Leaded Premium	300	1,366	Lubricating oils:		
Lead replacement petrol	300	1,362	White	249	1,133
			Greases	245	1,113
Middle distillate feedstock	286	1,301	Other	249	1,134
Kerosene:			Bitumen	214	973
Aviation turbine fuel	275	1,248	Petroleum coke	185	843
Burning oil	274	1,246	Petroleum waxes	261	1,187
			Industrial spirit	302	1,375
DERV fuel: all	260	1,183	White spirit	278	1,265
0.005% or less sulphur	260	1,180			
>0.005% sulphur	264	1,202			

Note: The above conversion factors, which for refined products have been compiled by the UK Petroleum Industry Association, apply to the year 1999, and are only approximate for other years.

A.1 Estimated average gross calorific values of fuels

	GJ per tonne		GJ per tonne
Coal:		**Renewable sources:**	
All consumers (weighted average) *(1)*	27.3	Domestic wood *(2)*	10.0
Power stations *(1)*	25.8	Industrial wood *(3)*	11.9
Coke ovens *(1)*	32.0	Straw	15.0
Low temperature carbonisation plants		Poultry litter	8.8
and manufactured fuel plants	30.1	General industrial waste	16.0
Collieries	29.3	Hospital waste	14.0
Agriculture	28.9	Municipal solid waste *(4)*	9.5
Iron and steel	30.7	Refuse derived waste *(4)*	18.7
Other industries (weighted average)	26.6	Tyres	32.0
Non-ferrous metals	25.1		
Food, beverages and tobacco	29.4	**Petroleum:**	
Chemicals	29.7	Crude oil (weighted average)	45.7
Textiles, clothing, leather etc.	30.1	Petroleum products (weighted average)	45.2
Paper, printing etc.	29.0	Ethane	50.7
Mineral products	26.7	Butane and propane (LPG)	49.4
Engineering (mechanical and		Light distillate feedstock for gasworks	47.7
electrical engineering and	29.3	Aviation spirit and wide cut gasoline	47.3
vehicles)		Aviation turbine fuel	46.2
Other industries	29.4	Motor spirit	47.1
		Burning oil	46.2
Domestic		Gas/diesel oil (DERV)	45.6
House coal	30.9	Fuel oil	43.2
Anthracite and dry steam coal	33.5	Power station oil	43.2
Other consumers	29.1	Non-fuel products (notional value)	43.4
Imported coal (weighted average)	29.2	Orimulsion	29.7
Exports (weighted average)	31.7		

	GJ per tonne		MJ per cubic metre
Coke (including low temperature	29.8	Natural gas *(5)*	39.5
carbonisation cokes)		Coke oven gas	20.2
Coke breeze	24.8	Blast furnace gas	3.0
Other manufactured solid fuel	30.9	Landfill gas	38.6
		Sewage gas	38.6

(1) Applicable to UK consumption - based on calorific value for home produced coal plus imports and, for "All consumers" net of exports.

(2) Based on a 50 per cent moisture content.

(3) Average figure covering a range of possible feedstock.

(4) Average figure based on survey returns.

(5) The gross calorific value of natural gas can also be expressed as 10.973 kWh per cubic metre. This value represents the average calorific value seen for gas when extracted. At this point it contains not just methane, but also some other hydrocarbon gases (ethane, butane, propane). These gases are removed before the gas enters the National Transmission System for sale to final consumers. As such, this calorific value will differ from that readers will see quoted on their gas bills.

Note: The above estimated average gross calorific values apply only to the year 1999. For calorific values of fuels in earlier years see Table A.2 and previous issues of this Digest. See the notes in Chapter 1, paragraph 1.72 regarding net calorific values. The calorific values for coal other than imported coal are based on estimates provided by the main coal producers. The calorific values for petroleum products have been calculated using the method described in Chapter 1, paragraph 1.48. The calorific values for coke oven gas and blast furnace gas are provided by the Iron and Steel Statistics Bureau (ISSB).

Data reported in this Digest in 'thousand tonnes of oil equivalent' have been prepared on the basis of 1 tonne of oil equivalent having an energy content of 41.87 gigajoules (GJ), (1 GJ = 9.478 therms) - see notes in Chapter 1, paragraphs 1.45 to 1.48.

A.2 Estimated average gross calorific values of fuels, 1970, 1980, 1990 and 1995 to 1999

GJ per tonne (gross)

	1970	1980	1990	1995	1996	1997	1998	1999
Coal								
All consumers (1)(2)	..	25.6	25.5	26.6	25.9	26.1r	26.1r	26.2
All consumers - home produced plus imports minus exports (1)	..	..	..	27.1	26.9	27.2r	27.2r	27.3
Power stations (2)	23.7	23.8	24.8	25.3	25.1	25.3	25.4	25.5
Power stations - home produced plus imports (1)	..	..	..	25.4	25.4	25.6	25.8	25.8
Coke ovens (2)	29.8	30.5	30.2	31.2	31.4	31.4	31.3	31.5
Coke ovens - home produced plus imports (1)	..	..	..	31.4	32.0	32.0	32.0	32.0
Low temperature carbonisation plants and manufactured fuel plants	29.8	19.1	29.2	29.9	30.0	30.4	30.5	30.1
Collieries	24.9	27.0	28.6	28.2	26.2	27.8	29.6	29.3
Agriculture	31.1	30.1	28.9	29.0	28.9	29.1	28.5	28.9
Iron and steel industry	29.1	29.1	28.9	31.5	31.3	31.3	31.3	30.7
Other industries (1)	27.0	27.1	27.8	27.7	27.3	27.0	26.9r	26.6
Non-ferrous metals	..	..	23.1	25.0	25.4	25.1	24.5	25.1
Food, beverages and tobacco	28.4	28.6	28.1	28.4	28.1	28.7	29.7	29.4
Chemicals	25.8	25.8	27.3	27.2	27.1	27.3	28.9	29.7
Textiles, clothing, leather & footwear	27.4	27.5	27.7	29.7	30.2	30.4	30.2	30.1
Pulp, paper, printing, etc.	26.5	26.5	27.9	28.4	27.7	27.4	29.0	29.0
Mineral products	..	..	28.2	27.1	27.4	27.0	26.6	26.7
Engineering (3).	27.7	27.7	28.3	28.1	29.3	29.6	29.4	29.3
Other industry (4)	28.4	28.4	28.5	28.7	26.6	27.5r	29.5	29.4
Unclassified	..	..	27.1	..	..	..	..	..
Domestic								
House coal	29.1	30.1	30.2	30.4	30.6	30.6	30.9	30.9
Anthracite and dry steam coal	33.8	33.3	33.6	34.0	33.9	33.9	34.1	33.5
Other consumers	29.1	27.5	27.5	28.9	30.4	29.3	29.2	29.1
Imported coal (1)	..	..	28.3	30.0	29.8	29.3r	29.2r	29.2
of which Steam coal					26.9	26.9	27.0	27.0
Coking coal					32.0	32.0	32.0	32.0
Anthracite					31.1	31.4r	32.0r	31.2
Exports (1)	..	..	29.0	32.0	29.5	30.7	30.8	31.7
of which Steam coal					28.5	30.4	30.1	32.1
Anthracite					31.0	30.9	31.4	31.5
Coke (5)	28.1	28.1	28.1	29.8	29.8	29.8	29.8	29.8
Coke breeze	22.9	24.4	24.8	24.8	24.8	24.8	24.8	24.8
Other manufactured solid fuels (1)(6)	28.1	27.6	27.6	27.9	30.2	30.4	30.7	30.9
Petroleum								
Crude oil (1)	..	45.2	45.6	45.7	45.7	45.7	45.7	45.7
Liquified petroleum gas	49.6	49.6	49.4	49.4	49.4	49.4	49.4	49.4
Ethane	52.3	52.3	50.6	50.7	50.7	50.7	50.7	50.7
LDF for gasworks/Naphtha	47.8	47.8	47.9	47.7	47.7	47.7	47.7	47.7
Aviation spirit and wide-cut gasoline (AVGAS & AVTAG)	47.2	47.2	47.3	47.3	47.3	47.3	47.3	47.3
Aviation turbine fuel (AVTUR)	46.4	46.4	46.2	46.2	46.2	46.2	46.2	46.2
Motor spirit	47.0	47.0	47.0	47.0	47.0	47.0	47.0	47.1
Burning oil	46.5	46.5	46.2	46.2	46.2	46.2	46.2	46.2
Vaporising oil	46.0	45.9	45.9	..	..	..	..	..
Gas/diesel oil (including DERV)	45.5	45.5	45.4	45.4	45.4	45.4	45.5	45.6
Fuel oil	43.0	42.8	43.2	43.2	43.2	43.3	43.2	43.2
Power station oil	43.5	42.8	43.2	43.2	43.2	43.3	43.2	43.2
Non-fuel products (notional value)	..	42.2	43.2	43.3	43.3	43.4	43.3	43.4
Petroleum coke	..	..	39.5	39.5	39.5	39.5	39.5	39.5
Orimulsion	..	..	29.7	29.7	29.7	29.7	29.7	29.7

(1) Weighted averages.
(2) Home produced coal only.
(3) Mechanical engineering and metal products, electrical and instrument engineering and vehicle manufacture.
(4) The low figure in 1997 for this sub-sector is due to the substantial disposal of low calorific value "discards" in this year.
(5) As of 1995 the source of these figures is the ISSB.
(6) These figures have been reviewed and revised from 1996 onwards.

Annex B
Energy and the environment

Introduction

B.1   The operations of the energy sector in the UK, as elsewhere, can affect the environment in many different ways. Detrimental effects can result from exploration, production, transportation, storage, conversion and distribution. The final use of the energy and the disposal of waste products can also damage the environment.

B.2   The particular areas of potential environmental concern related to the energy sector are:- ambient air quality; acid deposition; coal mining subsidence; major environmental accidents; water pollution; maritime pollution; land use and siting impact; radiation and radioactivity; solid waste disposal; hazardous air pollutants; stratospheric ozone depletion; and climate change.

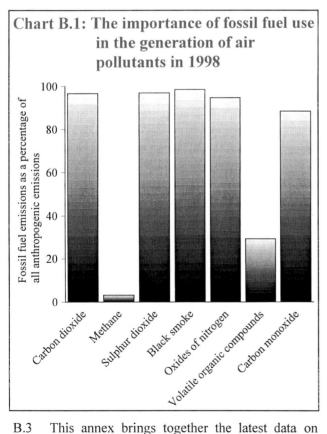

Chart B.1: The importance of fossil fuel use in the generation of air pollutants in 1998

B.3   This annex brings together the latest data on greenhouse gases and air pollution arising, at least in part, from energy related activities. Figures up to 1998 are based on data from the National Environmental Technology Centre, which are presented in further detail in a press release published by the Department of the Environment, Transport and the Regions on 30 March 2000. Provisional DTI estimates of carbon dioxide emissions are included for 1999. In addition, this Annex shows emissions from power stations in

relation to electricity generated, and emissions from road transport in relation to road traffic growth.

B.4   As Chart B.1 shows, fossil fuels are responsible for the majority of emissions of five of the seven pollutants shown. Only methane and VOCs were mainly caused by sources other than fossil fuels. The trend in emissions of each of these pollutants is downwards and the rest of this annex examines these in more detail.

Emission factors

B.5   The emissions estimates presented in this annex are largely based on detailed fuel use data from the *1999 Digest of United Kingdom Energy Statistics.* The 1999 edition of this publication is used due to the time scales involved in the production of an emissions inventory.   These data are combined with detailed emission factors, compiled at the National Environmental Technology Centre. Please note that due to ongoing refinements to the methodology the emissions data published up to 1998 in this edition are not directly comparable with data published up to 1997 in the *1999 Digest of United Kingdom Energy Statistics.*

Table B.1: Emission factors of fossil fuels in 1998

grams per GJ

	Natural gas	Oil	Coal
Carbon dioxide *(1)*	14,000	19,000	24,000
Methane	3.7	3	20
Sulphur dioxide	1	400	840
Black smoke	0	15	55
Nitrogen oxides	55	100	230
Volatile organic compounds	5	2	18
Carbon monoxide	7	8	210

(1) Expressed in terms of weight of carbon produced.

B.6   Table B.1 shows average emission factors for each pollutant by fuel, i.e. the amount of each pollutant released by the consumption of one unit of energy. These values are calculated from the estimated UK total emissions from stationary (i.e. excluding transport) combustion processes and the total energy consumed in these processes; more detailed emission factors are actually used in producing the emissions inventory. The figures presented in Table B.1 are for comparative purposes only. It can be seen that in all cases the combustion of gas and oil result in lower levels of emissions of all pollutants compared to coal. This explains the environmental benefits of the trend towards increased use of natural gas instead of coal for electricity generation.

Greenhouse gases

B.7 Naturally-occurring greenhouse gases maintain the earth's surface at a temperature 33°C higher than it would be in their absence. Water vapour is by far the most important greenhouse gas but there are also significant natural sources of carbon dioxide, methane, ozone and nitrous oxide. At present greenhouse gas concentrations in the atmosphere are increasing as a result of human activities. This includes trace gases such as some halogenated compounds (e.g. hydrofluorocarbons, HFCs) which are completely anthropogenic in origin. Increased greenhouse gas concentrations risk causing enhanced global warming and subsequent climate change.

B.8 Emissions data are available on two bases: the United Nations Economic Commission for Europe (UNECE) basis and the Intergovernmental Panel on Climate Change (IPCC) basis. The UK uses the methodology developed by the IPCC for reporting emissions of carbon dioxide and other greenhouse gases under the Framework Convention on Climate Change. The IPCC data agree with the UNECE data for carbon dioxide to about 5 per cent; the trend is almost identical. The main differences are that the IPCC definition includes land use change, domestic aviation and shipping, but excludes emissions from international marine and aviation bunker fuels. UNECE includes coastal shipping and aircraft movements below 1,000 metres. IPCC is the basis on which emission limitation and reduction targets are set for greenhouse gas emissions. The figures presented here show emissions which have been collected under IPCC[1] and UNECE[2] definitions.

---

[1] *Power stations here include public electricity and heat production which are part of the energy industries as defined by IPCC. Industrial combustion comprises the rest of the energy industries as well as industrial processes and manufacturing industries and construction. Domestic and commercial and public service are defined by the IPCC as residential and commercial/institutional respectively, and make up part of their 'other sectors' grouping. Land use change and forestry here include the IPCC grouping of land use change and forestry and fuel combustion for agriculture, forestry and fishery from the IPCC 'other sectors' grouping. Road transport and other transport, which comprises civil aviation, railways, navigation and aircraft support, make up the IPCC transport grouping. Other sectors here include military aircraft, naval vessels, fugitive emissions from fuel and waste related emissions.*

[2] *Industrial combustion here includes refineries, combustion in fuel extraction and transformation, iron and steel, other industrial combustion and production processes as defined by UNECE. Commercial and public service here includes the following UNECE groupings: commercial, public and agricultural combustion and waste treatment and disposal. Other transport here include off-road sources, military, railways, shipping and civil aircraft. Agriculture here relates to agriculture and managed forestry under UNECE.*

Carbon dioxide

B.9 The most important anthropogenic greenhouse gas is carbon dioxide. Carbon dioxide contributes about 80 per cent of the potential global warming effect of anthropogenic emissions of greenhouse gases. Although this gas is naturally emitted by living organisms, these emissions are balanced by the uptake of carbon dioxide by plants during photosynthesis; they therefore tend to have no net effect on atmospheric concentrations. The burning of fossil fuels, however, releases carbon dioxide fixed by plants many millions of years ago, and thus increases its concentration in the atmosphere.

Table B.2: Carbon dioxide emissions in 1998 collected under IPCC definitions

	Million tonnes of carbon	
Collected under IPCC definitions		
Power stations	41	(26%)
Domestic	23	(14%)
Commercial and public service	8	(6%)
Industrial combustion	40	(25%)
Land use change and forestry(1)	7	(5%)
Road transport	32	(20%)
Other transport	3	(2%)
Other sectors (2)	3	(2%)
Total	**156**	**(100%)**

(1) Includes fuel combustion for agriculture, forestry and fishing.
(2) Includes military aircraft and naval vessels, fugitive emissions from solid fuels, oil, natural gas and waste.

B.10 The UK contributes about 2½ per cent of global emissions from energy and industrial processes; the UK's emissions are broken down by fuel and by source in Tables B.2 and B.3. In 1998 about 92 per cent of carbon dioxide emissions came directly from energy consumption: petroleum products (34 per cent), coal and smokeless fuel (22 per cent) and natural gas (37 per cent); 3 per cent of emissions came from non-energy products and 5 per cent from land use change and forestry.

B.11 On an IPCC basis, the main sources of emissions were power stations (26 per cent), industry (25 per cent), road transport (21 per cent) and the domestic sector (14 per cent). In 1998, 156 million tonnes of carbon were emitted as carbon dioxide from the UK. Between 1997 and 1998 there was an overall increase of ½ million tonnes, with emissions from power stations increasing by almost 1 million tonnes. Between 1990 and 1998, emissions fell by 7 per cent or 10½ million tonnes, with emissions from power stations alone falling by 13½ million tonnes. Trends in carbon dioxide emissions on an IPCC basis are shown in Chart B.2. The figures provided for 1999 are provisional DTI estimates based on the difference between fuel consumption in 1998 and 1999.

Chart B.2: Carbon dioxide emissions[1] collected under IPCC definitions 1990 to 1999[2]

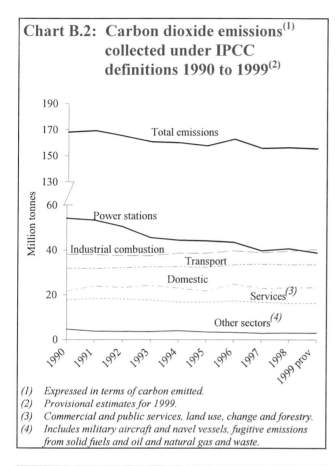

(1) Expressed in terms of carbon emitted.
(2) Provisional estimates for 1999.
(3) Commercial and public services, land use, change and forestry.
(4) Includes military aircraft and navel vessels, fugitive emissions from solid fuels and oil and natural gas and waste.

Chart B.3: Carbon dioxide emissions[1] collected under UNECE definitions 1970 to 1999[2]

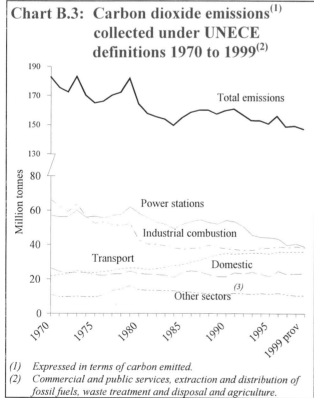

(1) Expressed in terms of carbon emitted.
(2) Commercial and public services, extraction and distribution of fossil fuels, waste treatment and disposal and agriculture.

B.12 On a UNECE basis, 149 million tonnes of carbon were emitted in 1998 as carbon dioxide from the UK. This represents a fall of 19 per cent from emissions in 1970, caused by a reduction in emissions from power stations and industry. Emission levels

fluctuated between 165 and 185 million tonnes through the 1970's. During the first half of the 1980's they reduced significantly following the recession in manufacturing industries and, later, the miners strike in 1984. From the low of 149½ million tonnes in 1984 emissions rose to 156 million tonnes in 1992. Since then there has been a downward trend with reduction in emissions due to the impact of power stations switching from coal to gas outweighing the upward pressure from rising levels of economic activity. Trends in carbon dioxide emissions since 1970 are shown in Chart B.3. The provisional figures provided for 1999 are DTI estimates based on the difference between fuel consumption in 1998 and 1999.

Table B.3: Carbon dioxide emissions in 1998 collected under UNECE definitions

By fuel	Million tonnes of carbon	
Coal	34	(23%)
Solid smokeless fuel	1	(-)
Petroleum:		
Motor spirit	19	(13%)
DERV	13	(9%)
Gas oil	6	(4%)
Fuel oil	5	(3%)
Burning oil	3	(2%)
Other petroleum	7	(5%)
Natural gas	57	(38%)
Other emissions	5	(3%)
Total	**149**	**(100%)**

Collected under UNECE definitions

Power stations	41	(27%)
Domestic	23	(15%)
Commercial and public service (1)	9	(7%)
Industrial combustion	39	(26%)
Extraction and distribution of fossil fuels	-	(-)
Road transport	32	(21%)
Other transport	4	(3%)
Agriculture	-	(-)
Total	**149**	**(100%)**

(1) Includes waste treatment and disposal.

B.13 Emissions from power stations, the largest single source, rose in 1998. Provisional 1999 estimates suggest a slight decrease due to decreased coal burn during 1999. As can be seen in Chart B.4, it is estimated that carbon dioxide emissions from power stations have fallen by 22 per cent between 1970 and 1999, whilst the amount of electricity generated has risen by 44 per cent. Carbon dioxide emissions per unit of electricity generated have declined steadily, falling by 45 per cent since 1970 level. This is due to the increasing contribution of nuclear power, general improvements in the efficiency of power stations and, in recent years, fuel switching from coal and oil, to gas.

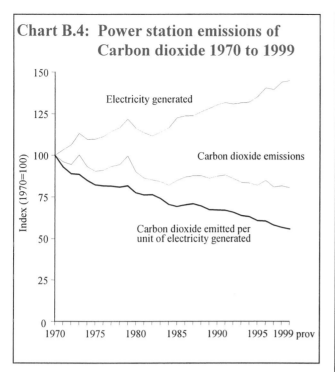

Chart B.4: Power station emissions of Carbon dioxide 1970 to 1999

Chart B.5: Methane emissions collected under IPCC definitions 1990 to 1998

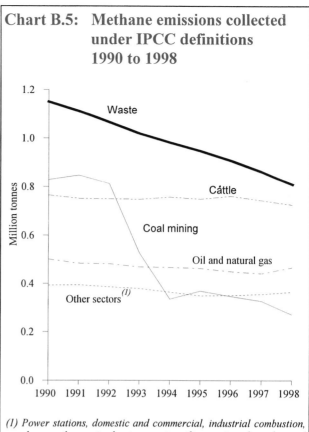

(1) Power stations, domestic and commercial, industrial combustion, sheep, other animals, transport, military aircraft and naval vessels, agriculture/ forestry/fishing.

Methane

B.14 Methane is the second most important anthropogenic greenhouse gas. Like carbon dioxide it also occurs naturally as a component of biological cycles. At present, world-wide methane emissions contribute around 20 per cent of anthropogenic greenhouse gas. Chart B.5 shows UK emissions of methane between 1990 and 1998 collected under IPCC definitions. The main sources are waste (32 per cent) and livestock (37 per cent). Only about 3 per cent of emissions come from combustion (i.e. from power stations, domestic and commercial combustion, industrial combustion, and transport) although emissions from coal mining and the extraction and distribution of oil and natural gas account for a further 28 per cent. In 1998, 2.7 million tonnes of methane were emitted, 25 per cent below 1990 levels.

Chart B.6: Methane emissions collected under UNECE definitions 1970 to 1998

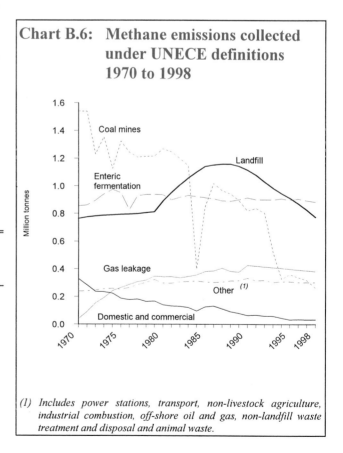

(1) Includes power stations, transport, non-livestock agriculture, industrial combustion, off-shore oil and gas, non-landfill waste treatment and disposal and animal waste.

Table B.4: Methane emissions in 1998 collected under IPCC definitions

Thousand tonnes

Collected under IPCC definitions		
Power stations	19	(1%)
Domestic and commercial	39	(1%)
Industrial combustion	18	(1%)
Cattle	656	(25%)
Sheep	208	(8%)
Other animals	129	(5%)
Road transport	19	(1%)
Waste	809	(31%)
Coal mining	264	(10%)
Oil and natural gas	444	(16%)
Other *(1)*	12	(-)
Total	**2,637**	**(100%)**

(1) Includes military aircraft and naval vessels, other transport and agriculture/forestry/fishing.

B.15 Under the UNECE definition, methane emissions were 30 per cent lower than in 1970. Since 1986, emissions have generally fallen steadily, with reductions from coal mining and landfill, and to a lesser extent from domestic and commercial combustion. Chart B.6 and Table B.5 present emissions on a UNECE basis.

dioxide emissions in 1998 in the UK by fuel and collected under UNECE definitions.

Table B.5: Methane emissions in 1998

Thousand tonnes

Collected under UNECE definitions

Power stations	19	(1%)
Domestic and commercial combustion	39	(1%)
Industrial combustion	24	(1%)
Coal mines	263	(10%)
Gas leakage	382	(14%)
Offshore oil and gas	62	(2%)
Road transport	19	(1%)
Other transport	3	(-)
Landfill	774	(29%)
Other waste disposal	57	(2%)
Enteric fermentation	883	(33%)
Animal wastes	112	(4%)
Non-livestock agriculture	-	(0%)
Total	**2,637**	**(100%)**

B.16 The figures used in this section are approximate and represent best current estimates; they will be revised as more accurate information becomes available. Although the estimates in Table B.5 are given to the nearest thousand tonnes, this does not reflect the accuracy of the estimates which is thought to be ±15 to 20 per cent overall[3]. Figures for landfill emissions are particularly uncertain.

Air pollution

B.17 Air pollution can have a wide range of environmental impacts, with excessively high levels potentially affecting soil, water, wildlife, crops, forests and buildings as well as damaging human health. This section gives information on the main air pollutants associated with fossil fuel combustion - sulphur dioxide, Black Smoke and PM_{10}, nitrogen oxides, volatile organic compounds and carbon monoxide.

Sulphur dioxide

B.18 Sulphur dioxide is a gas produced by the combustion of sulphur-containing fuels such as coal and oil. Most sulphur dioxide is generated by the combustion of coal. Trends on a UNECE basis are shown in Chart B.7. In 1998, 66 per cent of sulphur dioxide was emitted by power stations, with a further 25 per cent coming from refineries and other industrial combustion. In 1998 there were 1.6 million tonnes of sulphur dioxide emitted, a level 57 per cent lower than 1990 and 75 per cent lower than in 1970. The decrease is a result of lower coal and fuel oil consumption. Table B.6 shows figures for sulphur

[3] 'Treatment of Uncertainties for National Estimates of Greenhouse Gas Emissions', a report produced for Global Atmosphere Division, Department of the Environment, Transport and the Regions by AEA Technology Environment (1998).

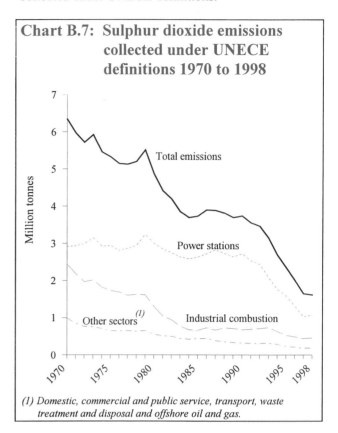

Chart B.7: Sulphur dioxide emissions collected under UNECE definitions 1970 to 1998

(1) Domestic, commercial and public service, transport, waste treatment and disposal and offshore oil and gas.

Table B.6: Sulphur dioxide emissions in 1998

Thousand tonnes

By fuel

Coal	1,158	(71%)
Solid smokeless fuel	11	(1%)
Petroleum:		
Motor spirit	11	(1%)
DERV	12	(1%)
Gas oil	42	(3%)
Fuel oil	227	(14%)
Burning oil	2	(-)
Other petroleum	75	(5%)
Other gases	30	(2%)
Other emissions	47	(3%)
Total	**1,615**	**(100%)**

Collected under UNECE definitions

Power stations	1,072	(66%)
Domestic	52	(3%)
Commercial and public service	34	(2%)
Refineries	172	(11%)
Other industrial combustion	219	(14%)
Non-combustion processes (1)	6	(-)
Road transport	23	(1%)
Other transport	38	(2%)
Waste treatment and disposal	1	(-)
Total	**1,615**	**(100%)**

(1) Offshore oil and gas (well testing).

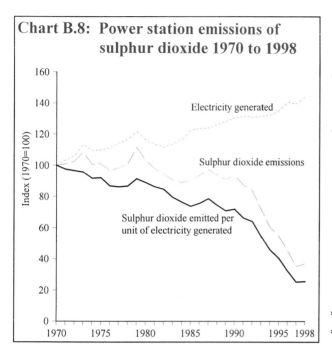

Chart B.8: Power station emissions of sulphur dioxide 1970 to 1998

Electricity generated

Sulphur dioxide emissions

Sulphur dioxide emitted per unit of electricity generated

Index (1970=100)

B.19 Chart B.8 shows that emissions of sulphur dioxide from power stations have fallen by 63 per cent since 1970. With electricity generation increasing by 44 per cent from 1970 levels, the amount of sulphur released per unit of electricity generated has fallen by 74 per cent since 1970. This fall reflects a gradual reduction in the quantity of coal used by power stations, as well as the increased output of the nuclear stations and the increasing contribution of gas-fired plant. In addition, since 1993, flue-gas desulphurisation has come progressively into operation on the 6 GWe of coal-fired generating capacity at the Drax and Ratcliffe-on-Soar power stations, helping to reduce emissions further.

Black smoke and PM_{10}

B.20 "Black Smoke" and "PM_{10}" refer to two different measurement methods for fine, suspended particles in the air. These particles may come from a wide range of man-made and natural sources, including incomplete fuel combustion, wind-blown soil, and dust generated by activities such as quarrying. The long-established British Standard Black Smoke measurement is based on the degree of blackening of a filter paper through which a known volume of air has been drawn. Thus the values do not represent the actual mass of particles but a combination of mass and blackness. A size-selective sampler is used, selecting mainly respirable particles with diameters below about 5 µm and some larger particles up to 10-15 µm. Estimates of black smoke emissions from non-combustion sources are not made.

B.21 Direct methods of measuring mass concentrations of particulate matter (PM) in air, independent of their colour, are now available, and the results are expressed in terms of the fineness of particles collected by the sampling head. For example, PM_{10} refers to the particle fraction sampled with an

aerodynamic diameter less than 10 µm (size at 50% sampling efficiency) and some larger particles up to about 40 µm.

Table B.7: Black smoke emissions[1] in 1998

		Thousand tonnes
By fuel		
Coal	76	(27%)
Solid smokeless fuel	3	(1%)
Petroleum:		
Motor spirit	5	(2%)
DERV	147	(53%)
Gas oil	34	(12%)
Fuel oil	5	(2%)
Burning oil	-	(- %)
Other petroleum	2	(1%)
Other emissions	4	(1%)
Total	**277**	**(100%)**
Collected under UNECE definitions		
Power stations	16	(6%)
Domestic	63	(23%)
Commercial and public service	2	(1%)
Industrial combustion	11	(4%)
Road transport	152	(55%)
Other transport	32	12%
Waste treatment and disposal	-	(- %)
Total	**277**	**(100%)**

(1) The figures expressed in this table are rounded to the nearest 5 thousand tonnes.

Table B.8: PM_{10} emissions in 1998

		Thousand tonnes
Collected under UNECE definitions *(1)*		
Power stations	23	(14%)
Domestic	26	(16%)
Commercial and public service	5	(3%)
Industrial combustion	25	(15%)
Non-combustion processes *(1)*	38	(23%)
Road transport	40	(24%)
Other transport	4	(3%)
Agriculture	-	(-%)
Total	**163**	**(100%)**

(1) Construction, quarrying, industrial processes and waste treatment and disposal.

B.22 Table B.7 shows estimates of Black Smoke emissions by fuel and source in the UK (rounded to the nearest five thousand tonnes). In 1998 estimated Black Smoke emissions were around 42 per cent lower than in 1990 and 74 per cent lower than in 1970. Emissions by the domestic sector have fallen by 92 per cent since 1970, while emissions from road transport have increased by 52 per cent over the same period. The method used to calculate the estimates of black smoke includes a weighting for the differing degrees of "blackness" of particles from various sources and hence the estimates represent a combination of both mass and blackness. The figures should not, therefore, be taken as estimates of the actual mass of black smoke emitted.

B.23 Table B.8 shows estimated anthropogenic emissions of PM_{10} in 1998. The major sources were non-combustion processes (construction, mining, quarrying and other industry: 23 per cent) and road transport (25 per cent). Emissions of PM_{10} are estimated to have fallen by 67 per cent since 1970, largely as a result of an 88 per cent fall in emissions from the domestic sector.

Table B.9: Emissions of nitrogen oxides in 1998

		Thousand tonnes
By fuel		
Coal	316	(18%)
Solid smokeless fuel	1	(-)
Petroleum:		
Motor spirit	495	(28%)
DERV	305	(17%)
Gas oil	171	(10%)
Fuel oil	43	(2%)
Burning oil	8	(-)
Other petroleum	71	(4%)
Natural gas	243	(14%)
Other emissions	101	(6%)
Total	**1,753**	**(100%)**
Collected under UNECE definitions		
Power stations	364	(21%)
Domestic	71	(4%)
Commercial and public service	34	(2%)
Refineries	52	(3%)
Industrial combustion	231	(13%)
Non-combustion processes	5	(-)
Extraction and distribution		
of fossil fuels	1	(-)
Road transport	799	(46%)
Other transport	195	(11%)
Waste treatment and disposal	3	(-)
Agriculture	-	(-)
Total	**1,753**	**(100%)**

Nitrogen oxides

B.24 A number of nitrogen compounds including nitrogen dioxide, nitric oxide and nitrous oxide (not included here) are formed in combustion processes when nitrogen in the air or the fuel combines with oxygen. These compounds can add to the natural acidity of rainfall. Chart B.9 shows trends on a UNECE basis. Road transport accounted for 46 per cent of emissions of nitrogen oxides, with a further 21 per cent coming from power stations. The total level of emissions in 1998 (at 1.8 million tonnes of nitrogen dioxide) was 37 per cent lower than in 1990, with substantial falls from both road transport and power

stations. Emissions rose steadily between 1984 and 1989, as increasing amounts were generated by road transport. Emissions from power stations have declined recently due to increased output from nuclear stations, combined cycle gas turbine stations replacing coal-fired plant, together with the effect of the installation of low NO_x burners at other coal fired power stations. Table B.9 shows figures for emissions of nitrogen oxides in 1998 by fuel and collected under UNECE definitions.

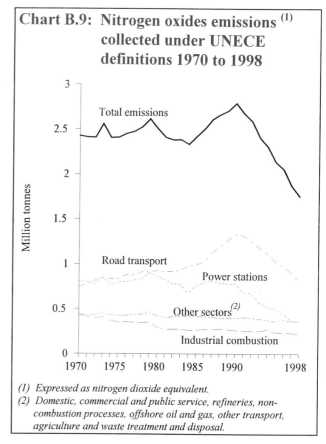

Chart B.9: Nitrogen oxides emissions [1] collected under UNECE definitions 1970 to 1998

(1) Expressed as nitrogen dioxide equivalent.
(2) Domestic, commercial and public service, refineries, non-combustion processes, offshore oil and gas, other transport, agriculture and waste treatment and disposal.

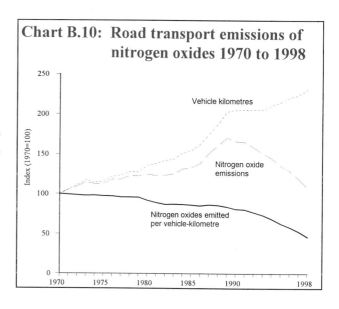

Chart B.10: Road transport emissions of nitrogen oxides 1970 to 1998

B.25 Chart B.10 shows estimates of the amount of nitrogen oxides emitted by road transport along with the total number of vehicle-kilometres travelled by all road vehicles. Both emissions and vehicle-kilometres rose substantially from 1970 to 1989, with emissions reaching a peak in that year, 70 per cent higher than their 1970 level. Since then emissions have declined, despite further small increases in the number of vehicle-kilometres travelled. This fall is mainly due to tighter emissions standards for passenger and goods vehicles, including the introduction of catalytic converters on all new cars since 1993. As a result of these recent improvements, the amount of nitrogen oxides emitted for each vehicle-kilometre travelled had fallen by 53 per cent between 1970 and 1998.

Volatile organic compounds

B.26 Volatile organic compounds consist of a wide range of chemicals, including hydrocarbons and oxygenated and halogenated organics, which are released from oil refining, petrol distribution, motor vehicles, industrial processes and solvent use. The main environmental impact of many of these compounds is through their involvement in the creation of ground level ozone. In addition, benzene is a known human carcinogen and 1.3 butadiene is a suspected human carcinogen.

Table B.10: Emissions of volatile organic compounds in 1998

	Thousand tonnes	
By fuel		
Coal	28	(1%)
Solid smokeless fuel	3	(-)
Petroleum:		
Motor spirit	361	(18%)
DERV	62	(3%)
Gas oil	15	(1%)
Fuel oil	1	(-)
Burning oil	-	(-)
Other petroleum	4	(-)
Natural gas	18	(1%)
Other emissions *(1)*	1,466	(69%)
Total	**1,958**	**(100%)**
Collected under UNECE definitions		
Power stations	6	(-)
Domestic	37	(2%)
Commercial and public service	4	(-)
Industrial combustion	8	(1%)
Non-combustion processes	289	(15%)
Extraction and distribution of fossil fuels	298	(15%)
Solvent use	532	(27%)
Road transport	527	(27%)
Other transport	48	(2%)
Waste treatment and disposal	29	(1%)
Forests	178	(9%)
Total	**1,958**	**(100%)**

(1) Includes industrial processes and solvents, petrol evaporation, offshore oil and gas activities, gas leakage, waste disposal and forests.

B.27 The largest sources of emissions on a UNECE basis are solvent use (27 per cent) and road transport (27 per cent). Non-combustion processes in the industrial sector (15 per cent) and extraction and distribution of fossil fuels (15 per cent) are both substantial non-combustion sources contributing to the total. Figures for 1998 show that emissions of volatile organic compounds were 25 per cent lower than in 1990. There were 2 million tonnes of volatile organic compounds emitted in 1998. Table B.10 shows emissions of non-methane volatile organic compounds in 1998 by fuel and collected under UNECE definitions.

B.28 Figures for 1998 show that emissions of volatile organic compounds were 14 per cent lower than in 1970. Through the 1970s and 1980s declining emissions from the domestic sector were more than offset by increases from road transport and offshore oil and gas activity. However, since 1989 emissions from road transport have been falling, resulting in an overall drop in emissions of volatile organic compounds in every year since.

Carbon monoxide

B.29 Carbon monoxide is derived from the incomplete combustion of fuel. On a UNECE basis, 4.8 million tonnes of carbon monoxide were emitted, a level 31 per cent lower than 1990 and 43 per cent lower than in 1970. Trends in carbon monoxide emissions are shown in Chart B.11. Table B.11 gives data for carbon monoxide emissions collected under UNECE definitions.

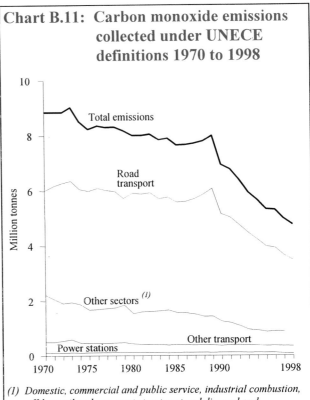

Chart B.11: Carbon monoxide emissions collected under UNECE definitions 1970 to 1998

(1) Domestic, commercial and public service, industrial combustion, offshore oil and gas, waste treatment and disposal and agriculture.

Table B.11: Carbon monoxide emissions in 1998

By fuel	Thousand tonnes	
Coal	297	(6%)
Solid smokeless fuel	27	(1%)
Petroleum:		
Motor spirit	3,630	(76%)
DERV	132	(3%)
Gas oil	41	(1%)
Fuel oil	4	(-)
Burning oil	1	(-)
Other petroleum	21	(-)
Natural gas	55	(1%)
Other emissions	551	(12%)
Total	**4,758**	**(100%)**

Collected under UNECE definitions		
Power stations	73	(2%)
Domestic	234	(5%)
Commercial and public service	18	(-)
Industrial combustion	594	(12%)
Extraction and distribution of fossil fuels	3	(-)
Road transport	3,479	(73%)
Other transport(1)	338	(7%)
Waste treatment and disposal	18	(-)
Agriculture	-	(-)
Total	**4,758**	**(100%)**

(1) Mainly off-road sources.

Chart B.12: Road transport emissions of carbon monoxide 1970 to 1998

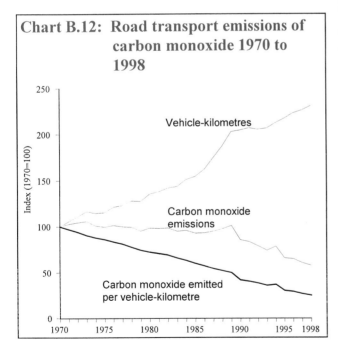

B.30 In 1998 road transport accounted for 73 per cent of carbon monoxide emissions. Emissions from road transport, shown in Chart B.12, fell whilst the number of vehicle-kilometres travelled increased. Hence the amount of carbon monoxide emitted per vehicle-kilometre shows a steady decline from 1970. In 1998, it was 74 per cent lower than in 1970. This decline is the result of the introduction of tighter emissions standards and the increasing penetration of diesel cars which emit less carbon monoxide than non-catalyst equipped petrol-fuelled cars. Also, real increases in road fuel duty will have helped restrain the demand for motor fuels, which in turn contributes to lower emissions of carbon monoxide from road transport. As with nitrogen oxides, the improvement in recent years should continue following the fitting of catalytic converters to all new petrol cars from 1993.

Lead and unleaded petrol

B.31 The main sources of lead in air are from lead in petrol, coal combustion and metal works. It is estimated that approximately three quarters of the lead in the air comes from petrol. Throughout the 1980s a number of steps have been taken to reduce lead emissions from petrol.

Chart B.13: Lead emissions from petrol engined vehicles and petrol consumption

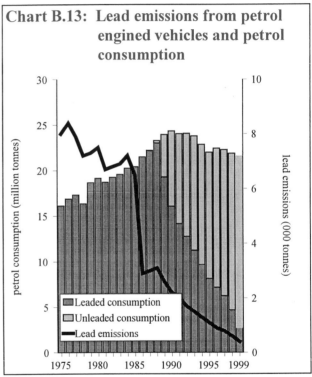

B.32 Emissions of lead from petrol-engined road vehicles fell after the reductions in 1981 and December 1985 in the maximum amount of lead permitted in petrol. A further reduction has occurred since the introduction of unleaded petrol and its increased uptake since 1988.

B.33 Table B.12 shows how emissions of lead have fallen by 95 per cent since 1975, despite the 33 per cent increase in petrol consumption.

B.34 Chart B.13 shows the increase in the percentage of unleaded in total deliveries of motor spirit, which reached 79 per cent in 1999. Super premium unleaded was introduced in 1990, and accounted for 11 per cent of unleaded deliveries in that year. In 1992 it accounted for nearly 13 per cent but fell sharply towards the end of 1994 and in 1998 accounted for 2 per cent of unleaded deliveries.

Oil pollution and oil spills

Offshore spills

B.35 Table B.13 shows the number of oil spills reported around the coasts of the United Kingdom and offshore (North Sea). Also shown is the amount of oil discharged in reported spills from offshore installations along with oil discharged with produced water and on drill cuttings.

B.36 The amounts are small (maximum about 0.02 per cent and 0.004 per cent in 1999) in relation to total oil production, with the amounts discharged on drill cuttings, and with produced water generally much larger than from offshore installation spills. The figures shown for the total number of spills reported are taken from the Advisory Committee on Protection of the Sea Annual Surveys of Oil Pollution around the Coasts of the United Kingdom.

B.37 Since 1986 the United Kingdom has carried out surveillance flights over offshore installations to check for oil spills, with an aircraft fitted with infra-red and ultra-violet detectors and side-looking airborne radar. In 1999, 300 hours were spent on 54 'dedicated' oil rig patrols (funded solely by DTI). The total amount of oil observed was just over 2 tonnes from 40 separate detections of spills, proving that in the main, Operators are quick to report spillages to the Department of Trade and Industry, re-enforcing the accuracy of these reports.

Table B.12: Consumption of petrol (inland deliveries of motor spirit) and estimated emissions of lead from petrol-engined road vehicles

	Consumption of petrol (million tonnes)			Estimated emissions of lead from petrol-engined road vehicles [1]	
	Total	Of which Leaded	Index (leaded) 1975 = 100	Thousand tonnes	Index 1975 = 100
1975	16.12	16.12	100	8.0	100
1976	16.88	16.88	105	8.4	105
1977	17.34	17.34	108	7.9	98
1978	18.35	18.35	114	7.2	90
1979	18.68	18.68	116	7.3	92
1980	19.15	19.15	119	7.5	94
1981	18.72	18.72	116	6.7	83
1982	19.25	19.25	119	6.8	85
1983	19.57	19.57	121	6.9	86
1984	20.23	20.23	125	7.2	90
1985	20.40	20.40	127	6.5	81
1986	21.47	21.47	133	2.9	36
1987	22.19	22.17	138	3.0	37
1988	23.25	22.99	143	3.1	39
1989	23.92	19.28	120	2.6	33
1990	24.31	16.06	100	2.2	27
1991	24.02	14.15	88	2.0	24
1992	24.04	12.77	79	1.7	22
1993	23.77	11.27	70	1.5	19
1994	22.84	9.68	60	1.3	16
1995	21.95	8.12	50	1.1	14
1996	22.41	7.18r	45	0.9	11
1997	22.25	6.25	39	0.8	10
1998	21.85	4.69	29	0.6	7
1999p	21.51	2.76	17	0.4	5

(1) These figures are based on lead contents for petrol published by the Institute of Petroleum. It has been assumed that only 70 per cent of this lead in petrol is emitted from vehicle exhausts, the remainder being retained in lubricating oil and exhaust systems.

Table B.13: Oil spills reported and discharge from offshore installations

	Number of spills reported	Of which Offshore (N. Sea)	Oil from offshore operations		
			Spills (Tonnes)	With produced water (Tonnes)	On drill cuttings (Tonnes)
1982	517	42	162	927	8,600
1983	396	62	186	1,700	14,500
1984	367	47	130	1,430	19,800
1985	366	87	310	2,150	20,200
1986	436	166	3,540	2,710	13,000
1987	500	254	516	3,330	12,400
1988	559	259	2,627	3,234	18,500
1989	764	291	517	3,423	13,400
1990	791	345	899	4,393	12,310
1991	705	234	192	5,490	11,230
1992	611	194	225	4,850	7,169
1993	676	183	224	4,232	4,588
1994	540	147	174	4,418	3,820
1995	585r	145	84	5,855	3,180
1996	678	300 *(1)*	127	5,706	3,826
1997	..	349 *(1)*	866	5,767	- *(2)*
1998	..	392 *(1)*	137	5,692	- *(2)*
1999	..	372 *(1)*	120	5,641	- *(2)*

(1) This figure now includes observations from the aerial surveillance programme.
(2) Since 1 January 1997, the UK has not discharged mineral oil contaminated cuttings into the sea in accordance with international obligations.

B.38 The average oil content of discharged water in 1991 was 0.0035 per cent, and in 1992 was 0.0036 per cent (4,850 tonnes of oil in 135 million tonnes of water in 1992). This was higher than in 1990 (0.0027 per cent) and earlier years when the average remained reasonably constant despite the increase in the amount of water being produced. The higher averages in 1991 and 1992 were due to problems with treating equipment on several installations. The ratios have improved as operators have replaced older equipment with more modern treatment facilities; the average oil content in 1999 was 0.00215 per cent, a slight reduction on the 1998 figure and the lowest level yet recorded.

B.39 Despite improved methods of water treatment, oil discharged with produced water has increased during the 1990s due to renewed growth in oil production and the increase in water produced, because of the increasing age of the North Sea reservoirs. The industry recognises that this will remain a problem in the future, and is investigating methods by which it can be alleviated. Techniques under investigation include re-injection of produced water and water shut-off technology. Despite this longer-term prognosis, there was a decrease of 51 tonnes in the total quantity of oil discharged with produced water in 1999 compared with 1998.

Other oil spills

B.40 Excluding offshore spills (North Sea), there has been an increase in the number of spills recorded, from 279 in 1985 to 378 in 1996. The marked increase in recent years in the total number of oil spills is thought to be partly due to more comprehensive reporting using the improved surveillance techniques referred to earlier. The above figures exclude the oil pollution caused by discharges from tankers outside United Kingdom territorial waters, and inland oil pollution.

B.41 In recent years there have been two major oil spills as a result of shipping accidents in UK waters; the Braer in January 1993 (84,000 tonnes) and the Sea Empress in February 1996 (about 70,000 tonnes). Most oil spills, however, are small and the oil is usually dispersed and degraded naturally.

Gas flaring in the UKCS

B.42 Under the terms of petroleum production licences, gas may be flared only with the consent of the Secretary of State. Flaring at onshore fields in 1999 was minimal, whilst 5.76 million cubic metres of gas a day was flared at offshore installations. Chart B.14 shows the amounts of oil produced and gas flared at offshore oil fields over the period 1988 to 1999.

B.43 Flaring at offshore installations in 1999 was 4 per cent higher than in 1998. The 1994 figure was abnormally high because of an increase in offshore loaders, oil production and associated gas flaring, which resulted in higher levels of gas flaring than during normal operations. The *Development of the oil and gas resources of the United Kingdom 2000* publication (the '*Brown Book*') gives more detail.

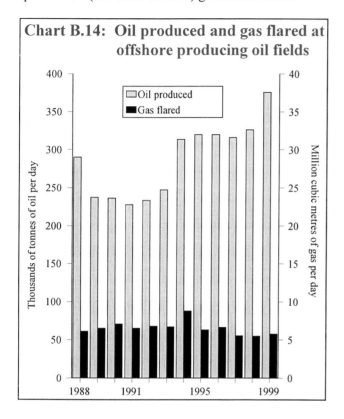

Chart B.14: Oil produced and gas flared at offshore producing oil fields

Contact: *Chris Bryant (Statistician, Emissions)*
              020 7215 0124
              Kevin Williamson (Statistician, Oil pollution and gas flaring)
              020 7215 5184

Annex C
United Kingdom oil and gas resources

Introduction

C.1 This section provides background information on the United Kingdom's resources of crude oil and gas: reserves, production (including a split by field), disposal and operations. This information is intended as a supplement to that in the commodity balances included in the main part of this chapter. Most of the data (including gas) are obtained from the Department of Trade and Industry's Petroleum Production Reporting System. Further information can be obtained from the DTI publication *Development of the Oil and Gas Resources of the United Kingdom 2000*, known as the *Brown Book*.

C.2 The annual statistics relate to the calendar years, or the end of calendar years, 1995 to 1999 and the data cover the United Kingdom Continental Shelf (onshore and offshore). Annual data for production, imports and exports of crude oil during the period 1960 to 1999 are given in Table 3.11.

Oil and gas reserves (Table C.1)

C.3 This table shows estimated reserves of oil and gas at the end of 1999. Estimates of initial recoverable reserves in present discoveries are given and, also, ranges of recoverable oil reserves (discovered and undiscovered) originally in place. The table also includes the range of potential additional reserves existing in discoveries that do not meet the criteria for inclusion as possible reserves defined below.

C.4 The terms **proven, probable** and **possible** are applied on a field by field basis and are given the following meanings in this context:

(i) **Proven** - those reserves which on the available evidence are virtually certain to be technically and economically producible (i.e. those reserves which have a better than 90 per cent chance of being produced).

(ii) **Probable** - those reserves which are not yet **proven** but which are estimated to have a better than 50 per cent chance of being technically and economically producible.

(iii) **Possible** - those reserves which at present cannot be regarded as **probable** but are estimated to have a significant but less than 50 per cent chance of being technically and economically producible.

C.5 Chart C.1 shows the changes in discovered recoverable oil and gas reserves and cumulative production since 1980, and shows that over this period the reserves (proven, probable and possible) have increased roughly in line with cumulative production, so that remaining reserves are broadly unchanged.

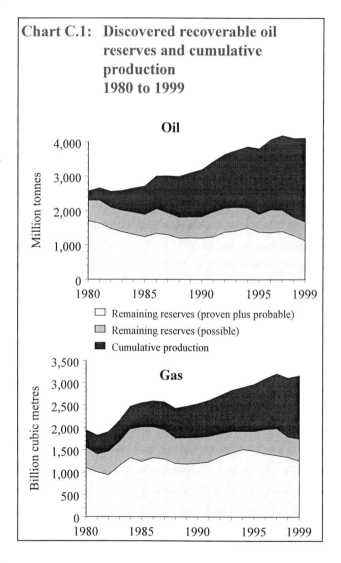

Chart C.1: Discovered recoverable oil reserves and cumulative production 1980 to 1999

□ Remaining reserves (proven plus probable)
▒ Remaining reserves (possible)
■ Cumulative production

C.6 Estimates for the size of reserves of oil and gas underneath the United Kingdom Continental Shelf are revised on an annual basis. This allows new discoveries of oil and gas reserves and reassessments of existing discoveries to be incorporated. These reassessments are needed so that overall estimates of reserves can take into account new information on existing fields as well as changes in the overall economic climate. Only if it is economically viable to exploit the discovery will oil and gas actually be produced from any discovered reservoir.

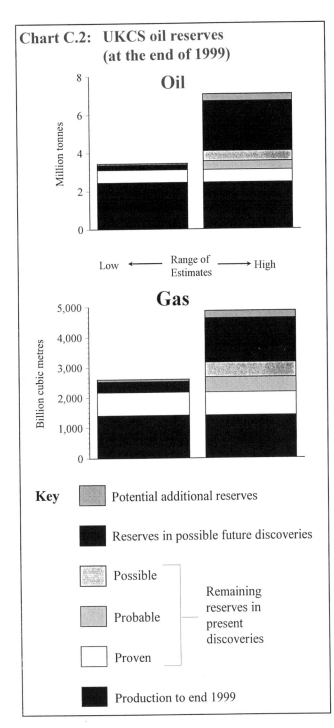

Chart C.2: UKCS oil reserves (at the end of 1999)

Oil

Low ←——— Range of Estimates ———→ High

Gas

Key

- Potential additional reserves
- Reserves in possible future discoveries
- Possible ⎱
- Probable ⎰ Remaining reserves in present discoveries
- Proven
- Production to end 1999

and possible reserves figure at the end of 1998. Actual production of 137 million tonnes in 1999 was very slightly offset by an upward revision of 5 million tonnes to the estimates of reserves in present discoveries. Remaining gas reserves on the same basis were 1,750 billion cubic metres at the end of 1999, 45 billion cubic metres lower than at the end of 1998. Production of gas was 105 billion cubic metres in 1999, accompanied by an upward revision of 55 billion cubic metres to the size of gas reserves in current discoveries.

C.9   These small revisions to the estimates of reserves are due to limited replacement of reserves from exploration successes or by fields previously held as only having potential additional reserves. This lack of replacement discoveries is due to the impact of low oil prices during 1998 and the first half of 1999 affecting the economics of field exploitation and also affecting company decisions with regards to exploration and development activities.

C.10   Table C.I contains simplified data on cumulative production of oil and gas compared to the latest estimates for the ranges of UK reserves of oil and gas, based on data for reserves in existing discoveries and adding in estimates for reserves yet to be discovered. The ranges of estimates for oil and gas reserves are illustrated in Chart C.2.

Table C.I: UK reserves of oil and gas

OIL	Million tonnes	Production as percentage of original reserves
Cumulative production to end 1999	2,444	
Original size of reserves in present discoveries:		
Lower	3,110	79%
Upper	4,110	59%
Total original size of reserves (present plus future discoveries):		
Lower	3,440	71%
Upper	7,080	35%

GAS	Billion cubic metres	Production as percentage of original reserves
Cumulative production to end 1999	1,410	
Original size of reserves in present discoveries:		
Lower	2,170	65%
Upper	3,160	45%
Total original size of reserves (present plus future discoveries):		
Lower	2,600	54%
Upper	4,870	29%

C.7   In recent years, advances in the technology used in the production of oil and gas, and economic situations, have been such that discoveries that were made sometimes several years ago have only recently become economically viable. As such, much of the increase in the estimates of the United Kingdom's reserves of oil and gas over recent years is made up of reassessments of known deposits of oil and gas rather than the discovery of new deposits. This is one of the reasons why the average size of oil and gas fields that have come into production in recent years is much smaller than during the 1970s and 1980s (see paragraph C.15 and Chart C.4).

C.8   At the end of 1999 remaining proven, probable and possible oil reserves stood at 1,670 million tonnes. This is a decrease of 130 million tonnes on the equivalent remaining proven, probable

C.11   Looking at the upper end of the ranges of estimates of the UK's recoverable reserves of oil, cumulative production to the end of 1999 is estimated to be nearly three-fifths (59 per cent) of the total expected (proven, probable and possible) from the recoverable reserves present in current discoveries. Taking the upper end of the ranges of total UK estimated recoverable reserves of oil (i.e. reserves in current discoveries plus the additional reserves anticipated to exist in future discoveries), a further 3 billion tonnes of oil may be found in the future compared with the 4 billion tonnes of reserves related to present discoveries (i.e. 3 billion tonnes on top of the total of production and reserves in Chart C.1). Taking the most pessimistic view of estimates of reserves of oil, the UK has already extracted over two-thirds (70 per cent) of the total reserves present, whereas if the upper end of total reserves is used the UK has extracted just over one-third (35 per cent) of the total recoverable reserves.

C.12   Similarly, looking at the upper end of the ranges of estimates of the UK's recoverable reserves of gas, cumulative production to the end of 1999 is estimated to be nearly two-thirds (65 per cent) of the recoverable reserves present in current discoveries. Taking the upper end of the ranges of total UK estimated recoverable reserves of gas, a further 1.7 billion cubic metres of gas may be found in the future compared with the 3.2 billion cubic metres of reserves related to present discoveries. Taking the most pessimistic view of estimates of reserves of gas, the UK has already extracted just over half (54 per cent) of the total reserves present, whereas if the upper end of total reserves is used the UK has extracted slightly under one-third (29 per cent) of the total recoverable reserves.

Offshore oil and gas fields and associated facilities (Table C.2)
C.13   Table C.2 (and Table C.9) provides operational details for the past five years. Table C.2 shows numbers of offshore oil and gas fields and of associated production facilities, including miles of operational pipelines, at the end of each year up to 1999.

C.14   This table also shows that the number of offshore oil fields in production or under development rose from 103 at the end of 1995 to 138 at the end of 1999. For offshore gas fields the equivalent increase has been from 71 to 90. These significant increases are shown in Chart C.3 (offshore fields in production).

C.15   The average size of fields commencing production has fallen during the 1990s (see Chart C.4). This reflects a decline in the size of fields discovered compared with the early development of the North Sea, and improved technology providing cost-effective means of extracting oil and gas from smaller fields and hitherto unpromising locations. The Cost Reduction in the New Era (CRINE) initiative has helped reduce industry production costs. Industry co-operation can allow the joint development of smaller fields that individually would not otherwise be economic.

C.16   Table C.2 also gives the numbers of production facilities installed, and length of pipelines operational. The length of offshore oil pipelines operational has increased by 55 per cent between 1995 and 1999, while the length of offshore gas pipelines has increased by 23 per cent in the period. For further information on the length of pipelines awaiting commissioning, or under construction, refer to *The Development of Oil and Gas Resources of the United Kingdom 2000.*

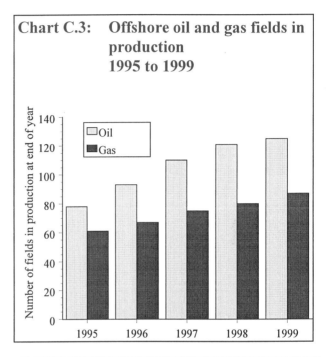

Chart C.3:   Offshore oil and gas fields in production 1995 to 1999

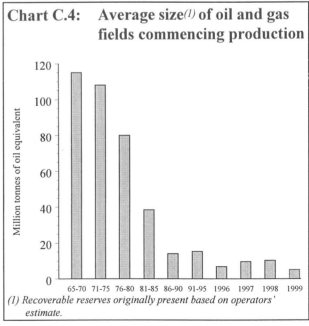

Chart C.4:   Average size[(1)] of oil and gas fields commencing production

(1) Recoverable reserves originally present based on operators' estimate.

Production (Table C.3, C.6, C.7 and C.8)
C.17   These tables show production of crude oil and methane (effectively, natural gas) onshore and offshore. Table C.3 shows totals for production of oil and gas, including condensates from gas and oil fields, and production of ethane, propane and butane. Table C.6 gives production of crude oil broken down by field and Table C.7 shows methane production by field. In

addition both tables show cumulative production to the end of 1999. Table C.8 contains information for some of the main gas producing systems in operation at the moment. Previously, details on production of the individual fields and areas that make up these systems were unavailable. However, recent developments have allowed estimates for individual field production to be produced covering the period 1995 to 1999.

Production of oil and gas (Table C.3)

C.18   Table C.3 shows gross production of crude oil and natural gas (well extraction less gas flared, vented or re-injected) from oil fields. Methane includes associated gas from oil fields, measured after initial separation of the gas from the oil on the fields. The gas is then further processed at onshore gas separation plants to yield methane [C_1], ethane [C_2], propane [C_3], butane [C_4], and condensates [C_{5+}]. Production of methane [C_1], and condensate [C_{5+}] from gas fields is measured at the field.

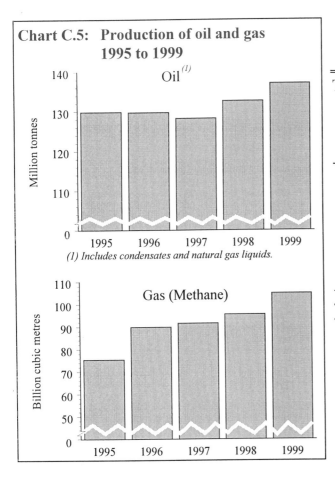

Chart C.5: Production of oil and gas 1995 to 1999

Oil [1]

(1) Includes condensates and natural gas liquids.

Gas (Methane)

C.19   The table also shows that while landward production of crude oil in 1999 was 15 per cent lower than in 1995, it still accounted for 3½ per cent of UK crude oil production in 1999.

C.20   Chart C.5 shows the recent growth of total oil production from 130 million tonnes in 1995 to 137 million tonnes in 1999, and the growth in methane production from 75 billion cubic metres in 1995 to 105 billion cubic metres in 1999.

Production of crude oil by field (Table C.6)

C.21   The figures show gross production (defined above). The first production of offshore oil from the UK sector of the Continental Shelf came in 1975, from the Argyll, Auk and Forties fields.

C.22   With the exceptions of Donan, Emerald and Medwin every field producing oil in 1995 was still producing in 1999. Production from the Cyrus field was suspended in 1992 and recommenced in 1996. 61 offshore fields have started production after 1994 (including the re-start of Cyrus). Production from those fields that have been established for some time has been dropping in recent years. This has been more than offset by the new fields that have commenced production. In 1999, the 61 fields that have started production since 1994 accounted for 44 per cent of total crude oil production. Of the top ten producing fields in 1999 that together accounted for 32 per cent of total crude oil production, four of these fields started production after 1994. A list of these top ten oil fields is given in Table C.II.

Table C.II: Top ten UK crude oil producing fields in 1999

Field	1999 production (million tonnes)	Production as percentage of annual total
Schiehallion	5.11	4.0
Brent	4.53	3.5
Nelson	4.51	3.5
Foinaven	4.33	3.4
Harding	4.28	3.3
Scott	4.01	3.1
Alba	3.99	3.1
Wytch Farm (onshore)	3.86	3.0
Forties	3.49	2.7
Andrew	3.29	2.6
Total top ten fields	41.40	32.3
Total crude oil production	128.26	

Production of methane by field (Table C.7)

C.23   The figures represent gross production from each field and include methane used for drilling, production and pumping operations, but excluding gas flared, vented or re-injected. The first production of offshore methane (natural gas) from the UK sector of the Continental Shelf came in 1967, from the West Sole field.

C.24   The table shows that methane production in 1999 was 39 per cent higher than in 1995, with production starting at 9 new gas fields during 1999. Associated gas was also produced for the first time from 9 oil and condensate producing fields during the year. Since 1995 the amount of associated gas produced from oil fields has increased, for example, with the expansions of the CATS, FLAGS and SAGE systems that have occurred. As mentioned above for older oil fields, the older gas fields that were discovered in the Southern North Sea have reduced the level of their production in recent years as the reserves originally present in the fields become depleted, for example, the Thames complex and Leman fields. Four gas fields ceased production during the period in question, Esmond and Gordon in 1995, Camelot North-East in June 1998 and Cleeton in February 1999. Production from Wensum was suspended in 1997 but restarted in 1999.

Production of methane by field for key production systems (Table C.8)

C.25   As stated above, prior to the last edition of this Digest, only aggregate production data was available for some gas production systems such as CATS, FLAGS and SAGE, although the systems themselves are made up of a number of individual producing fields. Recent developments have allowed the disaggregation of the system totals down to the individual fields involved, in some cases, or at least into smaller groups of fields. It has not been possible to carry this exercise back prior to 1995 for all systems, hence the cumulative data in the table does not include field detail for the FLAGS or Fulmar pipeline systems.

C.26   The availability of this data has allowed a more accurate assessment of the top producing gas fields to be made, given at Table C.III. Together these ten fields/systems accounted for 48 per cent of total production in 1999. Britannia is a new field that only started production in 1998, but already accounts for 7 per cent of UK gas production.

Table C.III: Top ten offshore UK gas producing fields in 1999

Field	1999 net production (billion cubic metres)	Production as percentage of annual total
Morecambe South	9.82	10.0
Britannia	6.86	7.0
Brent	6.15	6.2
Bruce	5.04	5.1
Brae Area	4.66	4.7
Armada	4.24	4.3
Beryl Area	3.09	3.1
Leman	2.85	2.9
ETAP	2.61	2.6
Viking	2.44	2.5
Total top ten fields	47.75	48.4
Total net gas production	98.71	

Transportation of crude oil production (Table C.4)

C.27   This table shows the mode of transportation of crude oil production, with separate figures for each main pipeline system and cumulative figures to the end of 1999.

C.28   The figures are compiled from the monthly returns sent to the Department of Trade and Industry by operators of oil fields and onshore terminals under the Petroleum Production Reporting System. Crude oil includes condensate and residual dissolved gases present in the disposals of crude oil by the industry. Heavier natural gases (ethane, propane and butane) and condensates produced in the treatment of liquid or gaseous hydrocarbons at the terminals are excluded. All pipeline systems terminate at one or other of the onshore terminals as indicated in the footnotes to the table. Crude oil is loaded directly offshore into adapted tankers by means of special facilities installed on certain fields. Until 1990 Brent and Thistle were served both by offshore loading and pipeline systems. In 1997, offshore loading from the Fulmar system ceased following the systems' connection to Teeside via the Norwegian pipeline from the Ekofisk system (Norpipe).

C.29   Between 1995 and 1999 the amount of oil loaded offshore rose by 52 per cent compared with a 6 per cent decrease in transportation by offshore pipeline systems. Transportation of crude oil produced from Joanna and Judy to Teeside via Norpipe commenced in 1995.

Disposals of crude oil (Table C.5)

C.30   Table C.5 and Chart C.6 show how crude oil is disposed of, split between amounts to UK refineries and exports (see technical notes, paragraphs C.44 to C.46) by country of destination (from which it may be transhipped elsewhere). The figures are obtained from returns made to the Department of Trade and Industry by operators of oil fields and onshore terminals under the Petroleum Production Reporting System (see paragraphs C.39 to C.41).

C.31   The exports figures in this table may differ from those compiled by the United Kingdom Petroleum Industry Association (UKPIA) and published in the main text of this chapter. UKPIA figures also include re-exports. These are products that might originally have been imported into the UK and stored before being exported back out of the UK, as opposed to actually having been produced in the UK.

C.32   The volume of exports of crude oil in 1999 was 2 per cent higher than in 1995, with disposals to UK refineries increasing by 8 per cent. Exports of crude oil in 1999 were 4 per cent higher than in 1998, partly due to the increased production in 1999 and also due to international markets achieving some degree of stability in oil prices after the major changes in 1998. During 1997 the price of crude oil fell by roughly 25 per cent. This was due to several factors involving increased supply (production by several of the larger members of OPEC exceeded their agreed quotas and oil started to flow from Iraq under the "Food for oil" arrangements) and reduced demand, particularly in Asia.

delays in the availability of oil from Iraq back onto international markets, price rises were seen during the latter part of 1998 and through 1999.

C.34   Chart C.7 illustrates this point by showing what proportion of the total amount of crude oil imported by the USA, France, the Netherlands and Germany is supplied by the UK. Whilst the USA represents a significant market for UK exports of crude oil, just over 4 per cent of total crude oil imports into the USA in 1999 were from the UK. The exports from the UK are a much more significant contributor to other European countries.

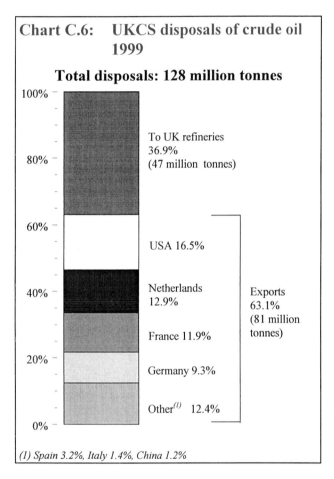

Chart C.6:   UKCS disposals of crude oil 1999

Total disposals: 128 million tonnes

To UK refineries 36.9% (47 million tonnes)

USA 16.5%

Netherlands 12.9%

France 11.9%

Germany 9.3%

Other[1] 12.4%

Exports 63.1% (81 million tonnes)

(1) Spain 3.2%, Italy 1.4%, China 1.2%

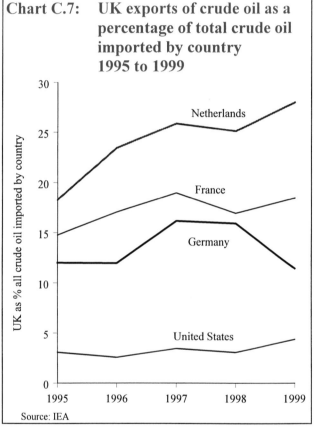

Chart C.7:   UK exports of crude oil as a percentage of total crude oil imported by country 1995 to 1999

UK as % all crude oil imported by country

Netherlands

France

Germany

United States

Source: IEA

C.33   The economic downturn in many of the countries of South East Asia in 1997 had a significant downward effect on global oil demand, since these countries had been exhibiting the strongest growth in demand in recent years. These factors led to a surplus of oil being available, thus reducing the price for crude oil on international markets. This had the effect of reducing the profit margins for exporting UK produced crude oil, given the extra transport costs involved in moving the oil from the UK to other areas in the world. Thus an increasing proportion of the crude oil extracted from the North Sea went for consumption in the UK, and the exports of the UK to more distant destinations such as the USA and Canada reduced while exports to European destinations increased. However, with the agreement of OPEC member countries to impose production cuts, and

Gas flaring at oil fields and terminals

C.35   Previous editions of this Digest have included tables showing the volumes of gas flared at each oil field and terminal. One of the changes that has been made to the format and content of the Digest is that this data no longer appears as a separate table. Detailed information by field can still be found in the *Development of the Oil and Gas Resources of the United Kingdom 1999,* known as the *Brown Book,* Appendix 9. Details are given in Table C.IV of the total amounts of gas flared in recent years.   The amounts of gas flared shown in this table are not included in Tables C.3, C.7 and C.8. Flaring at gas fields is minimal due to the fact that facilities exist for the gas to be put into the field transportation pipelines rather than being flared.

C.36   The amount of gas flared in 1999 was 2.2 billion cubic metres, 4 per cent higher than in 1998 and 8 per cent lower than in 1995.   This level of discharge is equivalent to 2 per cent of the production of gas from offshore gas fields.

Table C.IV: Gas flaring at oil fields and terminals, 1995 to 1999

			Million cubic metres
	Onshore	Offshore	Total
1995	95	2,293	2,388
1996	110	2,429	2,539
1997	100	2,022	2,122
1998	106	2,004	2,110
1999	99	2,102	2,201

Sales and expenditure by operators and other production licensees (Table C.9)

C.37   Table C.9 shows the value of sales, operating and exploration expenditure, gross trading profits and development expenditure, by operators and other production licensees.

C.38   Total receipts increased by 18 per cent in 1999, compared with 1998, with the recovery of oil prices from the low levels seen in 1998 increasing the value of sales.   While operating costs rose by 5 per cent, exploration expenditure fell by 40 per cent. As a result gross trading profits rose by 12 per cent in 1999, but were still below the levels seen in 1996 and 1997. Total development expenditure was down 38 per cent on the level in 1998, reflecting the decrease in drilling of development wells in 1999 (234 compared with 281 in 1998, and 244 in 1995).

Technical notes and definitions

Petroleum Products Reporting System

C.39   Licensees operating on the UK Continental Shelf are required to make monthly returns on their production of hydrocarbons to the Department of Trade and Industry. The DTI stores this information in the Petroleum Production Reporting System (PPRS). The PPRS is used to report flows, stocks and uses of hydrocarbon from the well-head through to final disposals from a pipeline or terminal and is the major source of the information presented in this chapter.

C.40   Returns are collected covering field, pipeline and terminal data compiled by relevant reporting units. Each type of return is provided by a single operator, but usually covers the production of a number of companies, since frequently operations carried out on the Continental Shelf involve several companies working together.

C.41   Every production system has one or more sets of certified meters to measure oil and gas or condensate production. The flows measured by the meters are used to check the consistency of returns, and are therefore used to assure the accuracy of the PPRS.

Sales and expenditure by operators and other production licensees

C.42   The data given in Table C.9 are compiled from the Quarterly Inquiry into Oil and Natural Gas carried out by the DTI. This inquiry collects information from operators and other production licence holders. The information collected covers all income and expenditure directly related to the production of oil and natural gas, including exploration, development and other capital expenditures together with operating costs and the value of sales. It covers all companies classified under Division 11 of the Standard Industrial Classification 1992, which includes production and exploration licence holders, together with service activities incidental to oil and gas extraction (other than surveying).

C.43   Information on the industry is published quarterly in *Energy Trends*, and in *The Development of Oil and Gas Resources of the United Kingdom 2000*. Annual information on the industry as a whole, which includes the expenditures and sales of non-exclusive exploration licence holders, drilling and other contractors, is available from the Office for National Statistics (ONS), published up to 1993 in the *Annual Census of Production* and in the *Annual Inquiries into Production* from 1994 onwards. For more details on the information available, contact the ONS library in Newport, Gwent on 01633-812399.

Exports

C.44   As stated in paragraph C.30 above, the term exports used in Table C.5 refers to figures recorded by producers of oil and gas for their exports. These figures may differ from the figures for exports compiled by HM Customs and Excise (HMCE) and given in Chapter 8. In addition HMCE now differentiate between EU and non-EU trade by using the term dispatches for trade going to other EU countries, with exports retained for trade going to non-EU countries. The differences can occur between results from the two sources of information because whilst the trader's figures are a record of actual shipments in the period, for non-EU trade HMCE figures show the trade as declared by exporters on documents received during the period stated.

C.45   In addition, trade in oil frequently involves a "string" of transactions, which can result in the actual destination of the exports changing several times even after the goods have been dispatched. As such, differences can arise between the final country of destination of the exports as recorded by the producers themselves and in the HMCE figures. The HMCE figures also include re-exports. These are products that might originally have been imported into the UK and stored before being exported back out of the UK, as opposed to actually having been produced in the UK.

C.46   In versions of the Digest before 1997, these exports were called "shipments" in an attempt to highlight their difference from the other sources of trade data. However, their name has now been changed to more clearly represent that the movements actually are movements out of the United Kingdom.

Units of measurement for gas

C.47   The basic unit of measurement for quantities of flows and stocks is volume in cubic metres at $15\,^\circ C$ temperature and 1.01325 bars pressure.

Monthly data

C.48   Monthly data on the production of crude oil and natural gas from the UKCS, along with details of imports and exports of oil, oil products and gas are published in Tables 11 and 14 of the monthly DTI publication *Energy Trends*. Quarterly data on the value of production and investment in UKCS oil and gas production are also published in Table 9 of *Energy Trends*, along with details on drilling activity in Table 8. See Annex F for more information on *Energy Trends*.

Contact: *Kevin Williamson (Statistician)*
020 7215 5184
Clive Evans
020 7215 5189
Ian Montague
020 7215 2711

C.1 Estimated oil and gas reserves on United Kingdom Continental Shelf [1]

(As at 31 December 1999)

Million tonnes

Oil reserves [2]

Initial recoverable oil reserves in present discoveries	Proven [3]	Probable [3]	Proven plus Probable	Possible [3]
Fields in production or under development	3,110	300	3,410	350
Other significant discoveries not yet fully appraised	-	155	155	190
Total initial reserves in present discoveries	**3,110**	**455**	**3,565**	545

Range of recoverable oil reserves (discovered and undiscovered) originally in place

	Range of oil reserves
Cumulative production to end of 1999	2,444
Remaining reserves in present discoveries	690-2,015
Reserves in potential future discoveries	
North and Central North Sea 56°N - 62°N	210-1,170
West of Shetlands	35-730
West of Scotland	0-520
Remainder of UK continental shelf	5-180

Total recoverable oil reserves originally in place on the UK continental shelf [4] — **3,385-7,060**

Potential additional reserves [5] — 85-370

Billion cubic metres

Gas reserves [6]

Initial recoverable gas reserves in present discoveries	Proven [3]	Probable [3]	Proven plus Probable	Possible [3]
Fields in production or under development				
Dry gas fields	1,480	105	1,585	105
Associated gas from oil fields	330	50	380	65
Gas from condensate fields	360	135	495	100
Other significant discoveries not yet fully appraised				
Dry gas fields	-	105	105	80
Associated gas from oil fields	-	10	10	25
Gas from condensate fields	-	100	100	115
Total initial reserves in present discoveries	**2,170**	**500**	**2,670**	**490**

Range of recoverable gas reserves (discovered and undiscovered) originally in place

	Range of gas reserves
Cumulative production to end of 1998 [7]	
Dry gas fields	1,108
Associated gas from oil fields	302
Remaining reserves in present discoveries	
Dry gas fields	370-765
Associated gas from gas condensate fields and oil fields	390-990
Gas reserves in potential future discoveries [8]	355-1,465

Total recoverable gas reserves originally in place on the UK continental shelf [4] — **2,525-4,630**

Potential additional reserves [5] — 75-245

(1) Includes onshore and offshore.

(2) With the exception of the production figures, entries are rounded to the nearest 5 million tonnes, and include gas liqiuds and liquefied products.

(3) The terms "proven", "probable" and "possible" are applied on a field by field basis and are given the internationally accepted meanings in this context. See Annex C, paragraph C.4.

(4) Rounded aggregate of the component ranges; the likelihood of the true figure lying outside this range is much smaller than for the individual component ranges.

(5) The potential additional reserves exist in discoveries which do not meet the criteria for inclusion as possible reserves.

(6) With the exception of the production figures, entries are rounded to the nearest 5 billion cubic metres.

(7) Excludes gas flared and gas used on platforms.

(8) Certain areas including the West of Scotland not assessed.

C.2 Offshore oil and gas fields and associated facilities

At end of year

	Unit	1995	1996	1997	1998	1999
Offshore oil fields						
Fields in production *(1)*	Number	78	93	110	121	125
Fields under development	"	25	21	19	17	13
Production facilities (installed):						
fixed production platforms *(2)*	"	57	68	68	70	70
floating production platforms	"	8	12	10	15	17
offshore loading systems *(2)*	"	15	15	15	15	15
flare towers	"	3	3	3	3	3
oil pipelines (operational) *(3)*	Kilometres	2,263	2,773	2,863	3,183	3,515
condensate pipelines (operational)	"	146	184	239	306	306
associated gas pipelines (operational)	"	3,213	3,290	3,299	3,343	3,406
onshore terminals oil (operational) *(3)*	Number	5	5	5	5	5
onshore terminals associated gas (operational)	"	4	4	4	4	4
Offshore gas fields						
Fields in production *(4)*	Number	61	67	75	80	87
Fields under development	"	10	5	5	8	3
Production facilities (installed):						
fixed production platforms *(5)*	"	146	149	152	154	158
flare towers *(5)*	"	1	1	1	1	1
dry gas pipelines (operational)*(6)*	Kilometres	2,176	2,275	2,320	2,603	3,211
onshore terminals (operational)	Number	9	9	9	9	9

(1) Includes Statfjord, Murchison fields and condensate fields.
(2) Fields connected to an offshore loading system. Excludes Statfjord 'A' 'B' and 'C' platforms and their offshore loading systems and the 2 pumping platforms on the Norwegian Ekofisk - Teeside oil pipeline.
(3) Excludes the 350 kilometre Ekofisk - Teeside oil pipeline and the 39 kilometre Heimdal-Brae Condensate pipeline.
(4) Includes Frigg field and Rough field which was granted approval in principle in 1982 for use as gas storage.
(5) Includes the 4 platforms and flare tower in the UK sector of Frigg field but excludes the 2 platforms in the Norwegian sector, platforms are counted if landed or operational.
(6) Includes the UK No.1 pipeline Frigg field to St. Fergus but excludes the parallel 350 kilometre Norwegian No.2 pipeline.

C.3 Production of oil and gas[(1)]

		Unit	1995	1996	1997	1998	1999
Crude oil	Land	Thousand tonnes	5,051	5,251	4,981	5,161	4,285
	Offshore	"	116,743	116,679	115,340	119,061	123,977
Condensates:	Land	"	-	-	-	-	-
	Offshore gas fields *(2)*	"	640	637	620	587	583
	Offshore oil fields *(3)*	"	891	913	925	1,147	1,374
Ethane *(3) (4)*		"	1,887	1,814	1,752	1,646	2,022
Propane *(3)*	Land	"	114	107	116	124	104
	Offshore	"	2,818	2,619	2,724	2,908	2,749
Butane *(3)*	Land	"	106	99	108	115	96
	Offshore	"	1,645	1,621	1,667	1,885	1,909
Total oil production		"	**129,894**	**129,742**	**128,234**	**132,633**	**137,099**
Methane *(5)*	Land *(6)*	Million cubic metres	322	382	388	335	310
	Offshore *(7)*	"	75,158	89,514	91,170	95,171r	104,584

(1) Monthly data on overall production of methane and crude oil are published in Tables 11 and 13 of Energy Trends - see Annex F.
(2) Includes condensate from the United Kingdom share of the Frigg field.
(3) Gaseous hydrocarbons associated with crude oil production.
(4) Excludes that part of the ethane from the Far- north Liquids and Associated Gas System (FLAGS) which is supplied to BG Trading Ltd as part of a predominantly methane mixture.
(5) Excludes gas flared, vented or re-injected.
(6) Excludes colliery methane.
(7) Includes the UK share of the Frigg field associated gas (notionally methane) produced and used on oil production platforms, and associated gas (notionally methane) delivered to St Fergus via the CATS, FLAGS, Frigg, Fulmar and SAGE pipeline systems. May include small quantities of other gases.

C.4 Transportation of crude oil production

Thousand tonnes

		1995	1996	1997	1998	1999	Total to end 1999
Offshore production:	Offshore loaded *(1)*	25,236	27,050	29,374	33,748	38,400	451,636
Pipeline systems:	Brent system *(2)*	23,300	23,094	19,575	17,394	16,240	568,264
	Ninian system *(3)*	14,043	14,026	11,900	11,308	10,321	308,823
	Flotta system *(4)*	12,064	10,653	9,575	10,061	9,559	277,997
	Forties system *(5)*	41,466	40,229	37,932	38,566	41,444	631,905
	Teeside system *(6)*	65	1,088	6,435	7,619	7,819	23,572
	Beatrice field	472	437	449	365	194	20,388
Total pipelines		91,410	89,527	85,866	85,313r	85,577	1,830,949
Onshore production:	Rail, road and pipeline *(7)*	5,051	5,251	4,981	5,161	4,285	47,436

(1) Production from Alba, Angus, Argyll, Auk (to 1996), Banff, Beryl, Bladon, Blenheim, Buckland, Captain, Clyde (to 1995), Crawford, Curlew, Cyrus (to 1993), Dauntless, Donan, Douglas, Duncan, Durward, Emerald, Fergus, Fife, Flora, Foinaven, Fulmar (to 1996), Gannet A, B, C, D (to 1996), Gryphon, Guillemot A, Harding, Hudson (to 1994), Innes, Kittiwake, Lennox, Leven (to 1996), Linnhe, Machar (to 1996), Mallard, Maureen, Medwin, Moira, Ness, Nevis, Pierce, Ross, Schiehallion, Statfjord (UK), Teal and Teal South.

(2) Brent, North Cormorant, South Cormorant, Deveron, Don, Dunlin, Dunlin SW, Eider, Hudson, Hutton, Hutton NW, Merlin, Murchison(UK), Osprey, Pelican, Tern and Thistle.

(3) Alwyn North, Columba B, D, E, Dunbar, Ellon, Grant, Heather,

Lyell, Magnus, Magnus South, Ninian, Staffa and Strathspey.

(4) Chanter, Claymore, Galley, Hamish, Highlander, Iona, Ivanhoe, MacCulloch, Petronella, Piper, Renee, Rob Roy, Rubie, Saltire, Scapa, and Tartan.

(5) Andrew, Arbroath, Arkwright, Balmoral, Beinn, Birch, Blair, Brae C, E, N, S, W, Brimmond, Britannia, Bruce, Buchan, Cyrus, Drake, Egret, Erskine, Everest, Fleming, Forties, Glamis, Heron, Kingfisher, Larch, Lomond, Machar, Marnock, Miller, Monan, Montrose, Mungo, Nelson, Scott, Sedgwick, Stirling, Telford, Thelma, Tiffany and Toni.

(6) Auk, Clyde, Fulmar, Gannet A, B, C, D, E and F, Janice, Joanne, Judy, Leven, Medwin and Orion.

(7) Wytch Farm including Kimmeridge, Stowborough, Wareham; and the West Midlands fields.

C.5 Disposals of crude oil[1]

Thousand tonnes

	1995	1996	1997	1998	1999
UK refineries	43,592	46,403	48,961	46,887	47,170
Exports:	79,154	75,462	71,933	77,322	80,613
Argentina	65	-	-	-	-
Bahamas *(2)*	139	-	-	257	143
Belgium	246	218	820	1,035	1,193
Canada	2,004	1,815	999	808	625
China	211	-	-	-	1,588
Croatia	1,720	428	-	-	-
Denmark	-	145	69	-	-
Finland	3,318	3,579	1,253	788	929
France	10,452	14,312	15,168	15,261	15,177
Germany	12,418	12,546	15,692	17,406	11,879
Greece	68	-	-	-	-
India	-	-	-	-	277
Italy	4,287	1,513	931	1,219	1,819
Japan	-	-	-	-	256
Madagascar	-	-	-	81	-
Martinique *(2)*	84	80	84	87	-
Morocco	3	-	-	-	-
Netherlands *(3)*	10,193	13,727	14,933	15,591	16,540
Norway	1,392	593	814	1,087	1,297
Poland	1,916	2,404	1,608	1,494	682
Portugal	217	379	867	1,157	1,394
Puerto Rico *(2)*	1,042	560	-	-	-
Republic of Ireland	340	302	-	82	69
Singapore	-	-	-	255	278
South Africa	1,555	271	-	1,028	-
South Korea	-	-	-	-	260
Spain	3,867	1,560	3,762	3,403	4,040
St. Lucia *(2)*	526	-	-	-	-
Sweden	1,022	1,380	1,025	1,266	1,024
Switzerland	134	-	-	-	-
Turkey	-	-	69	-	-
USA	21,962	19,650	13,478	15,017	21,142
Unknown	-	-	360	-	-
Total disposals *(4) (5)*	**122,746**	**121,865**	**120,893**	**124,209**	**127,783**

(1) Monthly data for aggregate disposals to refineries and exports are published in Table 13 of Energy Trends - see Annex F.

(2) Some of the exports to the Caribbean area may have been for transhipment to the USA.

(3) Exports to the Netherlands include oil for transhipment or in transit to other destinations (eg. Belgium and Germany).

(4) Includes disposals of onshore production. The difference between

disposals and production as shown in Table C.3 is accounted for by losses, platform and other field stock changes and by terminal and transit stock changes.

(5) Includes Norwegian crude from the Forties Terminal (382, 496, 330, 167 and 98 thousand tonnes in 1995 to 1999 respectively). Separate figures for UK and Norwegian disposals from Forties are not available.

C.6 Production of crude oil by field

						Thousand tonnes
	1995	1996	1997	1998	1999	Total to end 1999
Offshore						
Alba	3,767	3,804	4,844	4,376	3,989	23,080
Alwyn North	1,323	982	886	1,071	1,081	29,100
Andrew	-	855	2,794	3,239	3,293	10,181
Angus (1)	-	-	-	-	-	1,378
Arbroath	1,660	1,450	1,108	1,113	1,099	13,842
Argyll (2)	-	-	-	-	-	9,878
Arkwright	-	65	462	299	185	1,011
Auk	606	458	646	783	621	16,247
Balmoral	633	410	467	391	354	13,524
Banff (3)	-	380	278	-	1,110	1,768
Beatrice	472	437	449	365	194	20,388
Beinn	390	388	286	214	116	1,718
Beryl	4,377	4,229	3,744	2,957	2,293	97,461
Birch	285	1,024	767	499	226	2,801
Bladon	-	-	108	278	155	541
Blair (4)	-	-	-	-	-	83
Blenheim	1,096	845	398	229	142	2,710
Brae Central	486	405	383	474	288	5,326
Brae East	3,318	2,735	2,071	1,457	1,190	13,425
Brae North	540	468	361	412	334	16,095
Brae South	532	522	442	411	268	31,683
Brae West	-	-	107	1,128	1,546	2,781
Brent	9,009	9,077	6,255	6,046	4,530	246,520
Brimmond	-	20	60	80	48	208
Britannia	-	-	-	555	1,846	2,401
Bruce	1,710	1,703	1,287	897	1,842	10,382
Buchan	491	535	444	401	344	15,314
Buckland	-	-	-	-	474	474
Captain	-	-	1,451	2,833	2,522	6,806
Chanter	92	102	49	15	7	523
Claymore	2,255	2,151	2,094	1,816	1,656	66,401
Clyde	795	666	697	638	586	15,742
Columba B	-	146	177	150	154	627
Columba D	334	667	332	169	88	1,692
Columba E	-	-	-	217	170	387
Cormorant North	2,059	1,468	1,475	1,636	1,539	46,923
Cormorant South	811	967	1,011	807	858	23,861
Crawford (4)	-	-	-	-	-	539
Curlew	-	-	86	1,436	1,506	3,028
Cyrus (5)	-	202	603	540	402	2,357
Dauntless (6)	-	-	197	308	38	543
Deveron	54	58	26	52	40	2,127
Don	233	169	108	100	89	1,929
Donan (7)	357	282	193	-	-	2,021
Douglas	-	747	1,587	1,339	948	4,621
Drake	-	-	79	281	316	676
Dunbar	1,754	2,372	2,418	2,067	1,879	10,531
Duncan (2)	-	-	-	-	-	2,262
Dunlin	950	753	806	634	626	48,230
Dunlin South West	-	258	197	231	231	917
Durward	-	-	273	588	45	906
Egret	-	-	-	-	383	383
Eider	905	814	653	616	600	13,610
Ellon	89	138	395	281	129	1,038

(1) Production from Angus ceased in June 1993.
(2) Production from the Argyll, Duncan and Linnhe fields ceased during 1992.
(3) Production from Banff was suspended during 1997.
(4) Production from the Blair and Crawford fields ceased during 1990.
(5) Production from Cyrus was suspended during 1992. It resumed in 1996.
(6) Production from Dauntless, Durward, Maureen and Moira ceased during 1999.
(7) Production from Donan ceased during 1997.

C.6 Production of crude oil by field (continued)

Thousand tonnes

	1995	1996	1997	1998	1999	Total to end 1999
Offshore continued						
Emerald *(8)*	422	41	-	-	-	2,365
Erskine	-	-	35	1,141	881	2,057
Everest	262	276	312	285	235	1,740
Fergus	-	249	561	276	161	1,247
Fife	746	1,622	1,076	819	361	4,624
Fleming	-	-	93	506	476	1,075
Flora	-	-	-	152	505	657
Foinaven	-	-	221	3,591	4,330	8,142
Forties	5,444	5,141	4,104	3,993	3,491	323,996
Fulmar	1,240	1,039	546	468	373	70,902
Galley	-	-	-	945	1,358	2,303
Gannet A	997	1,313	1,191	1,013	865	5,985
Gannet B	196	97	58	35	29	591
Gannet C	1,615	1,638	1,150	918	687	8,731
Gannet D	362	408	436	466	359	2,682
Gannet E	-	-	-	644	366	1,010
Gannet F	-	-	326	463	327	1,116
Gannet G	-	-	-	-	268	268
Glamis	152	72	50	47	36	2,408
Grant	-	-	-	134	252	386
Gryphon	2,201	1,877	1,540	1,347	1,093	9,982
Guillemot A	-	249	1,025	687	419	2,380
Hamish	5	4	17	10	8	438
Harding	-	1,928	3,855	4,650	4,276	14,709
Heather	294	285	251	225	204	14,920
Heron	-	-	-	382	2,366	2,748
Highlander	306	272	149	188	102	9,394
Hudson	1,767	1,515	1,593	400	1,243	8,799
Hutton	1,185	900	786	581	557	25,775
Hutton North West	339	296	307	262	294	16,483
Innes *(9)*	-	-	-	-	-	755
Iona	-	-	28	13	77	118
Ivanhoe	618	519	400	282	237	8,381
Janice	-	-	-	-	1,710	1,710
Joanne	39	323	1,198	1,383	923	3,866
Judy	26	99	650	766	531	2,072
Kingfisher	-	-	211	914	675	1,800
Kittiwake	1,363	1,055	629	443	228	9,467
Larch	-	-	-	168	15	183
Lennox	-	105	453	886	855	2,299
Leven	83	59	83	42	37	753
Linnhe *(2)*	-	-	-	-	-	101
Lomond	152	181	197	206	182	1,179
Lyell	448	432	278	215	146	2,765
MacCulloch	-	-	583	1,998	1,752	4,333
Machar	837	443	-	396	1,730	4,031
Magnus	5,355	4,540	3,087	3,144	3,042	90,843
Magnus South	-	235	383	435	482	1,535
Mallard	-	-	-	148	700	848
Marnock	-	-	-	13	746	759
Maureen *(6)*	515	444	445	426	173	29,768
Medwin	53	7	-	-	-	144
Merlin	-	-	64	581	858	1,503
Miller	6,413	6,458	5,188	3,437	2,729	39,129
Moira *(6)*	38	29	16	12	3	560
Monan	-	-	-	75	341	416
Montrose	127	90	64	64	56	11,415
Mungo	-	-	-	706	1,873	2,579
Murchison (UK) *(10)*	585	680	805	791	743	36,783

(8) Production from Emerald ceased during 1996.
(9) Production from Innes ceased during 1991.
(10) UK share of field.

C.6 Production of crude oil by field (continued)

	1995	1996	1997	1998	1999	Total to end 1999
Offshore continued						
Nelson	6,639	7,072	5,595	4,689	4,509	33,627
Ness	91	80	171	104	123	4,035
Nevis	-	184	744	1,082	1,593	3,603
Ninian	2,760	2,419	2,364	2,195	2,051	146,607
Orion	-	-	-	-	137	137
Osprey	1,417	1,297	1,202	754	530	10,408
Pelican	-	1,530	1,268	1,257	1,075	5,130
Petronella	296	137	119	123	52	4,469
Pierce	-	-	-	-	1,414	1,414
Piper	4,023	3,144	2,413	1,949	1,488	130,603
Renee	-	-	-	-	706	706
Rob Roy	1,410	1,075	570	289	268	13,266
Ross	-	-	-	-	720	720
Rubie	-	-	-	-	182	182
Saltire	1,761	1,829	1,906	1,333	756	10,104
Scapa	846	946	914	769	638	13,520
Schiehallion (11)	-	-	-	1,196	5,107	6,303
Scott	8,757	7,028	5,562	4,525	4,012	39,479
Sedgwick (12)	-	-	52	496	-	548
Staffa (13)	-	-	-	-	-	511
Statfjord (UK) (10)	3,479	3,394	3,576	2,342	1,766	72,466
Stirling	61	42	37	9	16	165
Strathspey	1,686	1,810	1,329	1,005	643	7,881
Tartan	452	474	333	331	272	13,256
Teal	-	-	1,090	1,121	1,215	3,426
Teal South	-	44	267	122	136	569
Telford	-	104	1,517	1,519	1,012	4,152
Tern	3,322	2,777	2,590	2,284	2,122	28,800
Thelma	-	165	1,308	1,050	904	3,427
Thistle	664	535	429	362	305	52,747
Tiffany	1,952	1,762	1,203	761	425	8,051
Toni	1,462	1,056	683	793	654	5,265
Other offshore (14)	97	102	100	-	-	693
Total offshore	116,743	116,679	115,340	119,061	123,977	2,283,278
Wytch Farm	4,536	4,728	4,475	4,684	3,864	40,279
Other landward fields (15)	515	523	506	477	421	7,157
Total onshore	5,051	5,251	4,981	5,161	4,285	47,436
Total production	**121,794**	**121,930**	**120,321**	**124,222**	**128,262**	**2,330,714**

(11) Production figures include those for the 'Loyal' accumulation.
(12) From 1999 Sedgwick are included in Brae West figures.
(13) Production from Staffa ceased during 1994.
(14) Production from extended well tests other than those within established fields.
(15) Includes production from mining licence areas and extended well tests.

C.7 Production of methane by field

	1995	1996	1997	1998	1999	Total to end 1999
Alison	31	128	91	97	18	365
Alwyn North (1)	1,876	1,829	2,039	1,730	1,608	29,530
Amethyst East	991	1,416	848	870	724	11,028
Amethyst West	312	421	515	423	262	2,906
Anglia	615	439	284	391	296	3,693
Ann	399	428	270	140	166	2,044
Audrey	1,179	1,197	1,171	729	531	17,195
Baird	219	459	435	374	311	2,015
Barque	577	1,829	2,244	1,503	1,327	10,606
Barque South	6	-	8	2	-	16
Bessemer	139	777	812	735	692	3,155
Boulton	-	-	-	925	459	1,384
Bruce (1)	5,175	6,577	5,613	4,959	5,164	33,689
Bure	58	55	42	64	12	1,889
Bure West	-	-	-	22	124	146
Caister Bunter	388	295	343	235	315	1,861
Caister Carboniferous	745	649	642	364	390	3,615
Callisto (2)	102	254	254	199	45	854
Camelot Central and South	526	403	846	563	187	5,400
Camelot North	246	84	49	30	1	890
Camelot North East (3)	10	204	58	2	-	498
CATS (4)	1,941	2,334	4,429	10,126	13,605	35,361
Cleeton (5)	997	1,587	1,466	472	5	10,268
Clipper	621	1,190	1,152	669	508	8,174
Corvette	-	-	-	-	1,782	1,782
Dalton	-	-	-	-	267	267
Davy (6)	197	930	806	719	908	3,560
Dawn	1	170	92	94	103	460
Deben	-	-	-	66	240	306
Delilah	-	-	-	42	103	145
Dunbar (1)	954	1,371	1,359	1,121	1,133	5,961
Ellon (1)	337	521	791	448	162	2,285
Esmond (7)	36	-	-	-	-	8,866
Excalibur	811	876	599	681	552	3,751
FLAGS (8)	6,214	6,459	6,948	7,417	7,596	103,867
Forbes (9)	-	-	-	-	-	1,473
Frigg (10)	474	466	191	511	253	72,456
Fulmar (11)	1,854	1,716	1,505	1,890	2,104	14,654
Galahad	106	456	707	509	431	2,209
Galleon	518	1,398	1,501	1,493	1,168	6,348
Galley (1)	-	-	-	257	410	667
Ganymede (2)	532	1,708	1,655	947	1,148	5,990
Gawain	92	929	820	798	666	3,305
Gordon (7)	22	-	-	-	-	3,994
Grant (1)	-	-	-	322	672	994
Guinevere	358	243	271	227	232	1,786
Hamilton	-	-	1,176	1,752	1,416	4,344
Hamilton North	-	625	667	546	454	2,292
Hewett and Della	1,290	2,188	1,301	1,324	1,129	115,419
Hyde	346	357	284	291	259	2,123
Indefatigable (12)	1,133	2,139	1,507	2,055	1,345	125,685
Indefatigable South West(12)	63	242	210	179	198	892
Ivanhoe and Rob Roy (1)	159	152	79	38	48	1,754
Johnston	543	585	469	327	540	2,600
Ketch	-	-	-	-	297	297
KX	27	81	60	62	52	282
Lancelot	868	685	621	557	761	4,875
Leman	4,049	3,468	3,013	4,740	3,060	294,075
Malory	-	-	-	126	668	794
Markham (13)	933	807	663	514	485	4,945
Miller (14)	2,467	2,534	2,028	1,254	1,109	14,740
Millom	-	-	-	-	29	29
Mordred	-	26	82	17	39	164
Morecambe North	2,399	2,626	2,930	1,294	848	10,652

Million cubic metres

	1995	1996	1997	1998	1999	Total to end 1999
Morecambe South	7,675	7,099	6,170	7,993	9,971	77,906
Murdoch	1,110	1,127	1,150	1,376	836	6,950
Newsham	-	68	127	94	71	360
Orwell	1,470	789	720	832	667	5,838
Pickerill	1,790	1,345	1,288	879	626	10,052
Piper/Tartan Area *(1)*	1,037	950	633	452	421	12,843
Ravenspurn North	1,716	2,942	2,968	1,580	1,319	22,043
Ravenspurn South	852	1,253	1,433	1,186	1,006	11,817
Renee/Rubie	-	-	-	-	1	1
Ross	-	-	-	-	28	28
Rough *(15)*	-	-	-	-	-	4,370
SAGE *(16)*	6,829	7,321	8,035	10,398	15,307	55,458
Schooner	-	243	1,245	1,088	1,237	3,813
Sean East	501	512	301	227	253	1,859
Sean North and South	428	942	639	50	312	5,638
Thames	61	157	119	60	92	6,387
Trent	-	80	279	347	521	1,227
Tristan	206	27	18	7	90	969
Tyne North	-	-	76	130	255	461
Tyne South	-	109	539	435	479	1,562
Valiant North	144	277	295	334	172	4,106
Valiant South	177	349	391	397	298	6,300
Vampire	-	-	-	-	367	367
Vanguard	30	109	120	132	78	2,325
Victor	1,399	1,657	1,724	1,064	949	21,945
Viking	466	628	687	629	2,465	81,317
Vulcan	415	656	827	816	584	12,733
Waveney	-	-	-	137	741	878
Welland North West	411	358	386	629	326	5,030
Welland South	208	117	173	210	155	2,083
Wensum	4	3	3	-	2	52
West Sole	1,214	857	1,224	1,218	1,170	48,945
Windermere *(17)*	-	-	279	438	320	1,037
Yare	63	51	14	72	21	1,714
Others *(18)*	3,016	3,175	3,361	3,719r	3,937	49,743
Total offshore *(19)*	75,158	89,514	91,170	95,171r	104,584	1,485,827
Wytch Farm	182	245	242	156	149	1,611
Other landward fields	140	137	146	179	161e	1,172
Total onshore *(20)*	322	382	388	335	310	2,783
Total production *(21)*	**75,480**	**89,896**	**91,558**	**95,506r**	**104,894**	**1,488,610**

(1)   *Associated gas used offshore or delivered to land via the Frigg pipeline system.*
(2)   *With effect from August 1999 production from fields in the Jupiter area (Callisto, Ganymede and Sinope) and Bell have been reported with Ganymede.*
(3)   *Ceased production in June 1998.*
(4)   *Gas delivered to land via the Central Area Transmission System (CATS) from Andrew, Drake, Egret, Erskine, Everest, Fleming, Hawkins,   Heron, Janice, Joanne, Judy, Lomond, Machar, Marnock, Monan and Mungo.*
(5)   *Ceased production in February 1999.*
(6)   *With effect from December 1998 Davy includes the Brown field for which separate data are unavailable.*
(7)   *Ceased production in March 1995.*
(8)   *Gas delivered to land via the Far-north Liquids and Associated Gas System (FLAGS) from, Brent, North and South Cormorant, Magnus, Magnus South, Murchison (UK), Pelican, Statfjord (UK), Strathspey and Thistle. Gas used offshore is included in Others (see (18)).*
(9)   *Ceased production in February 1993.*
(10)   *UK share only.*
(11)   *Gas delivered to land via the Fulmar pipeline from Clyde, Curlew, Fulmar, Gannet A – G, Guillemot, Kittiwake, Leven, Mallard, Medwin, Nelson, Orion, Teal and Teal South. Gas used offshore is included in Others (see (18)).*
(12)   *Separate data for Indefatigable South West is only available from October 1995.*
(13)   *UK share only. Exported to Netherlands.*
(14)   *Gas delivered direct to Boddam (Peterhead) power station by dedicated pipeline.*
(15)   *Converted for use as an off-peak storage unit with effect from 1985.*
(16)   *Gas delivered to land via the Scottish Area Gas Evacuation (SAGE) system from the Beryl, Brae and Scott area and from Britannia.*
(17)   *Exported to Netherlands via Markham.*
(18)   *Associated gas, mainly methane, produced and used mainly on Northern Basin oil production platforms including those in the FLAGS and Fulmar systems.*
(19)   *Excludes Neptune and Mercury fields which started production in late December 1999 and for which data are unavailable.*
(20)   *Excludes colliery methane.*
(21)   *Gross production, i.e. includes own use for drilling purposes, production and pumping operations, but excludes gas flared and vented.*

C.8 Production of methane by field for key production systems

	1995	1996	1997	1998	1999	Total to end 1999 (1)
Central Area Transmission System (CATS):						
Andrew	-	58	174	286	338	856
Armada (2)	-	-	762	3,729	4,214	8,705
Banff	-	-	-	-	76	76
Erskine	-	-	4	812	1,026	1,842
ETAP (3)	-	-	-	179	2,679	2,858
Everest	1,109	1,181	1,264	1,362	1,471	7,970
J-Block (4)	-	66	1,130	2,424	2,317	5,937
Lomond	832	1,029	1,094	1,334	1,484	7,116
Total CATS fields	1,941	2,334	4,429	10,126	13,605	35,361
Far North Liquids and Associated Gas System (FLAGS):						
Brent	4,510	4,524	4,968	5,504	6,144	-
Cormorant N & S	33	90	129	106	74	-
Magnus	498	499	369	403	386	-
Murchison (UK)	17	26	42	23	17	-
Strathspey	636	724	856	771	618	-
Statfjord (UK)	520	596	584	610	357	-
Total FLAGS fields	6,214	6,459	6,948	7,417	7,596	103,867
Fulmar pipeline:						
Anasuria (5)	-	10	176	147	153	-
Clyde	21	10	8	3	7	-
Curlew	-	-	-	570	899	-
Fulmar	268	96	21	14	5	-
Gannet A to F	1,101	1,147	959	852	705	-
Kittiwake	52	34	20	21	60	-
Nelson	412	419	321	283	254	-
Orion	-	-	-	-	21	-
Total Fulmar fields	1,854	1,716	1,505	1,890	2,104	14,654
Scottish Area Gas Evacuation (SAGE)						
Beryl Area (6)	2,720	2,469	2,773	3,091	3,088	22,253
Brae Area (7)	3,260	4,172	4,314	4,560	4,661	21,917
Britannia	-	-	-	1,820	6,862	8,682
Scott Area (8)	849	680	948	926	696	4,605
Total SAGE fields	6,829	7,321	8,035	10,398	15,307	55,458

(1) Cumulative data for individual FLAGS and Fulmar fields/areas are not available.
(2) Gas from the Drake, Fleming and Hawkins fields.
(3) Gas from Eastern Trough Area Project comprising the Heron, Machar, Marnock, Monan and Mungo fields.
(4) Gas from the Janice, Joanne and Judy fields.
(5) Gas from the Guillemot, Teal and Teal South fields.
(6) Gas from the Beryl, Ness and Nevis fields.
(7) Gas from the Beinn, Brae Central, East, North, South and West, Kingfisher, Sedgwick, Thelma, Tiffany and Toni fields.
(8) Gas from the Scott and Telford fields.

C.9 Sales and expenditure by operators and other production licensees[1]

£ million

	1995	1996	1997	1998	1999
Sales *(2)*					
Crude oil	9,881	11,849	10,327	7,487	10,257
Natural gas liquids	614	748	700	551	712
Natural gas	4,141	5,295	5,254	5,313	5,075
Revenues from pipelines and terminals					
and other operators' revenues	1,166	1,243	1,279	1,453r	1,372
Total receipts	**15,802**	**19,135**	**17,560**	**14,804r**	**17,416**
Operating expenditure					
Oil fields	3,054	3,100	3,122	3,072	2,957
Gas fields	859	878	1,014	1,125r	1,287
Total operating expenditure *(3)*	3,950	4,009	4,184	4,299r	4,524
Gross trading profits *(4)*	**11,852**	**15,126**	**13,376**	**10,505r**	**12,892**
Development expenditure					
Platforms and modules	2,397	2,049	1,715	2,064	1,058
Offshore loading systems	142	247	316	154	66
Pipelines	163	171	133	268r	99
Terminals	144	157	167	85r	76
Production and appraisal wells	1,429	1,635	1,841	2,327	1,635
Other items	36	68	65	155	200
Total development expenditure	4,309	4,326	4,228	5,053	3,154
Exploration expenditure *(5)*	1,085	1,097	1,194	762	457
Other capital expenditure *(6)*	47	38	35	-58r	-39

(1) Quarterly data on the value of production and investment in the UKCS are published in Table 9 of Energy Trends - see Annex F.

(2) Deliveries on sale to third parties valued at the amount charged or appropriations by associated companies and other disposals at the landed c.i.f. value or post initial treatment valuation used for Petroleum Revenue Tax.

(3) Includes other costs not attributable to oil or gas fields.

(4) Gross trading profits are given as total receipts less total operating expenditure.

(5) Includes the cost of appraisal wells drilled prior to development approval.

(6) Other capital investment of operators and production licensees not classified to fields. This, together with development and exploration expenditure equals capital expenditure.

Annex D
Glossary

Advanced gas-cooled reactor (AGR)	A type of nuclear reactor cooled by carbon dioxide gas.
Anthracite	Within this publication, anthracite is coal classified as such by UK coal producers and importers of coal. Typically it has a high heat content making it particularly suitable for certain industrial processes and for use as a domestic fuel.
Anthropogenic	Produced by human activities.
Associated Gas	Natural gas found in association with crude oil in a reservoir, either dissolved in the oil or as a cap above the oil.
Autogeneration	Generation of electricity by companies whose main business is not electricity generation, the electricity being produced mainly for that company's own use.
Aviation spirit	A light hydrocarbon oil product used to power piston-engined aircraft power units.
Aviation turbine fuel	The main aviation fuel used for powering aviation gas-turbine power units (jet aircraft engines.
Benzole	A colourless liquid, flammable, aromatic hydrocarbon by-product of the iron and steel making process. It is used as a solvent in the manufacture of styrenes and phenols but is also used as a motor fuel.
Biogas	Energy produced from the anaerobic digestion of sewage and industrial waste.
Bitumen	The residue left after the production of lubricating oil distillates and vacuum gas oil for upgrading plant feedstock. Used mainly for road making and construction purposes.
Blast furnace gas	Mainly produced and consumed within the iron and steel industry. Obtained as a by-product of iron making in a blast furnace, it is recovered on leaving the furnace and used partly within the plant and partly in other steel industry processes or in power plants equipped to burn it. A similar gas is obtained when steel is made in basic oxygen steel converters, this gas is recovered and used in the same way.
Breeze	Breeze can generally be described as coke screened below 19 mm (¾ inch) with no fines removed, but the screen size may vary in different areas and to meet the requirements of particular markets.
BNFL	British Nuclear Fuels plc.
Burning oil	A refined petroleum product, with a volatility in between that of motor spirit and gas diesel oil primarily used for heating and lighting.
Butane	Hydrocarbon (C_4H_{10}), gaseous at normal temperature, but generally stored and transported as a liquid. Used as a component in Motor Spirit to improve combustion, and for cooking and heating (see LPG).
Calorific values (CVs)	The energy content of a fuel can be measured as the heat released on complete combustion. The SI (Système International - see note below) derived unit of energy and heat is the Joule. This is the energy per unit volume of the fuel and is often

measured in GJ per tonne. The energy content can be expressed as an upper (or gross) value and a lower (or net) value. The difference between the two values is due to the release of energy from the condensation of water in the products of combustion. Gross calorific values are used throughout this publication.

CO_2

Carbon dioxide. Carbon dioxide contributes about 60 per cent of the potential global warming effect of man-made emissions of greenhouse gases. Although this gas is naturally emitted by living organisms, these emissions are offset by the uptake of carbon dioxide by plants during photosynthesis; they therefore tend to have no net effect on atmospheric concentrations. The burning of fossil fuels, however, releases carbon dioxide fixed by plants many millions of years ago, and thus increases its concentration in the atmosphere.

Coke oven coke

The solid product obtained from carbonisation of coal, principally coking coal, at high temperature, it is low in moisture and volatile matter. Used mainly in iron and steel industry.

Coke oven gas

Gas produced as a by-product of solid fuel carbonisation and gasification at coke ovens, but not from low temperature carbonisation plants. Synthetic coke oven gas is mainly natural gas which is mixed with smaller amounts of blast furnace and basic oxygen steel furnace gas to produce a gas with almost the same quantities as coke oven gas.

Coking coal

Within this publication, coking coal is coal sold by producers for use in coke ovens and similar carbonising processes. The definition is not therefore determined by the calorific value or caking qualities of each batch of coal sold, although calorific values tend to be higher than for steam coal.

Colliery methane

Methane released from coal seams in deep mines which is piped to the surface and consumed at the colliery or transmitted by pipeline to consumers.

Combined cycle gas Turbine (CCGT)

Combined cycle gas turbine power stations combine gas turbines and steam turbines which are connected to one or more electrical generators in the same plant. The gas turbine (usually fuelled by natural gas or oil) produces mechanical power (to drive the generator) and heat in the form of hot exhaust gases. These gases are fed to a boiler, where steam is raised at pressure to drive a conventional steam turbine, which is also connected, to an electrical generator.

Combined Heat and Power (CHP)

CHP is the simultaneous generation of usable heat and power (usually electricity) in a single process. The term CHP is synonymous with cogeneration and total energy, which are terms often used in the United States or other Member States of the European Community. The basic elements of a CHP plant comprise one or more prime movers driving electrical generators, where the steam or hot water generated in the process is utilised via suitable heat recovery equipment for use either in industrial processes, or in community heating and space heating. For further information see paragraph 6.36.

Conventional thermal power stations

These are stations which generate electricity by burning fossil fuels to produce heat to convert water into steam, which then powers steam turbines.

Cracking/conversion

A refining process using combinations of temperature, pressure and in some cases a catalyst to produce petroleum products by changing the composition of a fraction of petroleum, either by splitting existing longer carbon chain or combining shorter carbon chain components of crude oil or other refinery feedstock's. Cracking allows refiners to selectively increase the yield of specific fractions from any given input petroleum mix depending on their requirements in terms of output products.

Crude oil

A mineral oil consisting of a mixture of hydrocarbons of natural origins, yellow to black in colour, of variable density and viscosity.

DERV

Diesel engined road vehicle fuel used in internal combustion engines that are compression-ignited (see gas diesel oil).

282

Distillation	A process of separation of the various components of crude oil and refinery feedstocks using the different temperatures of evaporation and condensation of the different components of the mix received at the refineries.
DNC	Declared net capacity is used to measure the maximum power available from generating stations that use renewable resources.
Downstream	Used in oil and gas processes to cover the part of the industry after the production of the oil and gas. For example, it covers refining, supply and trading, marketing and exporting.
Energy use	Energy use of fuel mainly comprises use for lighting, heating or cooling, motive power and power for appliances. See also non-energy use.
ESA	European System of National and Regional Accounts. An integrated system of economic accounts which is the European version of the System of National Accounts (SNA).
EESoPs	A review of the Energy Efficiency Standards of Performance Programme which currently supports the replacing of electric heating with Community Heating based CHP.
Ethane	A light hydrocarbon gas (C_2H_6) in natural gas and refinery gas streams (see LPG).
EUROSTAT	Statistical Office of the European Communities (SOEC).
Exports	For some parts of the energy industry, statistics on trade in energy related products can be derived from two separate sources. Firstly, figures can be reported by companies as part of systems for collecting data on specific parts of the energy industry (e.g. as part of the system for recording the production and disposals of oil from the UK continental shelf). Secondly, figures are also available from the general systems that exist for monitoring trade in all types of products operated by HM Customs & Excise.
	Before the 1997 edition of the Digest, the term "shipment" was used to distinguish figures derived from the former source from those export figures derived from the systems operated by HM Customs & Excise. To make it clearer for users, a single term is now being used for both these sources of figures (the term exports) as this more clearly states what the figures relate to, which is goods leaving the UK.
Feedstock	In the refining industry, a product or a combination of products derived from crude oil, destined for further processing other than blending. It is distinguished from use as a chemical feedstock etc. See non-energy use.
Final energy consumption	Energy consumption by final user - i.e. which is not being used for transformation into other forms of energy.
Fossil fuels	Coal, natural gas and fuels derived from crude oil (for example petrol and diesel) are called fossil fuels because they have been formed over long periods of time from ancient organic matter.
Fuel oils	The heavy oils from the refining process; used as fuel in furnaces and boilers of power stations, industry, in domestic and industrial heating, ships, locomotives, metallurgic operations, and industrial power plants etc.
Fuel oil - Light	Fuel oil made up of heavier straight-run or cracked distillates and used in commercial or industrial burner installations not equipped with pre-heating facilities.

Fuel oil - Medium	Other fuel oils, sometimes referred to as bunker fuels, which generally require pre-heating before being burned, but in certain climatic conditions do not require pre-heating.
Fuel oil - Heavy	Other heavier grade fuel oils which in all situations require some form of pre-heating before being burned.
Gas Diesel Oil	The medium oil from the refinery process; used as a fuel in diesel engines (i.e. internal combustion engines that are compression-ignited), burned in central heating systems and used as a feedstock for the chemical industry.
GDP	Gross domestic product.
GDP deflator	An index of the ratio of GDP at current prices to GDP at constant prices. It provides a measure of general price inflation within the whole economy.
Gigajoule (GJ)	A unit of energy equal to 10^9 joules (see note on joules below).
Gigawatt (GW)	A unit of electrical power, equal to 10^9 watts.
Gigawatt hour (GWh)	Unit of electrical energy, equal to 0.0036 TJ. A 1 GW power station running for one hour produces 1 GWh of electrical energy.
HMCE	HM Customs and Excise.
Imports	See the first paragraph of the entry for exports above. Before the 1997 edition of the Digest, the term "arrivals" was used to distinguish figures derived from the former source from those import figures derived from the systems operated by HM Customs & Excise. To make it clearer for users, a single term is now being used for both these sources of figures (the term imports) as this more clearly states what the figures relate to, which is goods entering the UK.
International Energy Agency (IEA)	The IEA is an autonomous body located in Paris which was established in November 1974 within the framework of the Organisation for Economic Co-operation and Development (OECD) to implement an international energy programme.
Indigenous production	For oil this includes production from the UK Continental Shelf both onshore and offshore.
Industrial spirit	Refined petroleum fractions with boiling ranges up to 200°C dependent on the use to which they are put – e.g. seed extraction, rubber solvents, perfume etc.
Joules	A joule is a generic unit of energy in the conventional SI system (see note on SI below). It is equal to the energy dissipated by an electrical current of 1 ampere driven by 1 volt for 1 second; it is also equal to twice the energy of motion in a mass of 1 kilogram moving at 1 metre per second.
Landfill gas	The methane-rich biogas formed from the decomposition of organic material in landfill.
LDF	Light distillate feedstock.
Liquefied petroleum gas (LPG)	Gas usually propane or butane, derived from oil and put under pressure so that it is in liquid form. Often used to power portable cooking stoves or heaters and to fuel some types of vehicle, e.g. some specially adapted road vehicles, fork-lift trucks.
Lead Replacement Petrol	An alternative to Leaded Petrol containing a different additive to lead (in the UK

(LRP)	usually Potassium based) to perform the lubrication functions of lead additives in reducing engine wear.
Lubricating oils	Refined heavy distillates obtained from the vacuum distillation of petroleum residues. Includes liquid and solid hydrocarbons sold by the lubricating oil trade, either alone or blended with fixed oils, metallic soaps and other organic and/or inorganic bodies.
Magnox	A type of gas-cooled nuclear fission reactor developed in the UK, so called because of the magnesium alloy used to clad the uranium fuel.
Major power producers	Companies whose prime purpose is the generation of electricity (paragraph 5.58 of Chapter 5 gives a full list of major power producers).
Motor spirit	Blended light petroleum product used as a fuel in spark-ignition internal combustion engines (other than aircraft engines).
Natural gas	Natural gas is a mixture of naturally occurring gases found either in isolation, or associated with crude oil, in underground reservoirs. The main component is methane; ethane, propane, butane, hydrogen sulphide and carbon dioxide may also be present, but these are mostly removed at or near the well head in gas processing plants.
Naphtha	(Light distillate feedstock) – Petroleum distillate boiling predominantly below 200°C.
Natural gas - compressed	Natural gas that has been compressed to reduce the volume it occupies to make it easier to transport other than in pipelines. Whilst other petroleum gases can be compressed such that they move into liquid form, the volatility of natural gas is such that liquefaction cannot be achieved without very high pressures and low temperatures being used. As such, the compressed form is more usually used as a "half-way house".
Natural gas liquids (NGL's)	A mixture of liquids derived from natural gas and crude oil during the production process, including propane, butane, ethane and gasoline components (pentanes plus).
Non-energy use	Includes fuel used for chemical feedstock, solvents, lubricants, and road making material.
NFFO	Non Fossil Fuel Obligation. The 1989 Electricity Act empowers the Secretary of State to make orders requiring the Regional Electricity Companies in England and Wales to secure specified amounts of electricity from renewable sources.
No$_x$	Nitrogen oxide. A number of nitrogen compounds including nitrogen dioxide are formed in combustion processes when nitrogen in the air or the fuel combines with oxygen. These compounds can add to the natural acidity of rainfall.
OFGEM	The regulatory office for gas and electricity markets.
Orimulsion	An emulsion of bitumen in water that can be used as a fuel in some power stations.
ONS	Office for National Statistics. Formerly the Central Statistical Office (CSO).
OTS	Overseas Trade Statistics of the United Kingdom.
Patent fuel	A composition fuel manufactured from coal fines by shaping with the addition of a binding agent (typically pitch). The term manufactured solid fuel is also used.

Petrochemical feedstock	All petroleum products intended for use in the manufacture of petroleum chemicals. This includes middle distillate feedstock of which there are several grades depending on viscosity. The boiling point ranges between $200°C$ and $400°C$.
Petroleum cokes	Carbonaceous material derived from hydrocarbon oils, uses for which include metallurgical electrode manufacture and in the manufacture of cement.
Petroleum wax	Includes paraffin wax, which is a white crystalline hydrocarbon material of low oil content normally obtained during the refining of lubricating oil distillate, paraffin scale, slack wax, microcrystalline wax and wax emulsions.
Photovoltaics	The direct conversion of solar radiation into electricity by the interaction of light with the electrons in a semiconductor device or cell.
Plant capacity	The maximum power available from a power station at one time (see also paragraph 5.62 of Chapter 5).
Plant loads, demands and efficiency	Measures of how intensively and efficiently power stations are being used. These terms are defined in paragraphs 5.64 and 5.65 of Chapter 5.
PPRS	Petroleum production reporting system. Licensees operating in the UK Continental Shelf are required to make monthly returns on their production of hydrocarbons (oil and gas) to the DTI. This information is recorded in the PPRS, which is used to report flows, stocks and uses of hydrocarbon from the well-head through to final disposal from a pipeline or terminal (see paragraphs C.39 to C.41 of Annex C).
Process oils	Partially processed feedstocks which require further processing before being classified as a finished product suitable for sale. They can also be used as a reaction medium in the production process.
Primary fuels	Fuels obtained directly from natural sources, e.g. coal, oil and natural gas.
Primary electricity	Electricity obtained other than from fossil fuel sources, e.g. nuclear, hydro and other non-thermal renewables. Imports of electricity are also included.
Propane	Hydrocarbon containing three carbon atoms (C_3H_8), gaseous at normal temperature, but generally stored and transported under pressure as a liquid.
PWR	Pressurised water reactor. A nuclear fission reactor cooled by ordinary water kept from boiling by containment under high pressure.
Reforming	Processes by which the molecular structure of different fractions of petroleum can be modified. It usually involves some form of catalyst, most often platinum, and allows the conversion of lower grades of petroleum product into higher grades, improving their octane rating. It is a generic term for processes such as cracking, cyclization, dehydrogenation and isomerisation. These process generally led to the production of hydrogen as a by-product which can be used in the refineries in some desulphurization procedures.
Refinery fuel	Petroleum products produced by the refining process that are used as fuel at refineries.
Renewable energy sources	Renewable energy includes solar power, wind, wave and tide, and hydroelectricity. Solid renewable energy sources consist of wood, straw and waste, whilst gaseous renewable consist of landfill gas and sewage gas.
Reserves	With oil and gas these relate to the quantities identified as being present in underground cavities. The actual amounts that can be recovered depend on the level of technology available and existing economic situations. These continually change, hence the level of the UK's reserves can change quite independently of whether or not new reserves have been identified.

| RPI | Retail Price Index (RPI) is published by the Office for National Statistics. RPI is calculated using prices collected on a day near the middle of the month. |

RPI Retail Price Index (RPI) is published by the Office for National Statistics. RPI is calculated using prices collected on a day near the middle of the month.

SI (Système International) Refers to the agreed conventions for the measurement of physical quantities.

SIC Standard Industrial Classification in the UK. Last revised in 1992 and known as SIC92, replaced previous classifications SIC80 and SIC68. Now compatible with European Union classification NACE Rev1 (nomenclature generale des activites economiques dans les Communautes europeennes as revised in October 1990).

Secondary fuels Fuels derived from natural primary sources of energy. For example electricity generated from burning coal, gas or oil is a secondary fuel, as are coke and coke oven gas.

Steam coal Within this publication, steam coal is coal classified as such by UK coal producers and by importers of coal. It tends to be coal having lower calorific values.

SO$_2$ Sulphur Dioxide. Sulphur dioxide is a gas produced by the combustion of sulphur-containing fuels such as coal and oil.

Synthetic coke oven gas Mainly a natural gas, which is mixed with smaller amounts of blast furnace, and BOS (basic oxygen steel furnace) gas to produce a gas with almost the same quantities as coke oven gas.

Temperature correction The temperature corrected series of total inland fuel consumption indicates what annual consumption might have been if the average temperature during the year had been the same as the average for the years 1961 to 1990.

Tonne of oil equivalent (toe) A common unit of measurement which enables different fuels to be compared and aggregated. (See paragraphs 1.45 to 1.46 of Chapter 1 for further information).

Tars Viscous materials usually derived from the destructive distillation of coal which are by-products of the coke and iron making processes.

Therm A common unit of measurement similar to a tonne of oil equivalent which enables different fuels to be compared and aggregated. (refer to Annex A).

Thermal efficiency The thermal efficiency of a power station is the efficiency with which heat energy contained in fuel is converted into electrical energy. It is calculated for fossil fuel burning stations by expressing electricity supplied as a percentage of the total energy content of the fuel consumed (based on average gross calorific values). For nuclear stations it is calculated using the quantity of heat released as a result of fission of the nuclear fuel inside the reactor.

UKCS United Kingdom Continental Shelf.

UKPIA UK Petroleum Industry Association. The trade association for the UK petroleum industry.

Ultra low sulphur Diesel (ULSD) A grade of diesel fuel which has a much lower sulphur content (less than 0.005 per cent or none at all) and of a slightly higher volatility than ordinary diesel fuels. As a result it produces fewer emissions when burned. As such it enjoys a lower rate of excise duty in the UK than ordinary diesel to promote its use.

Ultra low sulphur Petrol (ULSP) A grade of motor spirit with a similar level of sulphur to ULSD. In the March 2000 budget it was announced that a lower rate of duty than ordinary petrol for this fuel would be introduced during 2000, and it is anticipated that it will quickly replace premium grade unleaded petrol in the UK market place.

Upstream A term to cover the activities related to the exploration, production and delivery to a terminal or other facility of oil or gas for export or onward shipment within the UK.

VAT	Value added tax.
Watt (W)	The conventional unit to measure a rate of flow of energy. One watt amounts to 1 joule per second.
White spirit	A highly refined distillate with a boiling range of about 150°C to 200°C used as a paint solvent and for dry cleaning purposes etc.

Annex E
Major events in the Energy Industry since 1990

1990

Electricity

The new electricity licensing regime for electricity companies was established along with the post of Director General of Electricity Supply (DGES) by the 1989 Electricity Act. The Act also gave powers to the Secretary of State for Trade and Industry to replace existing public electricity boards by Plc's. Office of Electricity Regulation (Offer) now merged with OFGAS to form OFGEM set up in shadow form.

Provisions of the Act came into force in March. The Central Electricity Generating Board (CEGB) was split into four companies, National Power and PowerGen (fossil fuel generation), Nuclear Electric (nuclear generation) and the National Grid Company (NGC) (transmission). Twelve Regional Electricity Companies replaced the Regional Electricity Boards. At the same time South of Scotland Electricity Board and North of Scotland Hydro-Electric Board were replaced by Scottish Power and Scottish Hydro-Electric (generation, transmission, supply and distribution) and Scottish Nuclear (nuclear generation). The ordinary shares in the National Grid were transferred to the 12 Regional Electricity Companies (RECs).

At vesting, (31 March) price controls, put in place by the Government, came into being for the transmission business of NGC and the supply and distribution businesses of the RECs. There was not a generation price control as this sector was open to competition from the start.

The 12 Regional Electricity Companies in England and Wales were floated on the London Stock Exchange in December. The Government retained a special share in each of the privatised companies, known as the 'Special Share', which prevented any other investor from buying more than 15 per cent of the shares for a period of five years.

At the same time the market for customers with demand exceeding 1 MW was opened up to competition.

During 1990 the Electricity Pool was established. This is a trading mechanism, which allows electricity companies to purchase electricity from generators through a system of bidding.

New and Renewable Energy

The first Non Fossil Fuel Obligation (NFFO-1) Renewable Order for England and Wales made for 102 MW Declared Net Capacity (DNC).

1991

Electricity

60 per cent of the shares in National Power and PowerGen were floated on the London Stock Exchange in March. The Government retained the remaining 40 per cent of shares. Scottish Power and Scottish Hydro Electric were floated in June. The Government retained ownership of Nuclear Electric. In England and Wales the long term costs of generation from nuclear sources were funded from the proceeds of the "Fossil fuel Levy" on supplies of certain electricity. In Scotland contracts were established with the Scottish PESs for the output of the Scottish nuclear generating stations.

New and Renewable Energy

The second Non Fossil Fuel Obligation (NFFO-2) Renewables Order for England and Wales made for 457 MW Declared Net Capacity.

Oil

The Gulf War began in mid January and ended in late February, following the occupation of Kuwait by Iraq in August 1990. This provoked worldwide concern about the availability of oil. The annual average price of crude oil rose sharply, and took two years to settle back to its pre-war level.

One hundred oil and gas fields in production in the UK.

1992

Electricity

In March, generation was transferred from Northern Ireland Electricity to four independent generation companies. Northern Ireland Electricity plc became responsible for transmission and distribution. A special share was retained in Northern Ireland Electricity.

Gas

Following on from the withdrawal of BG legal monopoly relating to non-tariff or contract market for customers with demand greater than 25,000 therms per annum. British Gas in March was prompted by the Director General of Fair Trading to create the conditions whereby competing suppliers should be able to supply at least 60 per cent of the market for customers whose demand exceeded 25,000 therms. In August, the market sector open to competition was extended to include customers with an annual demand of between 2,500 and 25,000 therms. The agreement between British Gas and the Director General of Fair Trading was then redefined as 45 per cent of the market for demand greater than 2,500 therms per annum.

1993

Coal

Coal Review White Paper, "The Prospects for Coal", published on 23rd March. Main conclusions were:

- subsidy to be offered to bring extra tonnage down to world market prices,

- no pit to be closed without being offered to the private sector,

- no changes to the gas and nuclear sectors,

- increased investment in clean coal technology,

- regeneration package for mining areas increased to £200 million.

On 2nd December the Coal Industry Bill was published. Its main features were:

- to enable privatisation,

- to establish the Coal Authority,

- to protect the rights of third parties,

- to safeguard pension and concessionary fuel entitlements

- to retain HSE and HM Mines Inspectorate as bodies responsible for mine safety & inspection.

Electricity
In June, Northern Ireland Electricity plc was floated on the London Stock Exchange.

Gas
The Monopolies and Mergers commission published a report on competition in gas supply.

Oil
Unleaded petrol sales accounted for 50 per cent of the total UK market for motor spirits.

1994

Coal
Coal Authority brought into legal existence under Section 1 of the Coal Industry Act 1994 on 19th September.

The 31st October was the Coal Authority "Restructuring Date". Ownership of Britain's coal reserves transferred to the Authority and it assumed its full range of functions including powers to license coal operations.

In December, British Coal Corporation's mining activities were sold to the private sector.

Electricity
In April, competition in the electricity market was extended to include all customers whose demand exceeded 100 kW.

From 1 April the revised (tightened) supply price control took effect.

New and Renewable Energy
The third Non Fossil Fuel Obligation (NFFO-3) Renewables Order for England and Wales made for 627 MW Declared Net Capacity (DNC).

The first Scottish Renewables Order (SRO-1) for Scotland made for 76 MW Declared Net Capacity (DNC).

The first Northern Ireland NFFO (NI-NFFO-1) Renewables Order made for 16 MW Declared Net Capacity (DNC).

VAT
In April the Government introduced VAT on domestic fuel at a rate of 8 per cent.

1995

Coal
The Domestic Coal Consumer's Council was abolished.

Electricity
In March the Government's 'Special Share' in each of the Regional Electricity Companies expired, the companies were then exposed to the full disciplines and opportunities of the market, including acquisitions and mergers. In 1995 there were four bids involving Regional Electricity Companies, followed by a further seven successful bids in 1996 and two more in the first half of 1997.

In March, the Government also floated its remaining 40 per cent share in National Power and PowerGen on the London Stock Exchange. It did however retain its 'special share' in these companies.

Following a review, the distribution price control was revised from 1st April.

The DGES decided to review again the Distribution Price Control following Northern Electric's defence against a take-over bid from Trafalgar House.

The National Grid Company was floated on the London Stock Exchange in December. As a result, customers of the Regional Electricity Companies received a discount of £50 on their electricity bills in early 1996 as their share of the benefit from the sale. Before the company was floated, its Pumped Storage Business was transferred to a new company, First Hydro, which was then sold to a US generator, Mission Energy. The Government still holds a special share in the National Grid Company which is not time limited.

Gas

The Gas Act 1995 set out the Government's plans for the liberalisation of all gas markets, including the domestic sector. The Government and the industry put into place the licensing framework and the administrative/computer framework required to support the forthcoming gas pilot trials.

By the end of 1995, there were 40 independent gas marketing companies selling gas to UK end-users. They had captured 80 per cent of the firm industrial and commercial market, and 70 per cent of the market for "interruptible" sales.

The development of a "gas bubble" (an excess of supply over demand) in 1995 and into 1996 led to a sharp fall in the spot price of gas from 0.7p/KWh to around 0.4p/KWh (10p per therm), the main beneficiaries of this were customers on short term gas contracts.

Oil

Production from the UK sector of the North Sea and onshore sites reached a new record level of output, 130.3 million tonnes per annum.

Nuclear

In February electricity generation began at Sizewell B, the UK's only pressurised Water Reactor, completed the previous year.

In May the Government published a White Paper on the prospects for nuclear power in the UK. It concluded that nuclear power should continue to contribute to the mix of fuels used in electricity generation, provided it maintained its current high standards of safety and environmental protection; that building new nuclear power stations was not commercially attractive; and that there was no justification for any government intervention to support the construction of new nuclear stations.

1996   Coal

In January the company Coal Investments ceased trading, closing four pits and selling two to Midlands Mining Ltd.

In May the National Audit Office published its report into the privatisation of British Coal's mining activities.

Electricity

The revised Distribution Price control (further tightened as a result of the second review in 1995) took affect from 1st April.

In July 1996 Eastern Group leased a total of 6 GW of coal-fired electricity generation capacity from National Power (4 GW) and PowerGen (2 GW). As a result, the pool price cap was lifted.

Bids were made by National Power and Powergen for Southern Electric and Midlands Electricity respectively in 1995, which would have allowed significant vertical integration between generation and supply in the industry, were prohibited by the President of the Board of Trade in April following an investigation by the Monopolies and Mergers Commission.

Gas

On 6 February in response to the separation of licensing for gas distribution and supply in the Gas Act 1995, British Gas Plc announced it was demerging into two companies, one responsible for the transmission of gas (BG Transco) and one for the supply of gas (Centrica).

On 1st April, the first stage of the introduction of competition in the domestic market began - around 540,000 customers in South West England were enabled to purchase their gas from a variety of suppliers. By the end of the year just under 20 per cent of households switched to a new supplier, whose prices were on average 10-20 per cent less than those charged by British Gas.

Nuclear

The nuclear generating industry was formally restructured on 31 March 1996 in preparation for privatisation. A holding company, British Energy plc (BE) was created, together with two subsidiary companies - Nuclear Electric Ltd, which now operates the PWR and five AGR stations in England and Wales, and Scottish Nuclear Ltd, which operates two AGR stations in Scotland. In July 1996 British Energy, which operates the AGR/PWR nuclear electricity power stations in the UK, was floated on the London Stock Exchange by the Government. Magnox stations remain in the public sector under the ownership of Magnox Electric plc. Magnox Electric and BNFL merged early in 2000.

Because of the privatisation of British Energy, on 1st November, the Fossil Fuel Levy in England and Wales was reduced from 10 per cent to 3.7 per cent (on 1st April 1997 it was reduced further to 2.2 per cent).

The premium element of prices payable in Scotland under the Nuclear Energy agreement ended in July 1996. A new Fossil Fuel Levy was introduced at a rate of 0.5 per cent to support renewable energy.

In September, AEA Technology, the former commercial arm of the UK Atomic Energy Authority, was privatised.

Regulation

On 10th June, the Office for the Regulation of Electricity and Gas (OFREG) was formed to perform a similar role in Northern Ireland to that of OFFER and OFWAT in England and Wales. It was unique in that it was the only combined utility regulatory office in the UK.

Regulation

In June, the Government announced a review of utility regulation to cover in particular electricity, gas, telecommunications and water. It was aimed at ensuring that consumers get a fair deal from regulation, and at making regulation more consistent, transparent and accountable.

Electricity

Revised Transmission Price control took affect from 1st April reducing prices further over four years.

In October, the Government announced a review of the electricity trading arrangements including the Electricity Pool. OFFER were asked to report by July 1998.

In December, the Government announced a review of fuel sources for power stations. A consultation paper was issued in June 1998.

The Regulator announced a proposed timetable starting in April 1998 for the roll out of the final stage of supply competition using customer postcodes. The Regulator subsequently modified the timetable and put in place arrangements for the extensive testing of the systems necessary to make competition work.

Gas

Following the increased expenditure on exploration and development of gas fields in the North Sea in the early 1990s, gas production increased to the point where the UK became a net exporter of gas for the first time.

By March competition in the domestic market was extended to include another 0.5 million households in Avon and Dorset and 1.1 million households in the South East of England, bringing the total to two million customers. Over 20 per cent of these households switched to a new supplier, by the end of 1997

In November 1997, as part of the next stage in the liberalisation of the gas industry, competition was extended to another 2.5 million domestic customers in Scotland and North East England.

New and Renewable Energy

The fourth Non Fossil Fuel Obligation (NFFO-4) Renewables Order for England and Wales made for 873 MW Declared Net Capacity (DNC).

The second Scottish Renewables Order (SRO-1) made for 112 MW of capacity.

John Battle, Minister for Science, Energy and Industry announced a review of renewables energy policy on 6 June 1997.

VAT

On 1 September, VAT on domestic gas and electricity supplies was reduced to 5 per cent.

Windfall tax

In the July Budget, the Government announced that the privatised utilities would have to pay a one-off windfall tax on the excessive profits they had made, payable in two instalments - one in 1997 and the other in 1998. Altogether this was expected to raise £5.2 billion.

Coal

The 5-year contracts with the electricity generators ended in March 1998.

Silverdale Colliery closed in December 1998.

Electricity

The final stage of opening electricity supply markets began in September and was completed in May 1999.

The Government published a White Paper (CM 4071) on energy sources for power generation, and has adopted a more restrictive policy towards consents for new power stations, but with special provisions for CHP.

The Monopolies and Mergers Commission recommendation on revised transmission and distribution price control on Northern Ireland Electricity, which was subject to judicial review, was upheld by the Northern Ireland Court of Appeal.

Gas

Introduction of supply competition in Great Britain completed in May 1998.

In April the gas levy was reduced to zero.

The European Union Gas Liberalisation Directive entered into force in August 1998.

In October the UK - Belgium interconnector became operational, providing a path for UK gas exports to markets in Europe as well as another route for imports of gas into the UK.

Revision of Frigg Treaty with Norway was signed in August 1998.

New and Renewable Energy

The fifth Non-fossil fuel obligation (NFFO) Order was laid in September 1998 for 1,117 MW of capacity.

Oil

International agreement reached on decommissioning and disposal of offshore structures.

New Regulations require Environmental Impact Assessments for offshore projects.

200 oil and gas fields in production in the UK.

Nuclear

In January, the Government transferred its shareholding in Magnox Electric to BNFL as the first stage of a merger of the two companies. Full integration of the combined business of the two companies completed early in 2000.

The Health and Safety Executive and the Scottish Environment Protection Agency completed a full audit of safety at UKAEA Dounreay.

The acceptance of a small consignment of uranium from Georgia for non-proliferation reasons was subject to scrutiny by the Trade and Industry Committee, who approved of the Government's decision.

Utility Regulation

In March, following an inter-departmental review, the Government published a Green Paper entitled 'A fair deal for consumers' on utility regulation aimed at ensuring that consumers get a fair deal from regulation, and at making regulation more consistent, transparent and accountable. The Government's conclusions, in the light of consultation, were published in July, including confirmation that the regulators for gas and electricity should be merged and that the Electricity Act 1989 should be amended to require the distribution businesses of the PESs to be licensed separately from their supply businesses. Detailed proposals on energy and the creation of independent consumer councils were published in November.

Appointment of new gas regulator in November 1998 who took over as electricity regulator in January 1999.

1999

Nuclear

The Government announced that it was looking to introduce a Public Private Partnership (PPP) into BNFL, subject to the company's overall progress towards achieving targets on safety, health, environmental and business performance as well as further work undertaken by the DTI and its advisers. The Government's working assumption is that PPP would involve BNFL as a whole. Existing legislation provides for the sale of up to 49 per cent of the company.

BNFL, in partnership with US engineering group Morris Knudsen, has acquired the global nuclear business of the US company Westinghouse.

Following Government approval, BNFL commenced uranium commissioning of its Sellafield mixed oxide (MOX) fuel plant.

Electricity

Opening of supply market to full competition completed in May.

At the end of June, Ferrybridge and Fiddlers Ferry power stations were sold by PowerGen to Edison Mission Energy, and at the end of November National Power sold DRAX to AES.

The implementation date for the EU Electricity Liberalisation Directive was February 1999, this required an initial 25 per cent market opening to be implemented, with nearly all the Member States adhering to the timetable.

In October 1999 the Director General of Gas and Electricity supply (DGGES), published a report on Pool Prices. This concluded that the trading arrangements facilitated the exercise of market power. He proposed the introduction of a good market behaviour condition in the licences of the main generators.

New and Renewable Energy

In March 1999, a third Scottish Renewables Order (SRO) made for 150 MW (DNC) of capacity.

In March 1999 the Government published a consultation paper "New and Renewable Energy - Prospects for the 21st Century. The legal provision for a new renewables mechanism is the Utilities Bill announced in the Queen's speech in November 1999.

Coal

In March 1999, the Government issued mineral planning guidance with respect to opencast mining.

In April 1999, the Department published its policy paper Energy Paper 67 on research and development into cleaner coal technologies.

In July 1999 Calverton Colliery closed.

In October 1999, the Government's Coal Field Task Force published a progress report relating to the problem of those communities affected by pit closures.

Utility Regulation

New name for the combined OFFER and Ofgas regulating gas and electricity markets announced as Office of Gas and Electricity Markets - OFGEM.

In April, in its response to consultation, the Government confirmed its plans to establish independent consumer councils.

In November 1999 the Government announced a Utilities Reform Bill to provide a new framework for the regulation of the gas and electricity markets to provide a fair deal to consumers.

Fuel Poverty

Following discussions with expert organisations and across departments, the Government issued, in May, a consultation paper on a new Home Energy Efficiency Scheme (HEES) aimed at cutting the numbers of households in fuel poverty. The proposed scheme will concentrate on households in the private rented and owner occupier sectors and particularly those at greatest risk, e.g. the elderly. It will provide a greater range of energy efficiency measures, including new heating systems, than the existing HEES scheme. The new scheme will have a total budget of nearly £300 million in the first two years, (which includes the additional £150 million made available through the Comprehensive Spending Review).

BG Transco announced its Affordable Warmth programme which will help upgrade a million homes.

Gas

In June, BG announced its restructuring to separate Transco, the regulated pipeline company, from the rest of the business.

A consultation exercise into the Fundamental Review of Gas Safety was launched by the Health and Safety Executive.

In September 1999 the HSE issued a consultation document outlining the proposed amendments to "The Gas Safety (Management) Regulations (GS(M)R)", with a closing date for December 1999.

Environment

In the March 1999 Budget the Government announced its intention to introduce a climate change levy on the supply of energy to business, following up recommendations of the Marshall Report.

The Government has also followed up Lord Marshall's recommendations on Emissions Trading, by encouraging the launch of an industry-led project to design a pilot scheme for the UK. The scheme is intended to be operational by April 2001.

Oil

16th February 1999, the key Brent crude oil benchmark price touched $9 per barrel, a record low level. However, by December 1999 the oil price had recovered to over $25 a barrel.

Production from the UKCS reached a record level of 137 million tonnes of oil.

Drilling at BP Amoco's Wytch Farm onshore field achieved two world records - longest production well drilled and greatest horizontal drilling distance achieved.

Oil & Gas industry Task-Force report published in September 1999 sets a vision for the UKCS in 2010, aimed at increasing investment in UKCS activity, increasing employment in directly linked and related industries and prolonging UK self-sufficiency in oil and gas.

2000

Fuel Poverty

A cross Departmental Ministerial group was set up in January with the objective of developing and publishing a strategy setting out the Government's fuel poverty objectives, targets for achieving those objectives, the policies to deliver these and how progress should be monitored.

In the March 2000 Budget the Government announced a change in tax rules to facilitate the Affordable Warmth scheme (lease financing of central heating in social housing) being organised by Transco. This programme aims to install central heating in 850,000 local authority/registered social landlord homes and 150,000 pensioners private sector homes over the next seven years.

The first Social Action Plan was produced jointly in 1998 by Offer and OFGAS. It was thought to be inadequate, and in March 2000 OFGEM published a revised Social Action Plan following earlier publication and consultation on initial proposals. The Plan proposes licence amendments and new codes of practice to help the fuel poor; broader structural changes to help the fuel poor benefit more fully from competition; outlines areas where further research is needed to understand the issues better; and proposes a timetable for action and indicators for reviewing progress.

The new HEES scheme for England was formally launched in June 2000.

Gas

In March 2000 BG plc announced that Transco would become a separately listed company with the rest of the group forming BG International.

Electricity

In March National Power sold Eggborough power stations to British Energy and Killiongholme CCGT stations to NRG.

The Secretary of State for Trade and Industry announced on 17 April that he anticipated lifting the restrictions on the building of new gas-fired power stations, once new electricity arrangements are in place. This is expected to be in the autumn.

Energy Efficiency

In July 1999 the Regulator announced his intention to raise the level of Energy Efficiency Standards of Performance (EESOP), which obliges electricity supply companies to provide energy efficiency amongst domestic consumers, from £1.00 to £1.20 per customer with effect from April 2000. He also announced his intention to extend EESOP obligations to supply companies, also at a rate of £1.20 per customer.

The Utilities reform bill announced in November 1999 provides that future EESOP obligation is decided by the Government. The Government had previously indicated such an intention explaining that it was appropriate for the Government rather than the Regulator to decide social and environmental obligations that had significant financial cost.

In November 1999 the Government confirmed its intention to set a target of at least 10,000 MWe of CHP capacity by 2010: more than double current capacity. In December, the Government confirmed that electricity from CHP plants would be exempt from the Climate Change Levy where the electricity is used on site or sold direct to the consumers.

Coal

At the end of January 2000 Midlands Mining ceased to produce coal from its only remaining colliery, Annesley-Bentinck

On 17 April 2000, the Secretary of State announced a coal subsidy scheme designed to assist UK coal producers through a difficult transitional period arising from adverse market conditions and the imminent relaxation of the stricter gas consents policy. The Government's objective is to enable those elements of the industry with a viable future without aid to overcome short term market problems. The subsidy scheme is conditional upon approval by the European Commission and must meet the requirements of the ECSC Treaty and Commission Decision 3632/93/ECSC which establishes the rules for State aid to the coal industry. Payments will be made to coal producers who submit valid applications meeting the criteria devised by DTI to meet the Government's objective and fulfil Treaty obligations. Payments cannot be made until after Commission approval but will cover the period from the date of the announcement until the ECSC Treaty expires on 23 July 2002.

New and Renewable Energy

In February the Government published *Conclusions in Response to the Public Consultation* ("New and Renewable Energy: Prospects for the 21st Century"). The document summarises the Government's strategy to make progress to a target of 10% of UK electricity from renewables sources by 2001, subject to the cost to consumers being acceptable.

In May the European Commission adopted a proposal for a Directive on the promotion of electricity from renewable energy sources in the internal electricity market.

Nuclear

In February HSE published three reports into BNFL, covering

- The Storage of Liquid High Level Waste at Sellafield
- Falsification of Data at the Mox Demonstration Facility
- Team inspection of the Control and Supervision of Operations at Sellafield

In March the Government announced that it considered that the earliest date possible for the introduction of any PPP into BNFL could not be before the latter part of 2002.

In April BNFL published its response to two of the HSE reports, with the third response to follow in the summer as agreed with HSE.

In May BNFL acquired the nuclear business of ABB. BNFL also announced a strategy for managing lifetimes of its Magnox stations.

Annex F
Further sources of United Kingdom energy statistics

Some of the publications listed below give shorter term statistics, some provide further information about energy production and consumption in the United Kingdom and in other countries, and others provide more detail on a country or fuel industry basis. The list is not exhaustive and the titles of publications and publishers may alter. Unless otherwise stated, all titles are available from The Stationery Office and can be ordered through Government Bookshops.

Department of Trade & Industry publications on energy

Energy Trends

(monthly); (available on subscription only from EPTAC4f, Department of Trade and Industry, Bay 1128, 1 Victoria Street, London SW1E 5HE, tel. 020-7215 2697/2698).
Contains monthly and quarterly data on production and consumption of overall energy and of the individual fuels in the United Kingdom. Also includes data on foreign trade in fuels and on the prices of fuels.

Energy Sector Indicators 1999

This is a set of indicators grouped into 12 sections covering different aspects of the energy sector. The content is designed to show the extent to which secure, diverse and sustainable supplies of energy to UK Businesses and consumers at competitive prices are ensured.

Energy Report 1999

This is the sixth annual Energy Report, the purpose of which is to help competitive markets develop by setting out the key elements of energy policy and the main driving forces, both internal and external.

Development of the Oil and Gas Resources of the United Kingdom 2000

This is an annual report to parliament by Secretary of State for Trade and Industry on the activities to search for and exploit oil and gas in the UK sector of the Continental Shelf. Contains information about the upstream industry including information on the licensing and fiscal regimes governing it, production and remaining reserve levels.

Energy Consumption in the UK:- Energy Paper 66

A statistical review of delivered and primary energy consumption by sub-sector, end use and the factors effecting change.

Industrial Energy Markets: Energy markets in UK manufacturing industry 1973 to 1993 - Energy Paper 64

Using tables of data drawn from the 1989 Purchases Inquiry conducted by the Office for National Statistics, the report, which updates one produced in 1989, sets out the implications for the trends in industrial energy consumption over the period from 1973 to 1993.

Other publications including energy information

General

Basic Statistics of the Community (annual); *Statistical Office of the European Communities*
Digest of Environmental Statistics (annual); *Department of the Environment, Transport and the Regions*
Digest of Welsh Statistics (annual); Welsh *Office* (available from ESS Division, Welsh Office, Cathays Park, Cardiff)

Eurostatistics - Data for Short Term Analysis; *Statistical Office of the European Communities*
Monthly Digest of Statistics; *Office for National Statistics*
Northern Ireland Annual Abstract of Statistics (annual); *Department of Finance and Personnel,* (available from the Policy & Planning Unit, Department of Finance & Personnel, Stormont, Belfast BT4 3SW)
Overseas Trade Statistics of the United Kingdom; *H.M. Customs & Excise*
  - Business Monitor MM20 (monthly) (extra-EU trade only)
  - Business Monitor MM20A (monthly) (intra and extra EU trade data, relatively limited level of production detail)
  - Business Monitor MQ20 (quarterly) (intra-EU trade only)
  - Business Monitor MA20 (annual) (intra- and extra-EU trade);
Purchases Inquiry 1989, 1994-1998; Office *for National Statistics*
Rapid Reports - energy and industry (ad hoc); *Statistical Office of the European Communities*
Regional Trends (annual); *Office for National Statistics*
Scottish Abstract of Statistics (annual); *Scottish Office*

United Kingdom Minerals Yearbook (annual); *British Geological Survey* (available from the British Geological Survey, Keyworth, Nottingham, NG12 5GG)

Yearbook of Regional Statistics (annual); *Statistical Office of the European Communities*

Energy
Annual Bulletin of General Energy Statistics for Europe; *United Nations Economic Commission for Europe*

BP Statistical Review of World Energy (annual) ; (available from The Editor, BP Statistical Review, The British Petroleum Company plc, Corporate Communications Services, Britannic House, 1 Finsbury Circus, London EC2M 7BA)

Energy - Monthly Statistics; *Statistical Office of the European Communities*

Energy Balances of OECD Countries (annual); *OECD International Energy Agency*

Energy Statistics and Balances of OECD Countries (annual); *OECD International Energy Agency*

Energy Statistics and Balances of Non-OECD Countries (annual); *OECD International Energy Agency*

Energy - Yearly Statistics; *Statistical Office of the European Communities*

UN Energy Statistics Yearbook (annual); *United Nations Statistical Office*

Projections
Energy Projections for the UK, working paper (March 2000), EPTAC Directorate, DTI. Reports work in progress to update projections of the future UK energy demand and energy-related CO_2 emissions, last published in Energy Paper 65.

Coal
Annual Bulletin of Coal Statistics for Europe; *United Nations Economic Commission for Europe*

Annual Reports and Accounts of The Coal Authority and the private coal companies - *apply to the Headquarters of the company concerned.*

Business Monitor PA 130 Coal extraction and manufacture of solid fuels (annual, ceased publication in 1994); *Office for National Statistics*

Coal Information (annual); *OECD International Energy Agency*

Oil and gas
Annual Bulletin of Gas Statistics for Europe; *United Nations Economic Commission for Europe*

Annual Report of the Office of Gas Supply - OFGAS

BP Review of World Gas (annual); (available from British Petroleum Company plc, Corporate Communications Services, Britannic House, 1 Finsbury Circus, London EC2M 7BA)

Annual Reports and Accounts of British Gas Transco, Centrica and other independent gas supply companies - *apply to the Headquarters of the company concerned.*

Business Monitor PA 162 - Public gas supply (ceased publication in 1994); *Office for National Statistics*

Oil and Gas Information (annual); *OECD International Energy Agency*

Quarterly Oil Statistics and Energy Balances; *OECD International Energy Agency*

UK Petroleum Industry Statistics Consumption and Refinery Production (annual and quarterly); *Institute of Petroleum* (available from IP, 61 New Cavendish Street, London W1M 8AR).

Electricity
Annual Bulletin of Electric Energy Statistics for Europe; *United Nations Economic Commission for Europe*

Annual Reports and Accounts of the Electricity Companies and Generators - *apply to the Headquarters of the company concerned.*

Annual Report of the Office of Electricity Regulation - OFGEM

Business Monitor PA 161 Production and distribution of electricity (annual, ceased publication in 1994); *Office for National Statistics*

Electricity Supply in OECD Countries ; *OECD International Energy Agency*

National Grid Company - Seven Year Statement - (annual) *National Grid Company - For further details telephone 01203 423065*

Operation of Nuclear Power Stations (annual); *Statistical Office of the European Communities*

UK Electricity (annual); *Electricity Association plc* (available from the Electricity Association plc, 30 Millbank, London SW1P 4RD)

Electricity Information (Annual); *OECD International Energy Agency*

Prices
Energy Prices (annual); *Statistical Office of the European Communities* (summarises price information published in the European Commissions *Weekly Oil Price,* and half-yearly *Statistics in Focus* on *Gas Prices* and *Electricity Prices)*

Energy Prices and Taxes (quarterly); *OECD International Energy Agency*

Electricity prices; *Eurostat* (annual)

Gas prices; *Eurostat,* (annual)

Environment
Digest of Environmental Statistics (Annual); *Department of the Environment, Transport and the Regions*

Indicators of Sustainable Development for the United Kingdom; *Department of the Environment, Transport and the Regions*

Quality of life counts, *Indicators for a strategy for sustainable development for the United Kingdom: a baseline assessment.*

Environment Statistics; EUROSTAT (Annual).

UK Environment; DETR adhoc/one-off release.

Renewables

New and Renewable Energy, *Prospects for the 21ˢᵗ Century. This consultation paper reports on the outcome of the review conducted by the Government and the possible ways forward in implementing the Government's new drive for renewables.*

Useful energy related websites

The DTI website can be found at http://www.dti.gov.uk, the energy information and statistics website is at http://www.dti.gov.uk/epa

Other Government websites

Ofgem (at COI site)	http://www.ofgem.gov.uk
DETR	http://www.detr.gov.uk
Customs and Excise	http://www.hmce.gov.uk
ONS	http://www.ons.gov.uk
COI	http://www.nds.coi.gov.uk

Other useful energy related websites

International Energy Agency	http://www.iea.org
Eurostat	http://europa.eu.int/en/comm/eurostat/server/home.htm
UK Petroleum Industry Association	http://www.ukpia.com
Institute of Petroleum	http://www.petroleum.co.uk
US Energy Information Administration	http://www.eia.doe.gov/
ETSU	http://www.etsu.com
Electricity Association	http://www.electricity.org.uk
BP Amoco	http://www.bpamoco.com
World Energy Statistics	http://www.energyinfo.co.uk
UK Offshore Operators Association	http://www.ukooa.co.uk
NETCEN (Air quality estimates)	http://www.aeat.co.uk/netcen/airqual/welcome.html.
BRE	http://www.bre.co.uk
Coal Authority	http://www.coal.gov.uk/
Iron and Steel Statistics Bureau	http://www.issb.co.uk/
Europa	http://europa.eu.int/pol/ener/index_en.htm

The Stationery Office	http://www.the-stationery-office.co.uk/

NOTES

NOTES